ADVANCES IN MOLECULAR AND CELL BIOLOGY

ENZYMOLOGY *IN VIVO*

Series Editor: E. EDWARD BITTAR
Department of Physiology
University of Wisconsin
Madison, Wisconsin

Guest Editor: KEVIN M. BRINDLE
Department of Biochemistry
University of Cambridge
Cambridge, England

VOLUME 11 • 1995

JAI PRESS INC.

Greenwich, Connecticut *London, England*

55 Old Post Road, No. 2
Greenwich, Connecticut 06836

JAI PRESS LTD.
The Courtyard
28 High Street
Hampton Hill, Middlesex TW12 1PD
England

ISBN: 1-55938-844-7

Manufactured in the United States of America

CONTENTS

LIST OF CONTRIBUTORS

Hilary A. Berthon	Department of Biochemistry University of Sydney Sydney, Australia
Kevin M. Brindle	Department of Biochemistry University of Cambridge Cambridge, England
Athel Cornish-Bowden	Laboratoire de Chimie Bacterienne Centre National de la Recherche Scientifique Marseille, France
C. Doumen	Department of Cellular and Molecular Physiology Milton S. Hershey Medical Center Hershey, Pennsylvania
Alexandra M. Fulton	School of Biological Sciences University of Manchester The Medical School Manchester, England
Jessica M. Halow	Department of Biological Sciences and Pittsburgh NMR Center for Biological Research Carnegie Mellon University Pittsburgh, Pennsylvania
Douglas B. Kell	Department of Biological Sciences University of Wales Aberystwyth, Wales
Alan P. Koretsky	Department of Biological Sciences and Pittsburgh NMR Center for Biological Research Pittsburgh, Pennsylvania

Philip W. Kuchel — Department of Biochemistry, University of Sydney, Sydney, Australia

K. F. La Noue — Department of Cellular and Molecular Physiology, Milton S. Hershey Medical Center, Hershey, Pennsylvania

Craig R. Malloy — The University of Texas Southwestern Medical Center at Dallas, Rogers Magnetic Resonance Center, Dallas, Texas

Pedro Mendes — Department of Biological Sciences, University of Wales, Aberystwyth, Wales

Kenneth R. Miller — Department of Biological Sciences and Pittsburgh NMR Center for Biological Research, Carnegie Mellon University, Pittsburgh, Pennsylvania

Len Pagliaro — Center for Bioengineering, University of Washington, Seattle, Washington

A. Dean Sherry — Department of Chemistry, The University of Texas at Dallas, Richardson, Texas

Paul A. Srere — Department of Veterans Affairs Medical Center, Dallas, Texas

Balazs Sumegi — Department of Biochemistry, University Medical School, Pecs, Hungary

G. Rickey Welch — Department of Biological Sciences, University of New Orleans, New Orleans, Louisiana

Simon-Peter Williams — Department of Biochemistry, University of Cambridge, Cambridge, England

PREFACE

The idea for this volume came from a symposium that I organized for the Biochemical Society meeting held in Manchester in July 1991.

Much of our understanding of the control of metabolic processes in the cell is based on *in vitro* studies of the kinetic and allosteric properties of the enzymes involved. This reductionist approach has enabled us to formulate models of control of the primary metabolic pathways which can be found in most standard undergraduate biochemistry text books. But just how good are these models? They are based, after all, on the premise that we have a detailed knowledge of all the possible interactions of these enzymes with those molecules in the cell which may modulate their activity. These obviously would include their substrates as well as small molecule effectors and other proteins which may bind to them and affect their activity. Henrik Kacser illustrated the scale of this problem at the Manchester meeting by showing a series of slides which showed progressively more of that well known wall chart which depicts most of the known metabolic pathways in the cell. The level of any particular substrate or small molecule effector is a function of the system as a whole and changes in their levels can be expected to have a myriad of effects which will be communicated throughout the system. Superimposed on this complexity is cellular compartmentation. The problem is further compounded by the assumption that the map is complete. A casual reader of the biochemistry text books in the 1970s would have come away with the overwhelming impression that we had a thorough understanding of how the glycolytic pathway is controlled. Yet arguably one of the most important effectors in the control of this pathway, fructose-2,6-bisphosphate, was not discovered until 1980. This example illustrates graphically the limitations of the reductionist approach and shows why it is important that we study enzymes not only in the test tube but also in the intact cell i.e. we need to do our enzymology *in vivo*.

This volume describes some of the approaches that have been used to study enzymes *in vivo*. Metabolic control analysis provides a relatively simple framework with which to relate flux in a metabolic pathway to the kinetic properties of the component enzymes. More importantly it shows us how the importance of an enzyme in controlling flux in a pathway can be quantitated experimentally from measurements on the intact tissue. Fluorescence microscopy and NMR are two spectroscopic techniques which can be used to monitor, non-invasively, metabolite levels, metabolic fluxes and enzyme localization and mobility in intact biological systems. The potential of NMR for investigating the properties of enzymes *in vivo* has been greatly enhanced by using the technique in conjunction with molecular genetic methods for changing the levels and properties of specific enzymes in the intact cell.

Control of metabolism is regarded by some as a dead subject, with little new to learn. While it is true that the chemistry of the major metabolic pathways have been fully elucidated, our understanding of how they are controlled in the cell is still rather limited. Of particular interest is the emerging evidence for a high degree of spatial organization of the supposedly 'soluble' enzymes in the cytosol and the mitochondrial matrix. Much is still to be learnt on how this organization is effected and what influence it has on control of metabolic flux. If this volume excites some interest in this area of research and, furthermore, demonstrates that these problems are eminently addressable using the new techniques which are being developed, then it will have served a useful purpose.

Finally I would like to thank my co-contributors for writing such interesting chapters and for articulating, far more eloquently than I have here, the need to do enzymology *in vivo*.

Kevin M. Brindle
Guest Editor
December 1994

METABOLIC CHANNELING IN ORGANIZED ENZYME SYSTEMS: EXPERIMENTS AND MODELS

Pedro Mendes, Douglas B. Kell, and G. Rickey Welch

Advances in Molecular and Cell Biology
Volume 11, pages 1–19.

ISBN: 1-55938-844-7

ABSTRACT

The intracellular *milieu* is not a simple, homogeneous, aqueous state: protein concentration is high in eukaryotes, and even higher in prokaryotes and in organelles such as mitochondria, and membrane surfaces are clearly abundant. Evidence gathered with various techniques indicates that the cellular water does not have the same properties as water in dilute aqueous solutions. These findings support the view that classical enzymological studies may not provide sufficiently relevant information for generating a correct understanding of cellular physiology. Cellular organization exists at the molecular level: enzymes aggregate in clusters and in many cases this affects their catalytic activity. Consecutive enzymes in a number of metabolic pathways can channel their common intermediates without release to the "bulk" solution. This process can occur either via stable (static) multienzyme complexes or via short-lived (dynamic) enzyme–metabolite–enzyme complexes. Static complexes are found in anabolic pathways such as amino acid, nucleotide, and protein biosynthesis, where most of the intermediates have no other function or destination in the cell; dynamic complexes occur in amphibolic pathways where there are various flow-bifurcations. Channeling between dynamic complexes of enzymes is in some ways harder to demonstrate since the enzyme–enzyme complexes are not stable and are thus not isolatable. Theoretical developments, and simulations of existing metabolic channeling models, are not abundant. We review such studies and propose how modeling should evolve, the better to match the evolution of physiological experiments from *in vitro* to *in situ* to *in vivo*.

I. *IN VIVO* IS NOT THE SAME AS *IN VITRO*

Essentially since the beginning of modern biochemistry itself (Schlenk, 1985), enzymologists have studied enzymes *in vitro*. This (reductionist) approach to understanding cellular behavior is based on the belief that the phenomena observed in cells can be attributed solely to the properties of the cell components. The quest then has been that of isolating the thousands of different enzymes in the living world and studying their physicochemical properties *in vitro*, with the implicit assumption that after this knowledge had been attained one could somehow "reconstitute" the properties of the cell, if not in practice at least in principle. Throughout the last three decades (and some would say the whole century), a large amount of evidence has accumulated that suggests this approach is essentially flawed. Two main arguments are as follows: (1) following the pioneering studies of Kacser and Burns (1973) and of Heinrich and Rapoport (1974) it has been shown that the steady-state behavior of fluxes and metabolite concentrations within a cell are *systemic* properties not properly accountable in terms of the behavior of single enzymes, but instead by the concerted action of *all* of them (indeed even by noncatalyzed processes); and (2) the conditions used for *in vitro* assays are so far from those observed in cells (which are generally unknown) as to make extrapolations from *in vitro* to *in vivo* at best hazardous and at worst completely misleading. While the first point is very important (it is discussed in detail in the Cornish-Bowden chapter of this volume), we shall concentrate on the second. Since this book is about enzymology *in vivo*, it is worth discussing some of the findings that lead to the conclusion that a reductionistic analysis of cell biology is doomed to fail, an analysis which may be seen as an implication of the Humpty Dumpty effect (Kell and Welch, 1991). It is worth rehearsing the general argument.

The notion of "analytical reductionism" is intimately associated with the principles of irreversibility and boundary conditions. As Prigogine and Stengers (1984) point out, "Irreversibility is either true on all levels or on none: it cannot emerge as if out of nothing, on going from one level to another," but as nicely delineated by Coveney and Highfield (1990), irreversibility remains a philosophical enigma. Newtonian physics is time-reversible; if we watch a film of billiard balls colliding, we cannot tell whether the film is running forwards or backwards. By contrast, if we observe a film of a bull in a china shop, we may be fairly confident that the film is running in one (the "forward") direction; bulls do not normally reassemble broken crockery and emerge smiling from retail stores. Thus, as one sees with Humpty Dumpty, there are many ways of breaking things, but only one way of putting them together correctly. The key point is that the successively higher levels of the hierarchically organized, complex living cell are dependent, reductionistically, not so much on the *elements* at the lower levels, but on the nature and existence of boundary constraints. If one removes the constraints at a given level, the systemic (or holistic) properties of all higher levels potentially collapse. Thus, while individual protein molecules can be persuaded to refold to their "native" states (Anfinsen, 1973), though not reversibly in the sense of *microscopic* reversibility, no one has succeeded in making a cell do so, let alone an organism such as Humpty Dumpty, and there are straightforward combinatorial arguments why they are unlikely to succeed (Kell, 1988a,b; Kell and Welch, 1991).

In recent decades, we have come to appreciate some of the boundary constraints extant *in vivo*. To begin with, the intracellular medium is not a simple, homogeneous, aqueous state. Its protein content is extremely high (100–300 mg/ml in eukaryotes, and maybe double that in prokaryotes), and membrane surfaces are clearly abundant. Electron microscopy has revealed a complex and diverse particulate infrastructure in living cells, especially in the larger eukaryotic cells. This structure encompasses not only an extensive membranous reticulation but also a "ground substance" which is laced with a dense array of proteinaceous cytoskeletal elements. The protein density in association with these membranous and fibrous structures is akin to that in crystals (Sitte, 1980). In particular, the work of Porter and collaborators (see for example Porter and Anderson, 1982; Porter and Tucker, 1981) has revealed an intricate network structure in the cytoplasm of eukaryotic cells. This network has been named the microtrabecular lattice (MTL) and it is observable in high-voltage electron photomicrographs. The existence of the MTL does not, by itself, exclude the hypothesis that the enzymes found in the soluble fraction would also be in solution in the cytoplasm. The extra evidence needed can be found in the experiments of Kempner and Miller (1968a,b) with *Euglena gracilis*. Kempner and Miller found that due to their hard cellular wall, *E. gracilis* cells can be centrifuged at 100,000 × *g* for 1 hour without disruption, after which the various cellular components become stratified inside the cell. An important aspect is that the cells remained viable under these conditions. Kempner and Miller analyzed quickly frozen stratified *E. gracilis* cells by cytochemical methods for the presence of 19 different enzymes and found that none of these enzymatic activities were present in the ostensibly "soluble" aqueous phase, but rather in denser layers. However, if the cells were homogenized *before* the centrifugation, all of those enzymes were then found in the 100,000 × *g* supernatant. These experiments undoubtedly show that most of the "soluble" enzymes are in fact not in solution at all within *E. gracilis* cells. Similar experiments made with *Neurospora* (Zalokar, 1960) and ultracentrifugation and biochemical studies on *Artemia* cysts (Clegg, 1982) produced similar results. We have no reason to think that other eukaryotic cells would be much different from these. Another strong piece of evidence for the bound state of cytoplasmic proteins in the cell comes from studies with cells whose plasma membranes were made permeable (Kell and Walter, 1986; Clegg and Jackson, 1988; 1990). In some cases the pores in the plasma membrane were big enough to allow molecules of ~400 kDa to pass through them; nevertheless the loss of protein from these cells was small, indicating that most proteins are associated with some structure (or at least in complexes bigger than 400 kDa) (Clegg and Jackson, 1988; 1990).

Additional evidence for the cytoskeletal infrastructure comes from electron spin resonance (ESR) (Mastro and Hurley, 1987), fluorescence recovery after photobleaching (FRAP) (Luby-Phelps et al., 1988) and microfluorimetric (Fushimi and Verkman, 1991) studies *in situ*, which each show the interstititial voids (200–400 Å in diameter) to contain a medium akin to a dilute aqueous *milieu* of low macromolecular density.

It is widely understood that in order to reproduce *in vitro* the properties of enzymes that in their native cellular *milieu* are rigidly membrane-associated one must provide them with some sort of proteolipid environment, frequently by isolating them in fragments of the original membrane or otherwise by incorporating them into proteoliposomes. Unfortunately, the same belief is not so commonly held for the so-called soluble enzymes that are present in the 100,000 × *g* supernatant fraction. In not seeking to emulate more closely the native microenvironment *in vitro,* we take the risk of building models of cells which have little resemblance to reality. One immediate consequence of this extensive organization of enzymes in the cytoplasmic compartment (and others) is that the classic, bulk-phase, scalar concept of concentration is no longer very helpful. Instead we may have to start thinking in terms of "local concentrations" (Welch, 1977).

Available evidence from ESR (Mastro and Keith, 1981), nuclear magnetic resonance (NMR) (Seitz et al., 1981), quasi-elastic nuclear scattering (Trantham et al., 1984), and microwave dielectric measurements (Clegg et al., 1982; 1984; Bruni et al., 1989) indicate that the cellular water does not have the same properties as water in dilute aqueous solutions. In the late 1930s, Kopac and Chambers (reviewed in Chambers, 1940), using microinjection of oil droplets in echinoderm egg cells, observed an absence of the Devaux effect (the surface of the droplet crinkles due to adsorption of proteins), which should be observable if the intracellular proteins were in aqueous *solution*. These findings support the view that classical enzymological studies may not be terribly relevant to what goes on inside a cell.

II. ORGANIZATION LEADS TO CHANNELING

In parallel to these findings about the structure of the cytoplasmic matrix, there is a considerable amount of evidence for enzyme–enzyme interactions, both in this compartment and inside other cellular organelles such as the mitochondrion, and within prokaryotic cells. These interactions could of course simply be a consequence of proximity effects if enzymes are part of the MTL or are adsorbed to it in the cytoplasmic compartment, or because the enzyme concentration is high (as in mitochondria and prokaryotes). However, a very interesting and important set of enzyme–enzyme interactions, for the existence of which there is large body of evidence (e.g., Friedrich, 1984; Srivastava and Bernhard, 1986; Srere, 1987; Keleti, et al., 1989), is that in which two (or more) sequential enzymes in a pathway transfer their common intermediate (the product of the first and the substrate of the second) *directly*, i.e. without releasing it to the bulk solvent. The evidence for this phenomenon, known as *metabolic channeling*, has been gathered by several methods, from X-ray crystallography to kinetic experiments *in vitro* (Keleti et al., 1989).

III. STATIC VERSUS DYNAMIC CHANNELS

One obvious situation in which metabolic channeling can easily occur is that of the "static" multienzyme clusters, complexes of several polypeptides with different enzymatic activities. There are many well-known examples of such clusters, including the pyruvate dehydrogenase and fatty acid synthase complexes and the ribosome. In such cases it is common knowledge that intermediates of the reactions catalyzed by these clusters are not released to the solvent. Of particular interest is the case of tryptophan synthase, whose structure (from *Salmonella typhimurium*) has been determined by X-ray crystallography to 2.5 Å resolution (Hyde et al., 1988; Hyde and Miles, 1990). This cluster is an $\alpha_2\beta_2$ tetramer; the α chains convert indole 3-glycerol-phosphate to indole and D-glyceraldehyde 3-phosphate, and the β chains convert indole and L-serine to L-tryptophan. The images reconstructed by Hyde and co-workers show a tunnel going from the active center of one α polypeptide chain to the active center of one β chain (there are two of these tunnels per cluster). It is through this tunnel that indole undergoes "facilitated diffusion" from one active center, and is thus not released to the solvent. One obvious advantage of this mechanism is that if indole were to be released to solution it would be able, due to its hydrophobic character, to diffuse freely across the plasma membrane and thus escape (Manney, 1970) (as indeed it does in strains harboring the tryptophanase gene (Goodacre and Kell, 1993)). This former fact is particularly nicely illustrated in a recent paper by Murdock et al. (1993), who were interested in the biotechnological production of indigo from carbohydrates, via indole as a cellular intermediate within the cell. Only after site-directed mutagenesis of the tryptophan synthase, *to cause indole to leak from its channel*, were satisfactory yields achieved.

The channeling of metabolic intermediates, here known as *static* channeling or channeling via static complexes is not, however, restricted to stable multienzyme complexes; it can also take place via short-lived enzyme complexes referred to as *dynamic* channels (Friedrich, 1974; 1984; 1985). Channeling between dynamic complexes of enzymes is in some ways harder to demonstrate since the enzyme–enzyme complexes are not stable and are thus not isolatable. Of course this classification describes two extreme situations, and one may consider that in reality enzymatic complexes have properties that are a continuum between the two extremes (Ovádi, 1991). Ovádi (1991) proposed a general mechanism which can account for this continuum, and although not absolutely complete (see e.g., Ryazanov, 1988), this general mechanism is the best and most convenient description to date.

There is a remarkable correlation between the organizational state and the functionality of supramolecular enzyme complexes (Welch and Marmillot, 1991): static complexes are found in anabolic pathways such as amino acid, nucleotide and protein biosynthesis. In these pathways most of the intermediates have no other function or destination in the cell (Srere, 1987; Alberts et al., 1989). In some cases, as for indole in the tryptophan synthase reactions in wild-type strains, there is essentially no leakage of the metabolite to the solvent (perfect channeling). Dynamic complexes occur in amphibolic pathways like glycolysis and the Krebs tricarboxylic acid (TCA) cycle where there are various flow-bifurcations. In this case there are metabolites at branch-points, which are known (e.g. from NMR measurements) to exist in soluble pools. However this does not rule out the possibility that a proportion of these substances is bound to enzymes (either individually or in complexes); in some cases, e.g. for oxaloacetate in mitochondria (Atkinson, 1969), the number of binding sites exceeds the number of molecules extractable using perchloric acid. Indeed, there is evidence that the role of enzymes is not simply catalytic but also structural (Wistow et al., 1987). Based on the high concentration of enzymes and the relatively low concentration of some of their intermediates in glycolysis, it has been argued that an important role of the cytoplasmic enzymes may be that of buffering the concentrations of their ligands (Betts and Srivastava, 1991; Srivastava, 1991). In this case, enzyme–intermediate complexes may act as better substrates for the next enzymes in the pathway than are the soluble intermediates themselves (Srivastava and Bernhard, 1985).

IV. SOME CONTROVERSIES ABOUT DYNAMIC CHANNELING

In a series of elegant *in vitro* kinetic experiments, Srivastava and Bernhard observed channeling of NADH among some dehydrogenases (reviewed in Srivastava and Bernhard, 1986; see also Srivastava et al., 1989). The basis of their experiments is the buffering of NADH by the first dehydrogenase, such that if channeling occurs then the rate of the reaction of the second dehydrogenase calculated from the concentration of free NADH present will be different from the experimental rate. A requirement for these experiments is that the first dehydrogenase must be in large molar excess with respect to NADH which is in turn in molar excess over the second dehydrogenase; also the dissociation constant of the NADH from the first dehydrogenase must be known. Srivastava et al. (1985) and Srivastava and Bernhard (1986) also showed a remarkable correlation between the stereospecificity of NADH binding to dehydrogenases and the channeling of NADH between pairs of dehydrogenases. It turns out that in order for channeling to occur the two dehydrogenases must be of opposite chiral specificity to NADH (there are two possible stereospecificities, A and B). Chock and coworkers reported inconsistencies of this analysis, based in part on their own repetition of these experiments (Chock and Gutfreund, 1988; Wu et al., 1991), but it is now clear that this was due to the fact that they failed to reproduce key features of the original experiments, such as the ionic strength (Srivastava et al., 1989; Srivastava, 1991). The remarkable correlation between enhanced catalysis and opposite chiral specificity of the dehydrogenases (and no enhancement for equal specificities) is the major factor supporting the channeling model and this was never addressed by Chock's group.

There are a number of other cases of experimental results that point to channeling in dynamic complexes, but which have not universally been accepted. Ovádi and Keleti (1978) proposed channeling of glyceraldehyde-3-phosphate from aldolase to glyceraldehyde-3-phosphate dehydrogenase but Kvassman et al. (1988), although reproducing Ovádi and Keleti's results, argued that these were also consistent with a free diffusion mechanism. Vértessy and Ovádi (1987; see also Srivastava and Bernhard, 1986) showed data that suggested the channeling of dihydroxyacetone phosphate from aldolase to glycerol-3-phosphate dehydrogenase, although Chock and Gutfreund (1988) offered an alternative interpretation. Weber and Bernhard (1982) suggested direct transfer of 1,3-diphosphoglycerate between glyceraldehyde-3-phosphate dehydrogenase and 3-phosphoglycerate kinase; however Kvassman and Pettersson (1989a,b) argued that if autocatalytic effects of NAD^+ were taken into account then it would not be necessary to invoke channeling. Unfortunately, in most of these reports the data were not obtained under the same conditions by the different groups; there has also been a certain tendency among those who argue against channeling to do so despite the fact that only data which *also* support free diffusion are shown.

There are, however, cases of metabolic channeling which occur through transient enzyme complexes and whose existence is unchallenged, despite the fact that, as in the previous cases, the actual multienzyme complex has never been detected. One example is that of electron transfers between enzyme-bound flavin coenzymes (Steenkamp, 1987).

The Krebs tricarboxylic-acid cycle (TCA), has been extensively studied in terms of its supramolecular organization (Sumegi et al., 1992 and references therein; see also Srere et al., this volume). Srere has developed the concept of the *metabolon*, a complex of the enzymes of the TCA cycle which is anchored to the inner mitochondrial membrane (Srere, 1985). The *metabolon* can be obtained intact from sonicated mitochondria (Srere, 1985; 1987) and there is evidence that the intermediates of the TCA cycle are channeled in the *metabolon*. Data that support this view were obtained *in situ* with gently disrupted (sonicated) mitochondria (Robinson and Srere, 1985; Robinson et al., 1987; Sumegi et al., 1992) and *in vivo* in yeast cells using NMR techniques with [3-^{13}C]-propionate as the substrate (Sumegi et al., 1990; 1992).

Using chromatographic techniques, Gontero et al. (1988) isolated a complex of five enzymes of the Calvin cycle from spinach leaves while Persson and Johansson isolated a 6-enzyme complex from the same source, using partition in an aqueous two-phase system (Persson and Johansson, 1989). Interestingly enough, no interaction between purified enzymes of these complexes could be demonstrated *in vitro*. This strengthens the argument that one should study viable cells, and that if for technical reasons this is not possible, that one should try to emulate the relevant physiological environment as much as possible. The words of Henrik Kacser (1986) could not be more appropriate: "But one thing is certain: to understand the whole, you must look at the whole."

One of the reasons that it is arguable is dominant in determining the difficulty in demonstrating dynamic complexes *in vitro* is the possibility that, in the intact cell, these complexes might be stabilized by other cellular structures (Clegg and Jackson, 1988). The MTL is an attractive candidate for such interactions (Clegg, 1984) but membranes can also play an important role (Wilson, 1978). The interaction of muscle cell glycolytic enzymes with actin (Clarke and Masters, 1976; Knull and Walsh, 1992) is a remarkable example, but other specialized cell structures are certainly also involved. It may also be that enzyme–enzyme complexes can affect the strength of allosteric effects (or even create new ones) (Welch, 1977). It is also worth pointing out that the inability of a cloned enzyme to affect the flux through a pathway may be due either to the fact that they have a low flux-control coefficient or because they are unable to participate in channeling due to their expression at unsuitable concentrations or in an unsuitable location; to hope to distinguish these, one should clone the relevant genes down as well as up (Brindle, 1988; Kell and Westerhoff, 1990).

For pathways such as glycolysis there is a distinct structure–function duality. This is evidenced for example in the reversible, "ambiquitous" partitioning of glycolytic enzymes between cytomatrix-bound and aqueous cytoplasm-free forms (Masters, 1981; Knull and Walsh, 1992; Uyeda, 1992). The direct-transfer "channeling" scheme is probably immaterial for the *unbound* enzyme fraction in the bulk cytoplasmic solution. Depending on the cell type, the "whole cell" concentration of the individual glycolytic enzymes is in the range 1–100 μM. Considering the high, *bulk* "viscosity" of the cytoplasm relative to the translational motion of the macromolecles (Mastro and Hurley, 1987; Luby-Phelps et al., 1988), it is unlikely that the binary complex, consisting of enzyme$_1$ bound to its nascent product molecule, would last long enough to form the requisite ternary complex with enzyme$_2$ (Keizer and Smolen, 1992). However, for the *cytomatrix-associated* enzyme population, the microenvironmental situation is quite different. The "local" enzyme concentration is probably much higher than the averaged, "whole-cell" value. In addition, the total macromolecular density at the cytomatrix–"cytosol" interfaces (≤ 100 Å into the bulk phase) is extremely high (Sitte, 1980), probably similar to that within the mitochondrial matrix (viz. 250–500 mg/ml; Srere, 1985). Recent *in situ* measurements of the mitochondrial matrix bulk viscosity show the diffusion coefficient for small-metabolite-sized particles to be as much as 30-fold smaller than that for normal water (Abney et al., 1993). With such high "local" enzyme concentrations, along with the high microenvironmental viscosity (Siegbahn et al., 1985) and high activity coefficients for protein-based reactions in such a "crowded" medium (Minton, 1990), the efficacy of formation of channeling ternary complexes has been calculated to be greatly enhanced (Keizer and Smolen, 1992). The size of the free metabolite pools for the dynamically interacting (pairwise) enzyme systems will also depend on molar ratios of the metabolically consecutive enzymes, as well as on the respective enzyme kinetic mechanisms; moreover, the physiological grounds for channeling may not always entail a kinetic flux advantage for the multienzyme system (Srivastava, 1991; Welch and Easterby, 1994).

Meanwhile, one should be cognizant of the weight of the *in vivo* evidence for channeling in these pathways, obtained from the many whole-cell studies (Paul, 1989; Clegg and Jackson, 1990; Srere, 1992; Sumegi et al., 1992), as well as the correlation with *in vitro* indications of a "physiological" regulation of enzyme–cytomatrix associations (Masters, 1981; Knull and Walsh, 1992; Uyeda, 1992) and of the control of enzyme–enzyme interactions at pathway branchpoints (Tompa et al., 1987; Neuzil et al., 1990). The ambiquitous character of the enzyme organization—and more specifically the variability in the degree (and spatial locale) of channeling—in the branching amphibolic pathways may reflect the need to maintain adequate catabolite pools for proper responsiveness of such multifunctional processes to varying cellular demands (Easterby, 1991). It is within the localized cytomatrix microenvironments that channeling of amphibolic flow may therefore be expected.

V. MODELING STRATEGIES FOR STUDYING ENZYMOLOGY *IN VIVO*

All interpretations of experimental results are based on models, and metabolic channeling is no exception. Generally, enzyme models exhibit nonlinear behavior, and it is very difficult (and dangerous) to make predictions from those models by simple reasoning alone. For that purpose one has to formulate the model in mathematical terms and use the equations to calculate the values that the variables in question will take as a function of different starting conditions (parameters). These calculations are best carried out using computer programs to minimize errors and to process the calculations at an acceptable speed. A number of such programs are available (see e.g. Letellier et al., 1990; Kell et al., 1993; and Cornish-Bowden, this volume); in some of our own work on the simulation of metabolic channeling (Mendes et al., 1992), we have used the program GEPASI (Mendes, 1993).

Smolen and Keizer (1990) simulated a model of dynamic channeling with the direct transfer of NADH between dehydrogenases in mind. This group concluded from their simulations that the conditions for the ternary complex dehydrogenase$_1$–NADH–dehydrogenase$_2$ to form are in accordance with the estimates of K_m for the oxidation of the complex dehydrogenase$_1$–NADH by dehydrogenase$_2$. More recently these authors have extended their analysis to complexes of three dehydrogenases which would cycle the redox couple NAD^+/NADH. They concluded that there would be novel methods for regulation of the redox state of the couple which are consequent upon the channeling mechanism (Keizer and Smolen, 1992).

One consequence of metabolic channeling that has frequently been mentioned is that the operation of the catalytic path through the channel would be expected to reduce the size of the soluble pool of the intermediate compared to the case where the reaction proceeded exclusively via diffusion in the solvent (Srere and Mosbach, 1974; Kell, 1979; Heinrich et al., 1991; Ovádi, 1991). Cornish-Bowden (1991) showed that this is not always the case by simulating a model of metabolite channeling through a dynamic bi-enzyme complex. He then argued that this reduction of soluble pool size could *never* happen with a dynamic channel, but it was later shown that this generalization does not hold: channeling can decrease the pool size substantially (Mendes et al., 1992). It was also shown that if channeling operated through a static bi-enzyme complex then the size of the soluble pool could be decreased to an arbitrarily low level by increasing the proportion of flux through the channel (Mendes et al., 1992).

Brooks and Storey (1991) investigated the possibility of the existence of a complex of glycolytic enzymes in muscle cells. They used *in vitro* data for the association constants of the several complexes (enzyme–enzyme, enzyme–F-actin and enzyme–calmodulin) and relative activities of these complexes to conclude that complexes of enzymes on F-actin "may not exist;" complexes of *some* sequential glycolytic enzyme pairs could exist to a significant degree and increase the glycolytic flux, and the binding of phosphofructokinase with F-actin could be a regulatory mechanism to control glycolytic rate. However, this study was based solely on the equilibrium distribution of free and bound protein species. This is far from the reality of muscle cells where the glycolytic enzymes are responsible for very high net fluxes. Also not considered in this analysis was the effect of the intermediate metabolites, which is very important, as shown for example in the studies of Smolen and Keizer (1990; Keizer and Smolen, 1992)—the association constant for an E1–E2 enzyme complex is certain to be different from that when the intermediate is part of the complex (E1–M–E2).

Sauro and Kacser (1990), from a theoretical metabolic control analysis of a model of static channeling, predicted that the increase of the logarithmic flux of a binary–enzyme complex when the logarithmic concentration of both enzymes is increased simultaneously is not linear. Westerhoff and Kell (1988), Kell and Westerhoff (1990) and Welch and Keleti (1990) detail other properties of metabolic channels for which metabolic control analysis may be used to distinguish channeling from pool behavior.

All the above studies were based on catalysis in homogeneous solution, which we know is irrelevant for cellular compartments *in vivo* (Porter and Tucker, 1981; Clegg, 1984; Srere, 1987). Many of the supposed consequences of channeling through multienzyme complexes are specifically dependent on this feature. Perhaps the most frequently cited consequence of channeling is the reduction of the transient time of a reaction sequence, numerically equal to the sum of the metabolite concentrations divided by the pathway flux. A lower bound on the value of the transient time, τ, for the enzyme reaction (viewed as an intermediary metabolic process) can, under simplified (viz. pseudo-first order) steady-state conditions, be expressed as follows:

$$\tau = \frac{1}{\Omega \cdot D_{ES} \cdot R_{ES} \cdot f_e \cdot f_g \, [E]_T \cdot N_0} + \frac{k_{-s} + k_{cat}}{k_{+s} \cdot k_{cat} \cdot [E]_T} \tag{1}$$

$$= \tau_d + \tau_r$$

where Ω is a steric-orientation factor (e.g. a geometric solid angle within the range $0 < \Omega \leq 4\pi$) relating to the approach of the substrate to the enzyme active site, D_{ES} is the sum of the diffusion coefficients of E and S (where usually $D_{ES} \approx D_S$), R_{ES} is the "reaction distance" (viz. the sum of the radii of the substrate molecule and of the "recognition volume" of the enzyme active site), f_e is an electrostatic term arising from the possible influence of the net charge of the globular protein on an approaching, charged substrate molecule (with typical values of f_e ranging from 0.1 to 10, depending on whether the algebraic product of the net charge of the globular protein and that of the substrate molecule is positive or negative, respectively), f_g is a "gating" term due to the potential effect of local motions of proteinaceous lobes surrounding the active site (where typically $f_g \leq 0.5$ if slow protein-dynamical gating is identifiable), $[E]_T$ is the total enzyme concentration, N_0 is Avogadro's number, k_{+s}, k_{-s} and k_{cat} are the (intrinsic) unitary rate constants for the binding of substrate to the enzyme, the release of substrate, and catalytic turnover respectively. (For details of the derivation, see Westerhoff and Welch, 1992.)

Such effects can only be investigated if the spatial dimension is explicitly part of the model. Reaction–diffusion models are unfortunately more difficult to simulate than normal kinetic models and there are no software packages available for this purpose. We suggest that only when such studies are properly performed can the "physiological significance" of metabolic channeling be understood.

Marmillot et al. (1992) have recently studied the spatiotemporal organization of the reaction catalyzed by phosphofructokinase (PFK). They extended a model of Goldbeter and Lefever (1972) by allowing PFK to exist in both free and bound (to subcellular structures) forms. They observed sustained unidirectional wave propagation, a consequence of the nonuniform distribution of oscillation periods in the soluble phase (Marmillot et al., 1992). In this case, the segregation of a metabolite in more than one pool *in the aqueous phase* is temporal as well as spatial (see also Friedrich, 1984; 1985). In parallel with the arguments for doing enzymology *in vivo* rather than *in vitro*, this type of spatiotemporal modeling (as in Goodwin and Trainor, 1985) must become widely used to account for the inhomogeneity of the cellular compartments, and indeed of populations of cells generally (Kell, 1988a; Kell et al., 1991). Model studies of "well-stirred" reactions can give insight only into experiments *in vitro*. This is even more important in the case of metabolic channeling and only then can we really start discussing the consequences of channeling in cells. We can only be tempted to call this type of analysis "*in vivo* modeling".

VI. CONCLUDING REMARKS

Biology seems beset today with the same *fin de siècle* euphoria which affects the science of physics. Many sages are hailing the "end of physics," what with the apparent explanatory successes of modern-day relativistic quantum field theory, "string" theory, etc., leading to what some are calling a "Theory of Everything." Likewise, the discovery of the molecular basis of genetics in the 1950s (what many biologists have dubbed the "Secret of Life"), has sometimes seemed to have cast an air of finality on the study of living systems. As the 20th century comes to a close, the subject of "metabolism" has become rather *passé*. The great focus is on the isolation, cloning, sequencing and cutting/splicing of genetic elements. As we enter the 21st century, the United States government (with some contributions from other countries) stands poised to spend billions of dollars on the singular task of identifying all the loci within the human genome. One may argue that the science of biology has lost its philosophical view of life as a *process*, in favor of the perspective of *substance*. Hopefully, the import of a book with the title *Enzymology in vivo* will assert to the readership that the subject of metabolism is, in fact, far from being "solved."

The Humpty Dumpty effect (Kell and Welch, 1991) might appear to throw a shadow of nihilism on any analytical reductionist effort to understand the living state. In actuality, Humpty Dumpty serves as an abiding reminder that in hierarchically ordered systems one must seek to analyze the "parts" within the context of the "whole." It is only thus that an understanding of the emergent properties at each level of organization is attainable. In today's utilitarian science, the "understanding" of Nature has come to mean the "control" of Nature; in this vein one may indeed conclude that the organizational properties of the cellular metabolic machinery are crucial to (our understanding of) its control.

NOTE ADDED IN PROOF

H-K. Luh and S. L. Pimm (The assembly of ecological communities: a minimalist approach, J. Animal Ecol. 62, 749–765, 1993) have also used the phrase "The Humpty Dumpty Effect" to account for the inability to reassemble structured ecosystems from mixtures of their components.

Cornish-Bowden and Cárdenas (Channeling can affect concentrations of metabolic intermediates at constant net flux: artefact or reality? Eur. J. Biochem. 213, 87–92, 1993; hereafter CBC) have claimed that simulation results previously published by us (Mendes, P., Kell, D.B., and Westerhoff, H.V. Channeling can decrease pool size, Eur. J. Biochem. 204, 255–266, 1992) which had demonstrated that large reductions of intermediate pool sizes could be accompanied by increasing channel flux in a model metabolic pathway, were an artefact of changes in the pathway's overall flux of the order of 0.0075%, or of inappropriate alterations of enzyme activities. They also claimed to prove that "channeling of an intermediate cannot affect its free concentration at constant net flux".

In recent work (Mendes, P., Kell, D.B., & Westerhoff, H.V., in preparation) we consider the co-response of the intermediate metabolite concentration ("pool") and the channel flux to changes in kinetic (or thermodynamic) parameters. Both by analytical proofs and by numerical examples we show that this co-response can be positive, negative or null, depending on the parameter change. In particular we prove that there is always a number of ways of changing parameters such that the intermediate metabolite concentration decreases with increasing channel flux, whether the total flux varies or is constant. We also show that increased stability of the (dynamic) enzyme-intermediate-enzyme complex, as well as a single parameter change that similarly displays no cross-over effects, can lead to decreased intermediate metabolite concentration and increased channel flux at constant total flux.

More specifically:

(1) the algebraic analysis ("general proof") given in CBC contains the constraint that the elasticities of various steps to the modulation parameters which were used to vary the channel flux at constant net flux were unity. This is an unfortunate and unnecessary constraint which when lifted means that the concentration of the pool in the general case can indeed change at constant net flux. A "simplified proof" given in CBC also fails, due in addition to the consequent failure to include mass conservation relations for some of the enzymes.

(2) in the systems studied by CBC, flux is properly to be considered as a variable (since it varies during the transition to the steady state), and not a parameter, and as such cannot per se affect the magnitude of other variables in the steady state;

(3) by relaxing the constraint referred to in (1), above, and by making dual modulations (i.e. of more than on parameter at once) which are different from those carried out in CBC we find many instances in which channeling (described by a parameter p) does significantly affect the concentration of the pool intermediate C at constant total flux.

(4) in the same pathways, but in which the flux is held constant by setting it via a zero-order flux-generating reaction, the addition of a channel is also able significantly to modulate the size of the pool at constant total flux.

These results show that the effectiveness of channeling in decreasing a pool, even at constant flux, is very much a reality.

ACKNOWLEDGMENTS

P.M. thanks the J.N.I.C.T., Portugal for financial support (Grant BD-197/90-IF), and P.M. and D.B.K. are grateful to the Wellcome Trust and to the BBSRC Chemicals and Pharmaceuticals Directorate, for similar reasons.

REFERENCES

Abney, J. R., Scalettar, B. A., & Verkman, A. S. (1993). Rotational and translational dynamics of metabolite-sized molecules in the mitochondrial matrix. Biophys. J. 64, A163.

Alberts, B., Bray, D., Lewis, J., Raff, M., Roberts, K., & Watson, J. D. (1989). The Molecular Biology of the Cell, 2nd ed. Garland, New York.

Anfinsen, C.B. (1973). Principles that govern the folding of protein chains. Science 181, 223–230.

Atkinson, D. E. (1969). Limitation of metabolite concentrations and the conservation of solvent capacity in the living cell. Curr. Top. Cell. Reg. 1, 29–42.

Betts, G. F. & Srivastava, D. K. (1991). The rationalization of high enzyme concentration in metabolic pathways such as glycolysis. J. Theoret. Biol. 151, 155–167.

Brindle, K.M. (1988). ^{31}P NMR magnetization-transfer measurements of flux between inorganic phosphate and adenosine-5′-triphosphate in yeast cells genetically modified to overproduce phosphoglycerate kinase. Biochemistry 27, 6187–6196.

Brooks, S. P. J. & Storey, K. B. (1991). A quantitative evaluation of the effect of enzyme complexes on the glycolytic rate *in vivo*: mathematical modeling of the glycolytic complex. J. Theoret. Biol. 149, 361–375.

Bruni, F., Careri, G., & Clegg, J. S. (1989). Dielectric properties of *Artemia* cysts at low water contents. Evidence for a percolative transition. Biophys. J. 55, 331–338.

Chambers, R. (1940). In: The Cell and Protoplasm (Moulton, F. R., ed.), pp 20–30, Publ. 14 AAAS. Science Press, Lancaster, PA.

Chock, P. B. & Gutfreund, H. (1988). Reexamination of the kinetics of the transfer of NADH between its complexes with glycerol-3-phosphate dehydrogenase and with lactate dehydrogenase. Proc. Natl. Acad. Sci. USA 85, 8870–8874.

Clarke, F. M. & Masters, C. J. (1976). Interactions between muscle proteins and glycolytic enzymes. Int. J. Biochem. 7, 359–365.

Clegg, J. S. (1982). Interrelationships between water and cellular metabolism in *Artemia* cysts. IX. Evidence for organization of soluble cytoplasmic enzymes. Cold Spring Harbor Symp. Quant. Biol. 46, 23–37.

Clegg, J. S. (1984). Properties and metabolism of the aqueous cytoplasm and its boundaries. Am. J. Physiol. 246, R133–R151.

Clegg, J. S. & Jackson, S. A. (1988). Glycolysis in permeabilised L-929 cells. Biochem. J. 255, 335–344.

Clegg, J. S. & Jackson, S. A. (1990). Glucose metabolism and the channeling of glycolytic intermediates in permeabilised L-929 cells. Arch. Biochem. Biophys. 278, 452–460.

Clegg, J. S., Szwarnowwski, S., McClean, V. E. R., Sheppard, R. J., & Grant E. H. (1982). Interrelationships between water and cell metabolism in *Artemia* cysts. X. Microwave dielectrics. Biochem. Biophys. Acta 721, 458–468.

Clegg, J. S., McClean, V. E. R., Szwarnowwski, S., & Sheppard, R. J. (1984). Microwave dielectric measurements (0.8–70 GHz) on *Artemia* cysts at a variable water content. Phys. Med. Biol. 29, 1409–1419.

Cornish-Bowden, A. C. (1991). Failure of channeling to maintain low concentrations of metabolic intermediates. Eur. J. Biochem. 195, 103–108.

Coveney, P. & Highfield, R. (1990). The Arrow of Time. W. H. Allen, London.

Easterby, J. S. (1991). Homeostasis, flexibility and conflict in the kinetic advantage of channelling. J. Theoret. Biol. 152, 47–48.

Friedrich, P. (1974). Dynamic compartmentation in soluble enzyme systems. Acta Biochim. Biophys. Acad. Sci. Hung. 9, 159–173.

Friedrich, P. (1984). Supramolecular Enzyme Organization. Quaternary Structure and Beyond, pp. 194–195. Pergamon Press/Akadémiai Kiadó, Oxford/Budapest.

Friedrich, P. (1985). In: Organized Multienzyme Systems: Catalytic Properties (Welch, G. R., ed.), pp. 141–176. Academic Press, New York.

Fushimi, K. & Verkman, A. S. (1991). Low viscosity in the aqueous domain of cell cytoplasm measured by picosecond polarization microfluorimetry. J. Cell Biol. 112, 719–725.

Goldbeter, A. & Lefever, R. (1972). Dissipative structures for an allosteric model: application to glycolytic oscillations. Biophys. J. 12, 1302–1315.

Gontero, B., Cárdenas, M. L., & Ricard, J. (1988). A functional 5-enzyme complex of chloroplasts involved in the calvin cycle. Eur. J. Biochem. 173, 437–443.

Goodacre, R. & Kell, D. B. (1993). Rapid and quantitative analysis of bioprocesses using pyrolysis mass spectrometry and neural networks: application to indole production. Anal. Chim. Acta 279, 17–26.

Goodwin, B. C. & Trainor, L. E. H. (1985). Tip and whorl morphogenesis in acetabularia by calcium-regulated strain fields. J. Theoret. Biol. 117, 79–106.

Heinrich, R. & Rapoport, T. A. (1974). A linear steady-state treatment of enzymatic chains. General properties, control and effector strength. Eur. J. Biochem. 42, 89–95.

Heinrich, R., Schuster, S., & Holzhütter, H.-G. (1991). Mathematical analysis of enzymatic-reaction systems using optimization principles. Eur. J. Biochem. 201, 1–21.

Hyde, C. C. & Miles, E. W. (1990). The tryptophan synthase multienzyme complex: exploring the structure function relationships with X-ray crystallography. Bio/Technology 8, 27–32.

Hyde, C. C., Ahmed, S. A., Padlan, E. A., Miles, E. W., & Davies, D. R. (1988). Three-dimensional structure of the tryptophan synthase $\alpha_2\beta_2$ multienzyme complex from *Salmonella typhimurium*. J. Biol. Chem. 263, 17857–17871.

Kacser, H. (1986). In: The Organization of Cell Metabolism. (Welch, G. R. & Clegg, J. S., eds.), pp. 327–337. Plenum, New York.

Kacser, H. & Burns, J. A. (1973). The control of flux. Symp. Soc. Exp. Biol. 27, 65–104.

Keizer, J. & Smolen, P. (1992). Mechanisms of metabolite transfer between enzymes: diffusional versus direct transfer. Curr. Top. Cell. Reg. 33, 391–504.

Keleti, T., Ovádi, J., & Batke, J. (1989). Kinetic and physico-chemical analysis of enzyme complexes and their possible role in the control of metabolism. Progr. Biophys. Mol. Biol. 53, 105–152.

Kell, D. B. (1979). On the functional proton current pathway of electron transport phosphorylation: an electrodic view. Biochem. Biophys. Acta 549, 55–99.

Kell, D. B. (1988a). In: Bacterial Energy Transduction (C.J. Anthony, ed.), pp. 429–490. Academic Press, London.

Kell, D.B. (1988b) In: Biological Coherence and Response to External Stimuli (H. Fröhlich, ed.), pp. 233–241. Springer, Heidelberg.

Kell, D. B. & Walter, R. P. (1986). In: The Organization of Cell Metabolism, (Welch, G. R. & Clegg, J. S., eds.), pp 215–231. Plenum, New York.

Kell, D. B. & Westerhoff, H. V. (1990). In: Structural and Organizational Aspects of Metabolic Regulation, UCLA Symposia on Molecular and Cellular Biology, New Series, Vol. 134 (P. Srere, M.E. Jones & C. Mathews, eds.), pp. 273–289. Alan R. Liss, New York.

Kell, D. B. & Welch, G. R. (1991). No turning back. Times Higher Education Supplement, 9 August issue, p. 15.

Kell, D. B., Ryder, H. M., Kaprelyants, A. S., & Westerhoff, H. V. (1991). Quantifying heterogeneity: flow cytometry of bacterial cultures. Antonie van Leeuwenhoek 60, 145–158.

Kell, D. B., Westerhoff, H. V., Fell, D., Thomas, S., & Mendes, P. (1993). Demonstration programs illustrating the modeling of metabolic systems. Binary 5, 47–49.

Kempner, E. S. & Miller, J. H. (1968a). The molecular biology of *Euglena gracilis*. IV. Cellular stratification by centrifuging. Exp. Cell Res. 51, 141–149.

Kempner, E. S. & Miller, J. H. (1968b). The molecular biology of *Euglena gracilis*. V. Enzyme localization. Exp. Cell Res. 51, 150–156.

Knull, H. R. & Walsh, J. L. (1992). Association of glycolytic enzymes with the cytoskeleton. Curr. Top. Cell. Reg. 33, 15–30.

Kvassman, J. & Pettersson, G. (1989a). Evidence that 1,3-bisphosphoglycerate dissociation from phosphoglycerate kinase is an intrinsically rapid reaction step. Eur. J. Biochem. 186, 261–264.

Kvassman, J. & Pettersson, G. (1989b). Mechanism of 1,3-bisphosphoglycerate transfer from phosphoglycerate kinase to glyceraldehyde-3-phosphate dehydrogenase. Eur. J. Biochem. 186, 265–272.

Kvassman, J., Pettersson, G., & Ryde-Pettersson, U. (1988). Mechanism of glyceraldehyde-3-phosphate transfer from aldolase to glyceraldehyde-3-phosphate dehydrogenase. Eur. J. Biochem. 172, 427–431.

Letellier, T., Mazat, J.-P., Irvine, D. H., Savageau, M. A., Voit, E. O., Mendes, P., Hofmeyr, J.-H. S., Cornish-Bowden, A., & Atkinson, D. E. (1990). In: Control of Metabolic Processes, (Cornish-Bowden, A. & Cárdenas, M., eds.), pp. 433–436. Plenum Press, New York.

Luby-Phelps, K., Lanni, F., & Taylor, D. L. (1988). The submicroscopic properties of cytoplasm as a determinant of cellular function. Ann. Rev. Biophys. Biophys. Chem. 17, 369–396.

Manney, T.R. (1970). Physiological advantage of the mechanism of the tryptophan synthetase reaction. J. Bacteriol. 102, 483–488.

Marmillot, P., Hervagault, J.-F., & Welch, G. R. (1992). Patterns of spatiotemporal organization in an "ambiquitous" enzyme model. Proc. Natl. Acad. Sci. USA 89, 12103–12107.

Masters, C. J. (1981). Interactions between soluble enzymes and subcellular structure. CRC Crit. Rev. Biochem. 11, 105–143.

Mastro, A. M. & Keith, A. D. (1981). In: The Transformed Cell. (Cameron, I. L. & Pool, T. B., eds.), pp. 327–347. Academic Press, New York.

Mastro, A. M. & Hurley, D. J. (1987). In: Organization of Cell Metabolism (Welch, G. R. & Clegg, J. S., eds.), pp. 57–74, Plenum Press, New York.

Mendes, P. (1993). GEPASI: a software package for modeling the dynamics, steady states and control of biochemical and other systems. Comput. Applic. Biosci. 9, 563–571.

Mendes, P., Kell, D. B., & Westerhoff, H. V. (1992). Channeling can decrease pool size. Eur. J. Biochem. 204, 257–266.

Minton, A. P. (1990). In: Structural and Organizational Aspects of Metabolic Regulation (Srere, P. A., Jones, M. E., & Mathews, C. K., eds.), pp. 291–306, Wiley-Liss, New York.

Murdock, D., Ensley, B.D., Serdar, C., & Thalen, M. (1993). Construction of metabolic operons catalysing the *de novo* biosynthesis of indigo in *Escherichia coli*. Bio/Technology 11, 381–385.

Neuzil, J., Danielson, H., Welch, G. R., & Ovádi, J. (1990). Cooperative effect of fructose bisphosphate and glyceraldehyde-3-phosphate dehydrogenase on aldolase action. Biochim. Biophys. Acta 1037, 307–312.

Ovádi, J. (1991). Physiological significance of metabolic channeling. J. Theoret. Biol. 152, 1–22.

Ovádi, J. & Keleti, T. (1978). Kinetic evidence for interaction between aldolase and D-glyceraldehyde-3-phosphate dehydrogenase. Eur. J. Biochem. 85, 157–161.

Paul, R. J. (1989). Smooth-muscle energetics. Ann. Rev. Physiol. 51, 331–349.

Persson, L.-O. & Johansson, G. (1989). Studies of protein–protein interaction using countercurrent distribution in aqueous two-phase systems. Partition behaviour of six Calvin-cycle enzymes from a crude spinach (*Spinacia oleracea*) chloroplast extract. Biochem. J. 259, 863–870.

Porter, K. R. & Tucker, J. B. (1981). The ground substance of the living cell. Sci. Am. 244, 56–67.

Porter, K. R. & Anderson, K. L. (1982). The structure of the cytoplasmic matrix preserved by freeze-drying and freeze-substitution. Eur. J. Cell Biol. 29, 83–96.

Prigogine, I. & Stengers, I. (1984). Order out of Chaos, Heineman, London.

Robinson, J. B., Jr. & Srere, P. A. (1985). Organization of Krebs tricarboxylic-acid cycle enzymes in mitochondria. J. Biol. Chem. 260, 10800–10805.

Robinson, J. B., Jr., Inman, L., Sumegi, B., & Srere, P. A. (1987). Further characterization of the Krebs tricarboxylic acid cycle metabolon. J. Biol. Chem. 262, 1786–1790.

Ryazanov, A. G. (1988). Organization of soluble enzymes in the cell: relay at the surface. FEBS Lett. 237, 1–3.

Sauro, H. M. & Kacser, H. (1990). Enzyme–enzyme interactions and control analysis. 2. The case of non-independence: heterologous associations. Eur. J. Biochem. 187, 493–500.

Schlenk, F. (1985). Early research on fermentations—a story of missed opportunities. Trends Biochem. Sci. 10, 252–254.

Seitz, P. K., Chang, D. C., Hazlewood, C. F., Rorschach, H. E., & Clegg, J. S. (1981). The self-diffusion of water in *Artemia* cysts. Arch. Biochem. Biophys. 210, 517–524.

Siegbahn, N., Mosbach, K. & Welch, G. R. (1985). In: Organized Multienzyme Systems (Welch, G. R., ed.), pp. 271–301. Academic Press, New York.

Sitte, P. (1980). In: Cell Compartmentation and Metabolic Channeling (Nover, L., Lynen, F., & Mothes, K., eds.), pp. 17–47. Elsevier/North Holland, Amsterdam.

Smolen, P. & Keizer, J. (1990). Kinetics and thermodynamics of metabolite transfer between enzymes. Biophys. Chem. 38, 241–263.

Srere, P. A. (1985). The metabolon. Trends Biochem. Sci. 10, 109–110.

Srere, P. A. (1985). In: Organized Multienzyme Systems (Welch, G. R., ed.), pp. 1–61. Academic Press, New York.

Srere, P. A. (1987). Complexes of sequential metabolic enzymes. Ann. Rev. Biochem. 56, 21–56.

Srere, P. A. (1992). The molecular physiology of citrate. Curr. Top. Cell. Reg. 33, 261–275.

Srere, P. A. & Mosbach, K. (1974). Metabolic compartmentation: symbiotic, organellar, multienzymic and microenvironmental. Annu. Rev. Microbiol. 28, 61–83.

Srivastava, D. K. (1991). Physiological constraints on evolution of enzymes for cellular metabolic pathways. J. Theoret. Biol. 152, 93–100.

Srivastava, D. K. & Bernhard, S. A. (1985). Mechanism of transfer of reduced nicotinamide adenine dinucleotide among dehydrogenases. Biochemistry 24, 623–628.

Srivastava, D. K. & Bernhard, S. A. (1986). Enzyme–enzyme interactions and the regulation of metabolic reaction pathways. Curr. Top. Cell. Reg. 28, 1–68.

Srivastava, D. K., Bernhard, S. A., Langridge, R., & McClarin, J.A. (1985). Molecular basis for the transfer of nicotinamide adenine dinucleotide among dehydrogenases. Biochemistry 24, 629–635.

Srivastava, D. K., Smolen, P., Betts, G. F., Fukushima, T., Spivey, H. O., & Bernhard, S. A. (1989). Direct transfer of NADH between α-glycerol phosphate dehydrogenase and lactate dehydrogenase: fact or misinterpretation? Proc. Natl. Acad. Sci. USA 86, 6464–6468.

Steenkamp, D. J. (1987). Preferential cross-linking of the small subunit of the electron-transfer flavoprotein to general acyl-CoA dehydrogenase. Biochem. J. 243, 519–524.

Sumegi, B., Sherry, A. D., & Malloy, C. R. (1990). Channeling of TCA cycle intermediates in cultured *Saccharomyces cerevisiae*. Biochemistry 29, 9106–9110.

Sumegi, B., Porpaczy, Z., McCammon, M. T., Sherry, A. D., Malloy, C. R., & Srere, P. A. (1992). Regulatory consequences of organization of citric acid cycle enzymes. Curr. Top. Cell. Reg. 33, 249–260.

Tompa, P., Batke, J., Ovádi, J., Welch, G. R., & Srere, P.A. (1987). Quantitation of the interaction between citrate synthase and malate-dehydrogenase. J. Biol. Chem. 262, 6089–6093.

Trantham, E. C., Rorschach, H. E., Clegg, J. S., Hazlewood, C. F., Nicklow, R. M., & Wakabayashi, N. (1984). The diffusive properties of water in *Artemia* cysts as determined from quasi-elastic neutron scattering spectra. Biophys. J. 45, 927–938.

Uyeda, K. (1992). Interactions of glycolytic enzymes with cellular membranes. Curr. Top. Cell. Reg. 33, 31–46.

Vértessy, B. & Ovádi, J. (1987). A simple approach to detect active-site-directed enzyme–enzyme interactions. The aldolase/glycerol-phosphate-dehydrogenase enzyme system. Eur. J. Biochem. 164, 655–659.

Weber, J. & Bernhard, S. A. (1982). Transfer of 1,3-diphosphoglycerate between glyceraldehyde-3-phosphate dehydrogenase and 3-phosphoglycerate kinase via an enzyme-substrate–enzyme complex. Biochemistry 21, 4189–4194.

Welch, G. R. (1977). On the role of organized multienzyme systems in cellular metabolism: a general synthesis. Prog. Biophys. Molec. Biol. 32, 103–191.

Welch, G. R. & Keleti, T. (1990). In: Structural and Organizational Aspects of Metabolic Regulation, UCLA Symposia on Molecular and Cellular Biology, New Series, Vol. 134 (P. Srere, M.E. Jones, & C. Mathews, eds.), pp. 321–330. Alan R. Liss, New York.

Welch, G. R. & Marmillot, P. R. (1991). Metabolic "channeling" and cellular physiology. J. Theoret. Biol. 152, 29–33.

Welch, G. R. & Easterby, J. S. (1994). Metabolic channeling versus free diffusion—transition-time analysis. Trends Biochem. Sci. 19, 193–197.

Westerhoff, H. V. & Kell, D. B. (1988). A control theoretical analysis of inhibitor titrations of metabolic channeling. Comments Mol. Cell. Biophys. 5, 57–107.

Westerhoff, H.V. & Welch, G.R. (1992). Enzyme organization and direction of metabolic flow: physicochemical considerations. Curr. Top. Cellul. Reg. 33, 361–390.

Wilson, J. E. (1978). Ambiquitous enzymes: variation in intracellular distribution as a regulatory mechanism. Trends Biochem. Sci. 3, 124–125.

Wistow, G. J., Mulders, J. W. M., & de Jong, W. W. (1987). The enzyme lactate dehydrogenase as a structural protein in avian and crocodilian lenses. Nature (London) 326, 622–624.

Wu, X., Gutfreund, H., Lakatos, S., & Chock, P. B. (1991). Substrate channeling in glycolysis: a phantom phenomenon. Proc. Natl. Acad. Sci. USA 88, 497–501.

Zalokar, M. (1960). Cytochemistry of centrifuged hyphae of *Neurospora*. Exp. Cell Res. 19, 14–22.

METABOLIC CONTROL ANALYSIS IN THEORY AND PRACTICE

Athel Cornish-Bowden

Advances in Molecular and Cell Biology
Volume 11, pages 21–64.

ISBN: 1-55938-844-7

ABSTRACT

Metabolic control analysis is the most widely used modern theoretical framework for analyzing the kinetic behavior of metabolic pathways and other multienzyme systems. It considers the responses of individual enzymes to changes in concentrations of their substrates and other metabolites in terms of elasticities, which correspond approximately to the traditional idea of order of reaction. These are local properties, because they treat each enzyme in isolation from its pathway; there are also systemic properties, which relate the behavior of a system variable such as metabolic flux or the concentration of an intermediate to changes in the external parameters, such as enzyme activities or concentrations of pool or sink metabolites. The most important of these systemic properties are the control coefficients, which express the contributions of the individual enzymes to the total effects on fluxes or metabolite concentrations. They obey summation relationships, implying for example, that control of flux is shared among the enzymes composing the system. Metabolic control analysis allows the classical ideas of metabolic regulation to be placed on a more secure theoretical base, because it allows one to dispense with any assumption that metabolic systems have to be studied in relation to certain regulatory enzymes whose properties define the behavior of the whole system. Instead of assuming the existence of "key enzymes," metabolic control analysis allows the contribution of each enzyme to be quantified.

I. INTRODUCTION

Enzymology has existed as a recognizable discipline for most of this century, and a great deal has been learned about individual enzymes and the kinetics of the reactions they catalyze. In nearly all cases the metabolic contexts of these reactions are also known: we know where each substrate has come from and where the product will go. Nonetheless, for most of the history of enzymology there has been little to connect the sorts of kinetic measurements people make with the physiological roles of the enzymes they study. An enzyme is typically first identified from some physiological observation, but the first thing an enzymologist does is purify it, or at least separate it from its physiological neighbors. Nearly all kinetic studies of enzymes are thus made on enzymes that have been deliberately taken out of physiological context. This is necessary, of course, if one is interested in understanding enzymes at the level of the chemistry of catalysis, but one can hardly understand how enzymes fulfill their roles in metabolic pathways if one only examines them under conditions where all other aspects of the pathway are suppressed.

Oddly enough, the discovery in the 1950s and 1960s of feedback inhibition and the associated properties of cooperative and allosteric interactions reinforced rather than removed this separation between the practice of enzymology and the physiology of enzymes. Certain enzymes, such as phosphofructokinase, display complex kinetics, having interactions with numerous effectors, and many of these interactions can be rationalized in terms of fairly simple ideas, such as regulation of the first committed step of a pathway by the end-product of the pathway. Once this was realized, it seemed natural to think that a few enzymes could be classified as "regulatory enzymes" and the rest could be largely ignored in discussions of physiological regulation. The most extreme form of this idea is to think that all that one needs to do to understand the regulation of a pathway is identify the regulatory step, usually assumed to be unique, and study all the interactions of the enzyme catalyzing it.

The classical approach to metabolic regulation has led to so many important discoveries that it cannot be wholly misconceived; it is difficult to imagine how so many examples of allosteric and cooperative enzymes would have been discovered in a few years at the end of the 1960s if the concept of feedback regulation of the first committed step by the end-product of a pathway had had no merit. Nonetheless, it includes too many undefined terms and unexplored questions to be wholly acceptable either. All biochemists talk about regulation and control, but there is almost no consensus about what these words mean: for some people they are interchangeable; for others they are quite different. All biochemists talk about pathways, but what is a pathway? Where does one pathway begin and another end? What is an end-product? If it is something lost to the environment, like water or carbon dioxide, then most of the metabolites referred to as end-products in the classical literature on regulation are not end-products at all but rather pathway products that serve as substrates for other pathways. If we really mean water and carbon dioxide when we refer to end-products, where are the regulatory enzymes that they inhibit?

Even such an apparently unambiguous term as "branch-point" is less clear than one might think from reading the classical literature. If one studies a chart of metabolic pathways, one does not find that it consists mainly of long straight pathways with occasional connections between them. On the contrary, one finds that it contains a great many branch-points, especially in the parts of metabolism that are regarded as its "core" e.g., glycolysis, the tricarboxylate cycle, oxidative phosphorylation, etc. Srere (1987) has estimated that ~80 percent of metabolites "have just one use in the cell," in the sense that they are substrates or products of no more than two enzymes each. However, even 20 percent of branch-point metabolites implies, at a rough estimate, ~40 percent of enzymes catalyzing "first committed steps," very far from a trivial proportion, and even then any implication that only a minority of enzymes are candidates for "regulatory enzymes" would be seriously misleading, because Srere's estimate is based on a representation of metabolism in Alberts et al. (1983) that largely ignores the fact that most enzymes have two or more substrates in each direction, the "extra" substrates being in most cases metabolite pairs like ATP/ADP, NAD^+/NADH, etc. that generate additional connections between pathways all over metabolism. Although it may be helpful to the human observer, at least in some contexts, to regard glucose as "the" substrate of hexokinase and ATP as just a coenzyme, such a classification is just a mental construct that has no reality in the cell: neither of the two necessary components in the reaction is any more "the" substrate than the other.

Even if all these problems of definition and terminology could be resolved to general satisfaction, there would remain an important unanswered question in the classical approach to regulation: how do we know that an effect on the activity of any enzyme will be translated into an effect on the flux of metabolites through a pathway? Analysis of the steady-state kinetic behavior of an individual isolated enzyme is now essentially a solved problem, trivial in theory and almost trivial in practice. This should not lead us to suppose that the same is true of multi-enzyme systems. Even two-enzyme systems generate serious problems, and these become rapidly more complex as more enzymes are added. There is nothing in the standard textbook treatment of enzyme kinetics that would justify an assumption about how even complete knowledge of the rate equation for a particular "regulatory enzyme" would allow any quantitative prediction of the effect that a change in its activity would have on the flux through the pathway in which it is embedded.

II. MODERN THEORIES OF METABOLIC CONTROL

Analysis of metabolic regulation at the systemic level, i.e. treating systems of enzymes as systems of enzymes and not as collections of individual enzymes, is not yet widely taught in biochemistry courses, and is hardly mentioned in biochemistry textbooks. Nonetheless, it had its origins 30 years ago in the efforts of Higgins (1965) to apply sensitivity analysis to metabolic systems, has been developing for more than 20 years, and has begun to attract wide attention in the past ten years. At present three main theoretical approaches are competing for attention: metabolic control analysis, biochemical systems theory, and flux-orientated theory.

Metabolic control analysis originated independently in work of Kacser and Burns (1973) and Heinrich and Rapoport (1974), but it is now recognized that there are no important contradictions between these two schools and there is also a general agreement on the principal terminology and symbols (Burns et al., 1985).

Biochemical systems theory originated in three papers by Savageau (1969a,b; 1970), and became more widely known as a result of his book (Savageau, 1976). It constitutes a major contribution to the subject, but in my opinion (not shared by Savageau and his school), it is an "engineering" approach rather than a "scientific" one (Cornish-Bowden, 1989): this is not meant to disparage it, but only to emphasize that it is more concerned with predicting how a system will respond to a particular stress than with understanding why it does so. Central to biochemical systems theory are factors with dimensions that are neither integral nor fixed: these present no particular difficulty if one is only concerned with prediction, but few biochemists concerned to understand regulation can feel entirely comfortable with quantities that have values like 0.73 $mol^{0.19}$ $L^{-0.19}$ s^{-1}.

The flux-orientated theory of Crabtree and Newsholme (1987a,b) is the only one of the three to have its roots in the classical tradition, in that it is the only one that derives from a group that has been and remains very active in the experimental study of metabolic regulation. Unfortunately, Crabtree and Newsholme presented it in terms and symbols that tended to obscure rather than illuminate its relationship with metabolic control analysis, and they introduced novel mathematical symbols even though conventional ones were available (for example writing dln X as $\overset{r}{X}$). Perhaps most serious for workers familiar with metabolic control analysis, they reject the classification of effectors into external effectors (whose concentrations are defined independently of the system under study) and internal effectors (with concentrations set by the system itself), introducing a new and confusing category of "partially external" effector.

There has been almost no attempt to compare these approaches objectively, and much of the considerable argument about which of them, if any, is most appropriate for studying real systems has been rhetorical. However, Groen and Westerhoff (1990) did make a detailed comparison, applying each of them (as well as a variant of metabolic control analysis given by Heinrich et al., 1977) to experimental results for gluconeogenesis. They found that all methods gave identical values for the same quantities, and that there was considerable, but not complete, overlap between the particular sets of numerical values given. They concluded (Groen and Westerhoff, 1990): "We are satisfied that the four calculation methods are simply somewhat different ways of achieving the same end. Suggestions that there are inherent conflicts between the methods appear to be void."

However, Groen and Westerhoff may have gone too far here in their efforts to be fair to flux-orientated theory. In effect they have studied not the whole theoretical edifice, but the particular parts of it that allow meaningful calculations. They mention, but do not discuss, the confusing idea of partially external regulators (see below), and they ignore assumptions in the theory that appear contrary to everyday experience. For example, flux-orientated theory insists that the first enzyme in the pathway under consideration must approach saturation by its substrate, the step catalyzed by such an enzyme being a "flux-generating step." In likening this idea to the battery in an electric circuit, Crabtree and Newsholme (1987b) appear to confuse the constant potential provided by such a battery with the constant flux generated by a saturated enzyme. The need for a constant potential is satisfied in metabolic control analysis by the requirement that at least two external metabolites exist. The mechanism by which the constant concentration is achieved is not ignored in metabolic control analysis, but it is considered to be external to the system under discussion.

If we accept Groen and Westerhoff's subset of flux-orientated theory as a valid theory, it follows from their study that the choice between the various theoretical approaches to metabolic control is not a matter of truth and error, but of convenience and preference. Fell (1992) noted that in the previous years about two-thirds of literature citations to theories of metabolic regulation were to metabolic control analysis. At least at present, therefore, it would appear that the preference of most biochemists is for metabolic control analysis, and in the remainder of this chapter I shall largely confine discussion to it.

III. ESSENTIALS OF METABOLIC CONTROL ANALYSIS

A. Scope of Metabolic Control Analysis

In its simplest form, metabolic control analysis is concerned with the steady states of systems of enzymes that connect a series of metabolites, with two or more reservoirs of metabolites whose concentrations are fixed independently of the enzymes in the system, and can thus be considered as "external" to it. The reservoirs include at least one source, from which metabolites flow, and at least one sink, into

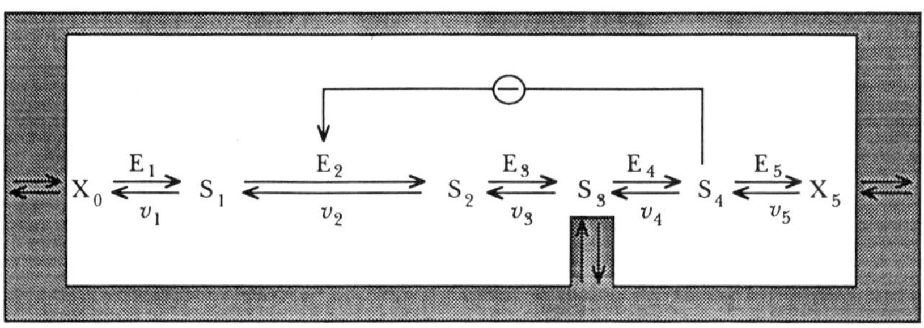

Figure 1. Definition of a pathway as a system. The entire figure represents metabolism in an entire organism. However, because such a whole system is too complicated to analyze, one must normally restrict attention to a subsystem, such as the pathway shown from X_0 to X_5, the remainder of the real system (the shaded region) being considered as outside the system under study. All connections of the pathway metabolites to the exterior (represented in this example by the shaded arrows to and from X_0, S_3 and X_5) are considered to be blocked. The metabolites at the beginning and end of the pathway (X_0 and X_5) are "external," which means that their concentrations are taken as fixed independently of the properties of the enzymes in the pathway. The arrow from S_4 to E_2 indicates that this enzyme is assumed to be subject to feedback inhibition by S_4. It is usually convenient to use different symbols for external and internal metabolites, which are distinguished here as X and S respectively.

which they flow. Neither of these flows need be irreversible, and the classification into sources and sinks is not absolute: both might well be considered as internal metabolites of a larger system under different circumstances. In the example shown in Figure 1, the system is considered to be the pathway from a source X_0 to a sink X_5: the heavily shaded part of the scheme, including the external connections to these metabolites and to S_3, is considered to be outside the system.

The other metabolites, S_1, S_2, S_3, and S_4, have concentrations that are determined by the steady-state properties of the enzymes, and are thus "internal." In addition to the metabolites connected by the enzymes, there can be any number of external effectors with fixed concentrations. In a living organism, of course, very few reactants are external, but there are so many reactions to be considered that the entire system is difficult to comprehend. To make metabolism manageable for analysis, therefore, one must define the system as just a part of the whole organism, and one must define the metabolites at the interfaces with the rest of the organism as external.

The real effectors studied in classical accounts of feedback inhibition are not external, of course, because the classical interpretation of such inhibition is that it serves to regulate the biosynthetic flux to the effector. If its concentration were fixed externally it would hardly need regulation by the system. It must therefore be treated as internal if one is to have any possibility of analyzing the effect of the feedback. It was probably to avoid this sort of unappealing necessity that Crabtree and Newsholme introduced the category of "partially external regulators," as an example of which they give ATP in glycolysis, whose concentration is, they say (Crabtree and Newsholme, 1987b) "*partially* determined by glycolysis" [italics in the original].

Unfortunately, this is probably worse than the disease it tries to cure, because there is no mathematical meaning one can attach to a quantity that is partially variable. This may readily be confirmed by trying to set up a computer model in which one of the concentrations is simultaneously variable and constant. The solution of the dilemma is to recognize that the limits of the system may have to be defined differently depending on the particular questions one wants to ask. If one is interested in the regulation of glycolysis by ATP there is no alternative but to treat ATP as internal and define the system so that there is a flux through ATP. If one is concerned with some other aspect of glycolysis one may be willing to treat the ATP concentration as fixed, and hence external; but there is no halfway house.

At present metabolic control analysis is largely concerned with steady states, though a start has been made in extending it to transient phenomena (Kohn and Chiang, 1982; Acerenza et al., 1989; Meléndez-Hevia et al., 1990; Heinrich and Reder, 1991). Moreover, it assumes that a steady state exists and that it is stable. As it is relatively easy to construct models, even for systems of only two enzymes, that have no steady state (e.g., by coupling two enzymes in series such that the first enzyme has an irreversible rate that exceeds the limiting rate of the second), one should not take the existence of a steady state as automatic. However, this rarely creates difficulties in experimental practice nor in computer models unless one over-simplifies to the point of treating as irreversible reactions that can behave properly only if their reversibility is allowed for. Stability of a steady state means that the system responds to a small perturbation by returning to the same state, not by moving to a quite different onc. This is less of a problem in practice than one might predict, because as Fell (1992) remarks, the fact that biochemists normally succeed in working under conditions that yield reproducible data itself implies that they normally succeed in finding conditions where steady states are stable.

The simplest versions of metabolic control analysis include some additional restrictions, but these are not absolute as they can be relaxed if one is willing to accept some additional complexity in the analysis. All reactions should be first-order in the enzymes that catalyze them, and zero-order in all other enzymes: in other words, each rate should be proportional to the concentration of exactly one enzyme; and no enzyme should act on more than one reaction in the system. Thus enzyme–enzyme interactions require special treatment, as do pathways such as fatty acid oxidation where the same enzymes are involved with numerous different reactions. Compartmentation might also seem to pose a problem, but in fact metabolic control analysis can readily accommodate multiple compartments as long as there are well defined connections between them and the components are distributed homogeneously within each compartment.

Two serious misconceptions about the scope of metabolic control analysis need to be cleared up, though both have proved so hardy that the task may be hopeless. The first is that metabolic control analysis is concerned only with effects brought about by changes in enzyme concentration. If this were true its scope would be restricted indeed, but fortunately it is not true. Although in their original paper Kacser and Burns (1973) defined the sensitivity coefficient (now called the control coefficient) in relation to changes in enzyme concentration, they made it clear that this was done only as a convenience: they recognized that whatever change in enzyme activity might occur, brought about by whatever mechanism, this could always be treated as equivalent to a change in enzyme concentration. For example, referring to the effect of an inhibitor, they wrote (Kacser and Burns, 1973):

> For this no reference to the immediate cause of the change is called for. The change in rate, caused by the change in *inhibitor*, can always be thought of as equivalent to some change in the *concentration of the enzyme*. If, for example, some inhibitor change causes a 1 percent reduction in rate, this is equivalent to a 1 percent reduction in enzyme concentration.

Unfortunately not everyone has had the same capacity as Kacser and Burns to recognize equivalence of effects, and the idea that metabolic control analysis is concerned exclusively with changes in enzyme concentration remains widespread. The diluted version of this misconception, that it is concerned with effects on the limiting rates (or "maximum velocities") of enzymes, is equally mistaken, and the remark that "control by modulation of V_{max} is rare," made as a criticism of metabolic control analysis (Atkinson, 1990), may be true, but it is irrelevant to metabolic control analysis.

The second misconception about metabolic control analysis is that it is restricted to unbranched pathways in which each enzyme rate is proportional to the concentration of its substrate. Both components of this idea are false: branched pathways are not excluded; there are no assumptions about the kinetic expressions or about the magnitudes of substrate concentrations in relation to Michaelis (or other) constants. Unbranched pathways with each enzyme obeying first-order kinetics have the important advantage that exact kinetic expressions for steady-state fluxes and concentrations can be derived. When Kacser and Burns (1973) included an appendix in which such an exact analysis was presented, they were simply following the common pedagogical practice of illustrating a general theory with an example simple enough to be understood in complete detail. But just as no one (I hope) would conclude from textbook discussions of the hydrogen atom that wave mechanics has nothing to say about the rest of chemistry, no one should confuse an exact example of metabolic control analysis with the whole subject.

B. Elasticity Coefficients or Kinetic Orders

Enzyme kinetic behavior has conventionally been expressed in terms of rate equations such as the Michaelis–Menten equation:

$$v = \frac{\dfrac{k_{0\mathrm{A}}}{K_{\mathrm{mA}}} e_0 a - \dfrac{k_{0\mathrm{P}}}{K_{\mathrm{mP}}} e_0 p}{1 + \dfrac{a}{K_{\mathrm{mA}}} + \dfrac{p}{K_{\mathrm{mP}}} + \dfrac{i}{K_{\mathrm{i}}}} \tag{1}$$

It is shown here in its less familiar reversible form, for interconversion of metabolites A and P, at concentrations a and p, in the presence of an inhibitor I that is not a reactant and has concentration i, and a total enzyme concentration e_0. It contains five kinetic constants in this equation: the forward and reverse catalytic constants, $k_{0\mathrm{A}}$ and $k_{0\mathrm{P}}$, which are second-order rate constants; and the forward and reverse Michaelis constants, K_{mA} and K_{mP}, and inhibition constant, K_{i}, all of which have the dimensions of concentrations.

The reversible form is normally always preferable in considering physiological states, because products are normally always present. (In metabolic simulations one should be extremely cautious before writing irreversible equations, because doing so can generate entirely false results about the behavior of a pathway, and can suggest that no steady state is possible in conditions where more realistic, albeit more complex, equations yield a perfectly stable steady state). This contrasts with the usual case in the cuvette, where it is easy to create conditions of irreversibility.

Investigation of enzyme mechanisms by studying steady-state kinetics is mainly a matter of determining how the apparent values of the Michaelis–Menten parameters vary with the concentrations of inhibitors, activators, additional substrates, etc., or with the pH or other environmental conditions. For this purpose it is clearly necessary to express the kinetic behavior in terms of an equation that resembles Eq. (1). However, in metabolic control analysis one is not much interested in enzyme mechanisms, not because they are not important, but because one is concerned with studying a different aspect of the system. The sort of question that one asks is not "what mechanism accounts for the variation of v with a?" but "how much will v change if there is a small change in a?". Equation (1) can certainly answer this question, but it does so in a way that is inconveniently indirect. Thus, partial differentiation of Eq. (1) with respect to a gives the following expression:

$$\frac{\partial v}{\partial a}=\frac{\dfrac{k_{0A}e_0}{K_{mA}}\left[1+\dfrac{p}{K_{mP}}\left(1+\dfrac{k_{0P}}{k_{0A}}\right)+\dfrac{i}{K_i}\right]}{\left(1+\dfrac{a}{K_{mA}}+\dfrac{p}{K_{mP}}+\dfrac{i}{K_i}\right)^2} \tag{2}$$

As it stands this derivative has the dimensions of reciprocal time, and as one is usually more interested in relative derivatives than absolute ones, it is usual to convert it to a relative form by multiplying by a/v:

$$\frac{\partial \ln v}{\partial \ln a}=\frac{a\partial v}{v\partial a}$$

$$=\frac{1+\dfrac{p}{K_{mP}}\left(1+\dfrac{k_{0P}}{k_{0A}}\right)+\dfrac{i}{K_i}}{\left(1-\dfrac{k_{0P}K_{mA}p}{k_{0A}K_{mP}a}\right)\left(1+\dfrac{a}{K_{mA}}+\dfrac{p}{K_{mp}}+\dfrac{i}{K_i}\right)}$$

$$=\frac{1}{1-\dfrac{k_{0P}K_{mA}p}{k_{0A}K_{mP}a}}-\frac{\dfrac{a}{K_{mA}}}{1+\dfrac{a}{K_{mA}}+\dfrac{p}{K_{mP}}+\dfrac{i}{K_i}}$$

$$=\frac{1}{1-\Gamma/K_{eqm}}-\frac{\alpha}{1+\alpha+\pi+\iota} \tag{3}$$

in which $\Gamma=p/a$, is the "mass action ratio," $K_{eqm}=k_{0A}K_{mP}/k_{0P}K_{mA}$ is the equilibrium constant by virtue of the Haldane relationship (see, e.g., Cornish-Bowden, 1979), and $\alpha=a/K_{mA}$, $\pi=p/K_{mP}$ and $\iota=i/K_i$ are the concentrations scaled by the appropriate Michaelis or inhibition constants. This equation may appear complicated, but when it is rearranged as the difference between two fractions, as in the last forms shown, it is seen that both of the these fractions have simple interpretations: the first measures the "disequilibrium," i.e. the departure of the system from equilibrium; the second measures the degree of saturation of the enzyme with the reactant considered. However, complicated or not, in metabolic control analysis one is not usually concerned with the algebraic form of this derivative, but with its numerical value: if it is zero, the rate does not vary with a; if it is positive, the rate increases as a increases; if it is negative, the rate decreases as a increases. Accordingly it is given a name, the *elasticity coefficient*, and symbol ε, to express its central importance in metabolic control analysis:

$$\varepsilon_a^v \equiv \frac{\partial \ln v}{\partial \ln a} \equiv \frac{a \partial v}{v \partial a} \qquad (4)$$

Although the superscript v in the symbol may seem sufficiently obvious to be superfluous, one should remember that in metabolic control analysis one is always concerned with systems of more than one enzyme; accordingly one needs a superscript to specify *which* rate is being considered.

Kacser and Burns (1973) did not explain why they chose the name elasticity for this property of an enzyme, and it has unfortunately remained an obscure way of referring to an idea that is both simple and, under the name "order of reaction," familiar. Wright and Albe (1990) for example, regret the use of "such passive and mysterious words as 'elasticity'", and one can hardly doubt that their feelings are shared by many who read the literature on metabolic control analysis for the first time, as it is difficult to discern much relationship with the everyday use of the term "elastic."

By contrast, the idea of order of reaction is very familiar to anyone who has studied elementary chemical kinetics. A reaction is said to be of first-order with respect to a if its rate is proportional to a, of second-order if it is proportional to a^2, etc. It is a matter of simple calculus to show that these values of 1 and 2 for the order of reaction correspond exactly to the values of what has been called the elasticity coefficient. The only difference is that whereas the IUBMB recommendations (International Union of Biochemistry, 1982) discourage the use of the term "order of reaction" for cases where its value is not an integer, suggesting "apparent order" instead, there is no suggestion in Eq. (4) that the quantity defined should be an integer. However, this recommendation, which was intended to avoid conflict with the IUPAC recommendations that were then newly revised (International Union of Pure and Applied Chemistry, 1981), has not been widely followed and few biochemists see any objection to nonintegral orders: students are commonly taught, for example, that an enzyme obeying irreversible Michaelis–Menten kinetics displays an order of 1 with respect to substrate concentration when the concentration is very small, and decreases as the substrate concentration increases, with a value of 0.5 at half-saturation and approaching zero at saturation. It is easy to confirm, by putting $p = 0$ in Eq. (3), that the elasticity coefficient behaves in exactly the same way; it is, in fact, identical to the quantity normally understood by biochemists as the order of reaction.

In biochemical systems theory the local properties of enzymes are expressed in terms of *kinetic orders* (Savageau, 1976), which are sometimes considered to be identical to what metabolic control analysis calls elasticity coefficients. In reality, this equivalence is less clear than it may appear at first sight, because kinetic orders are normally treated in biochemical systems theory as being constant over the concentration range considered, whereas metabolic control analysis never assumes that elasticities are constant (see Cornish-Bowden, 1989). Moreover, the current tendency in biochemical systems theory to "aggregate" enzymes implies that individual enzymes are not necessarily considered at all. Nonetheless, the case could be made that the term kinetic order could also be adopted in metabolic control analysis in the interests of greater clarity. Until this happens, however, ordinary readers can find the idea of an elasticity coefficient less obscure by remembering that apart from lacking any implication of constancy it is the same as what they normally understand by order of reaction.

Returning to Eq. (1), we can differentiate it with respect to each concentration in turn to obtain the following complete set of elasticity coefficients, which will now be written as differences between disequilibrium and saturation terms [as in the last form of Eq. (3)], as they are easiest to understand in this form:

$$\varepsilon_a^v = \frac{1}{1 - \Gamma/K_{\mathrm{eqm}}} - \frac{\alpha}{1 + \alpha + \pi + \iota} \tag{5}$$

$$\varepsilon_p^v = \frac{-\Gamma/K_{\mathrm{eqm}}}{1 - \Gamma/K_{\mathrm{eqm}}} - \frac{\pi}{1 + \alpha + \pi + \iota} \tag{6}$$

$$\varepsilon_{e_0}^v = 1 \tag{7}$$

$$\varepsilon_i^v = -\frac{\iota}{1 + \alpha + \pi + \iota} \tag{8}$$

The symbols on the right-hand sides of these equations have the same meanings as in Eq. (3). Although the expressions are derived from a specific model, the reversible Michaelis–Menten equation, and their exact forms are dependent on this model, they illustrate a number of points that apply fairly generally, in some cases universally: (1) Reactant elasticities are normally positive when the direction of disequilibrium is such that the reactant is a substrate, negative when it is a product. ("normally" here means that substrate inhibition and product activation lead to exceptions to this generalization). Note, however, that the passage from positive to negative as the reaction passes from one side of equilibrium to the other is not via zero, as one might naively guess, but via infinity: *reactant elasticities are infinite at equilibrium!* (For a curve illustrating this behavior, see Figure 2 of Westerhoff et al., 1984). This characteristic underlines the danger of writing irreversible rate equations in computer simulation. With irreversible reactions substrate elasticities are normally in the range 0 to 1 (ignoring such complications as cooperativity and substrate inhibition): values close to zero are characteristic of high substrate concentrations, and infinite elasticities are impossible. It follows that the numerical values of elasticities for such reactions are entirely different from those likely to be found in a living cell. (2) Enzymes have unit elasticities for their own reactions [and zero elasticities for other reactions, though this is not illustrated by Eq. (7)]. These generalizations are not universally true, as they depend on the assumption that each rate is proportional to the total concentration of one enzyme only. They fail if an enzyme associates (with itself or with other enzymes in the system) to produce species with altered kinetic properties. Much of metabolic control analysis assumes the truth of these generalizations, and the equations become considerably more complicated when they fail. (3) Elasticities for nonreactant inhibitors are always negative. Conversely, elasticities for nonreactant activators are always positive. The qualification "nonreactant" can be ignored as long as one remembers that a product inhibitor is transformed into a substrate when the direction of flux changes. In addition, elasticities for nonreactant (the qualification is now indispensable) inhibitors and activators are independent of the degree of disequilibrium.

C. Enzyme Kinetics Viewed From Control Analysis

From the point of view of metabolic control analysis, measuring elasticities is what enzymologists have been doing since the time of Michaelis and Menten, however unfamiliar the term itself may appear. Nonetheless, there are important differences in emphasis, with the result that the measurements made in traditional experiments may not be very useful for metabolic control analysis. In ordinary studies of enzymes, experiments are usually designed to reveal information about the mechanism of action. (Even if the experimenters' interests are primarily physiological, they usually follow procedures that have their origins in mechanistic studies). Because different mechanisms of action often predict patterns of behavior that differ only very slightly, if at all, one is often forced to design experiments carefully to illuminate any small deviations from expected behavior that may exist, and the experiments themselves must be done with great attention to accuracy. As any textbook of enzyme kinetics will illustrate, kinetic analysis frequently involves extrapolation of observations to infinite or zero concentrations. Moreover, experiments are rarely done with anything approaching a complete system, i.e. it is rare for an enzyme in a cuvette to encounter even half of the metabolites that might influence its activity in the cell; if any additional enzymes are present they are either trace contaminants that are believed to have no effect on the enzyme of interest, or they are coupling enzymes deliberately added in quantities designed to be optimal for the assay, without any relation to the concentrations that may exist in the cell.

All of these characteristics are quite inappropriate for metabolic control analysis. Although one is still interested in describing the kinetic behavior of an enzyme, the objective is not to understand the mechanism but to integrate the kinetic description into a description of the kinetic behavior of a system—at the simplest level a system of a few enzymes constituting a pathway, but ultimately a complete organ or organism. To a good approximation, properties that are at the limits of accuracy of one's equipment, and consequently are difficult to measure, are not important in the behavior of the system: if mechanistic differences don't produce major differences in kinetic behavior they don't matter.

On the other hand, one can no longer afford to simplify the experiment by omitting metabolites that affect the kinetics: all reactants and effectors should be present at concentrations as close as possible to those that occur in the cell. This includes products, of course, and implies that reactions need to be studied under reversible conditions. Even if the equilibrium constant strongly favors reaction in one direction, the conditions should be at least in principle reversible; apart from anything else product inhibition may be significant even if the complete reverse reaction is not.

Despite this emphasis on a complete realistic reaction mixture, elasticity measurements remain unnatural in one respect: they refer to an enzyme isolated from its pathway, i.e. they treat all concentrations of metabolites that influence the enzyme's activity as constants, ignoring the effects that other enzymes in the pathway might have on these concentrations. This isolation is illustrated by the unshaded region in Figure 2. In effect, elasticity measurements imply the existence of "walls" separating E_2 and the metabolites that affect it (S_1, S_2 and S_4) from E_1, E_3, E_4 and E_5.

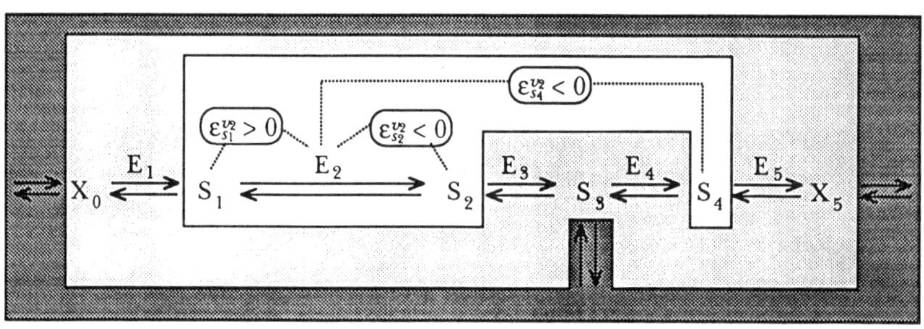

Figure 2. Isolation of an enzyme from its pathway. The pathway is the same as in Figure 1, but additional shading is used to isolate one enzyme, E_2, from the remainder of the pathway. Its kinetic behavior is expressed in terms of elasticities, which are nonzero for each metabolite that affects it, S_1, S_2 and S_4 in this example. The elasticities are defined at the concentrations of these metabolites (considered to be simultaneously present) that occur in the whole system. Typically substrates (S_1) have positive elasticities, whereas products (S_2) and feedback inhibitors (S_4) have negative elasticities.

D. Control Coefficients

To this point we have only been discussing the ordinary kinetic behavior of isolated enzymes, albeit in terminology rather different from that used in mechanistic studies. The objective of metabolic control analysis is now to determine how the kinetic behavior of a sequence of enzymes composing a pathway can be explained in terms of the properties of the individual isolated enzymes. If a system such as the one defined in Figure 1 is set up, the concentrations of the reservoirs X_0 and X_5 are constant, as are the kinetic properties of the enzymes, but the individual enzyme rates v_i and the concentrations of the internal metabolites S_j are free to vary. Even if these concentrations are initially arbitrary, they will tend to vary in such a way that each approaches a steady state. (Note that no steady state necessarily exists, and if one does exist it is not necessarily unique: for simplicity, however, we shall make the common assumption that a steady state does exist and that it is unique). If we consider S_1, for example, it is obvious that a steady state implies that the rate v_1 at which it is supplied must be equal to the rate v_2 at which it is consumed. A steady state in S_2 likewise implies $v_2 = v_3$ and so on; when all the metabolites are in steady state all the enzyme rates must be equal to one another, with a value J that is called the *flux* through the pathway. This equality of all the rates follows from the fact that Figure 1 defines an unbranched pathway. With branched pathways the relationships are more complex, but the principles are straightforward and obvious: the total flux into each branch-point metabolite is equal to the total flux out (Figure 3).

Enzyme rates are *local* properties, because they refer to enzymes isolated from the system. Steady-state fluxes and metabolite concentrations, by contrast, are *systemic* properties. Elasticities are also local properties, but there are analogous

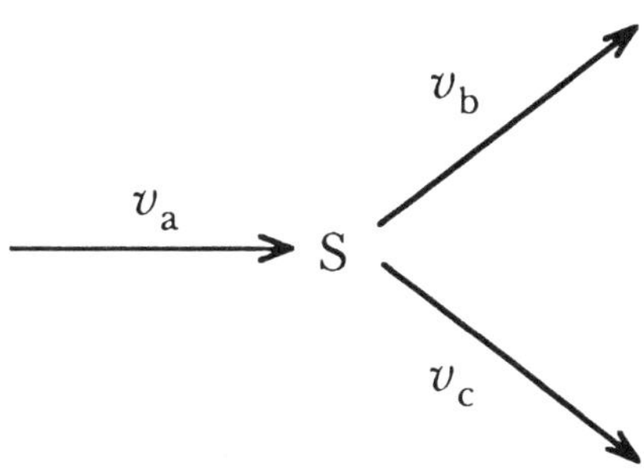

Figure 3. Steady state at a branch point. For a metabolite S at a branch point, steady state implies that the total rate at which it is supplied is equal to the total rate at which it removed, i.e. $v_a = v_b + v_c$. If the whole pathway is in a steady state these rates are identical to the corresponding fluxes, i.e. $J_a = J_b + J_c$. The apparent asymmetry in these expressions derives from the arbitrary directions of the fluxes assumed, but in reality any of the rates can be negative.

systemic properties called *control coefficients*. Suppose that some change in an external parameter p (undefined for the moment) brings about a change in a local rate v_i when the enzyme E_i is isolated, what is the corresponding effect on the system flux J when E_i is embedded in the system? This is not known *a priori*, and the ith *flux control coefficient* is defined by the following ratio of derivatives:

$$C_i^J = \frac{\partial \ln J}{\partial \ln p} \Big/ \frac{\partial \ln v_i}{\partial \ln p} = \frac{\partial \ln J}{\partial \ln v_i} \tag{9}$$

[The simpler form shown at the right is not strictly correct, because v_i is not a true independent variable in the system, but it is acceptable as long as it is remembered that there is always an implied external parameter p even if it is not shown explicitly. For fuller explanation of this point, see the discussion by Schuster and Heinrich (1992) of their Eqs. (3–5)] This definition corresponds to the way Heinrich and Rapoport (1974) defined their "control strength;" in apparent contrast, the "sensitivity coefficient" of Kacser and Burns (1973) was defined in terms of the effect of changes in enzyme concentration on flux:

$$C_i^J = \frac{\partial \ln J}{\partial \ln e_i} \tag{10}$$

These definitions might appear to be different, but provided that Eq. (7) is true, i.e. provided that each enzyme rate is proportional to the total enzyme concentration, Eqs. (9) and (10) are equivalent. Equation (9) has the advantage of avoiding the regrettably widespread misunderstanding that metabolic control analysis is limited to effects brought about by changes in enzyme concentration. As noted above, Kacser and Burns (1973) took the view that it would be simpler to write the definitions in terms of an effect that could easily be visualized, even though it is a rather abstract one in relation to the way in which metabolic control is usually exerted in short time scales (though covalent modification may alter concentrations of enzymically active proteins in times scales of seconds). The usual practice in metabolic control analysis has been to follow Kacser and Burns in using definitions similar to Eq. (10). However, there is an increasing tendency to consider that control coefficients ought not to be defined in terms of any specific parameter, and that Eq. (9) should be regarded as the fundamental definition of a control coefficient, as recently discussed by Schuster and Heinrich (1992). The quantity defined by Eq. (10) is then better regarded as an example of a *response coefficient*, which happens to be numerically equal to the corresponding control coefficient only because the connecting elasticity is assumed to be unity [see Eqs. (25–27) below].

A *concentration control coefficient* is the corresponding quantity that defines effects on metabolite concentrations, e.g., for a metabolite S_j with concentration s_j:

$$C_i^{s_j} = \frac{\partial \ln s_j}{\partial \ln p} \bigg/ \frac{\partial \ln v_i}{\partial \ln p} = \frac{\partial \ln s_j}{\partial \ln v_i} \qquad (11)$$

In this equation the simpler form at the right is subject to the same reservations as the corresponding form in Eq. (9).

E. Summation Relationships

The fundamental properties of the control coefficients are expressed in two *summation relationships*, of which the first, due to Kacser and Burns (1973) defines the sum of flux control coefficients:

$$\sum_{i=1}^{n} C_i^J = 1 \qquad (12)$$

and the second, due to Heinrich and Rapoport (1975), defines the sum of concentration control coefficients:

$$\sum_{i=1}^{n} C_i^{s_j} = 0 \qquad (13)$$

in which n is the number of enzymes in the system, and s_j is the concentration of any internal metabolite. If the pathway is branched there will be more than one flux: in this case Eq. (12) holds with J defined as any of these.

There are various ways of proving these relationships (Kacser and Burns, 1973; Reder, 1988; Giersch, 1988a,b; Cascante et al., 1989a,b), which are all ultimately equivalent, though they may appear different. Kacser and Burns (1973) used a "thought-experiment" approach, which can be understood with very little mathematical sophistication, at least as applied to any particular pathway. Readers with a thorough knowledge of linear algebra may regard the summation relationships as obvious, though arguing that results are obvious is often no more than an excuse for not demonstrating them; it is safer to say that they follow from the properties of homogeneous functions of degree 1 that were established more than 200 years ago by Leonhard Euler.

I shall not give any proof of the summation relationships here—there is now no doubt that they are correct, as any of the references cited in the preceding paragraph shows—but will simply indicate briefly the idea of the thought experiment of Kacser and Burns. It involves consideration of the effect of changing *all* enzyme concentrations by the same factor α. It is intuitively clear that this must cause all steady-state fluxes to change by the same factor α. For any metabolite S, this means that if the supply and removal fluxes were initially balanced for some concentration of S, they will still be balanced when they are both altered by the same factor;

consequently the steady-state concentration of S does not change. Although we started by assuming all enzyme concentrations to be changed simultaneously, Kacser and Burns showed that consideration of the changes as the net effect of n separate changes in the enzyme concentrations leads to Eqs. (12–13).

The essence of Eq. (12) is that control of flux through a pathway is shared by all the enzymes in the system. If all flux control coefficients are positive, the idea of sharing control is completely straightforward: no enzyme can have a control coefficient greater than one, and if any enzyme has one approaching one, those of the others must be correspondingly small. This is normally the case for unbranched pathways, though exceptions can occur if substrate inhibition or product activation dominate the behavior of some enzymes. With branched pathways the idea of sharing is less clear, because flux control coefficients are often then negative and they may also be greater than one. However, one often (not universally) finds that the following generalizations apply: any enzyme has a positive flux control coefficient for the flux through its own reaction; numerically significant negative flux control coefficients are not very common, occurring mainly for enzymes and fluxes that occur in different branches immediately after a branch point.

To the extent that these are correct, it follows that the sum of flux control coefficients for all the enzymes in a linear pathway will be approximately one even if the pathway is only a part of the entire system under study. Thus, the idea that control of flux through a pathway is shared among the enzymes catalyzing the reactions in the pathway retains enough meaning to be useful even in complicated branched systems. This interpretation is not universally accepted, however, and more work on the behavior of complicated branched pathways needs to be done to know whether it is useful. Savageau (1992), for example, argues that for any idea of sharing to be meaningful one should take the sum of *absolute* values of flux control coefficients, and points out that the value of such a sum can differ greatly from unity if there is feedback activation or feedforward inhibition. However, although he describes such phenomena as common they are actually very rare in biochemistry, and in general his argument is unconvincing.

F. Relationship between Elasticities and Control Coefficients

One of the major objectives of metabolic control analysis is to reveal how the properties of a whole system, i.e. its control coefficients, can be explained in terms of the properties of its components, i.e. in terms of the elasticities of the component enzymes. This would be a useful exercise even if both control coefficients and elasticities were equally easy to measure, but its importance is increased by the fact that the control coefficients are not usually easy to measure. Thus one is trying to assess the systemic behavior that is of fundamental interest but difficult to measure in terms of quantities that can be measured very easily by the methods developed by enzymologists over many years. It turns out that for any pathway there are exact

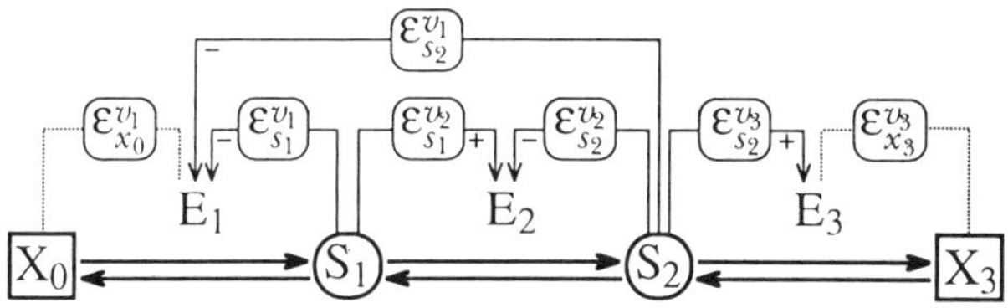

Figure 4. A three-step pathway with feedback inhibition. All nonzero elasticities are shown as connections from a metabolite to the enzyme it acts on. In all cases substrates and products are assumed to have nonzero elasticities, and in addition S_2 is assumed to have a feedback effect on E_1. The + and – signs represent the fact that substrate elasticities are usually positive, whereas product (and feedback) elasticities are usually negative; however neither of these generalizations is universal. From the point of view of enzymology, external metabolites (X_0 and X_3) have elasticities exactly like internal metabolites, and these are shown in the scheme. However, the concentrations of these metabolites must be held constant in metabolic control analysis, and consequently their elasticities do not enter into any summation relations. For this reason external elasticities are connected by dotted lines.

relationships between the control coefficients and the elasticities, and for the simple pathway shown in Figure 4 the flux control coefficients are as follows:

$$C_1^J = \frac{\varepsilon_{s_1}^{v_2}\varepsilon_{s_2}^{v_3}}{\varepsilon_{s_1}^{v_2}\varepsilon_{s_2}^{v_3} - \varepsilon_{s_1}^{v_1}\varepsilon_{s_2}^{v_3} + \varepsilon_{s_1}^{v_1}\varepsilon_{s_2}^{v_2} - \varepsilon_{s_1}^{v_2}\varepsilon_{s_2}^{v_1}} \tag{14}$$

$$C_2^J = \frac{-\varepsilon_{s_1}^{v_1}\varepsilon_{s_2}^{v_3}}{\varepsilon_{s_1}^{v_2}\varepsilon_{s_2}^{v_3} - \varepsilon_{s_1}^{v_1}\varepsilon_{s_2}^{v_3} + \varepsilon_{s_1}^{v_1}\varepsilon_{s_2}^{v_2} - \varepsilon_{s_1}^{v_2}\varepsilon_{s_2}^{v_1}} \tag{15}$$

$$C_3^J = \frac{\varepsilon_{s_1}^{v_1}\varepsilon_{s_2}^{v_2} - \varepsilon_{s_1}^{v_2}\varepsilon_{s_2}^{v_1}}{\varepsilon_{s_1}^{v_2}\varepsilon_{s_2}^{v_3} - \varepsilon_{s_1}^{v_1}\varepsilon_{s_2}^{v_3} + \varepsilon_{s_1}^{v_1}\varepsilon_{s_2}^{v_2} - \varepsilon_{s_1}^{v_2}\varepsilon_{s_2}^{v_1}} \tag{16}$$

in which the numerators are placed over the corresponding terms in the denominators in order to emphasize that not only is the denominator identical in the three expressions, but also that it consists of the sum of the three numerators, in accordance with the summation relationship, Eq. (12). Each term in the denominator consists of a product of elasticities, one for each internal metabolite in the systems; for a linear pathway there is one elasticity for each enzyme in the system except the one being modulated, i.e. the one whose control coefficient is being expressed.

Corresponding relationships exist for each of the metabolite concentrations, e.g., for s_1:

$$C_1^{s_1} = \frac{\varepsilon_{s_2}^{v_3} - \varepsilon_{s_2}^{v_2}}{\varepsilon_{s_1}^{v_2}\varepsilon_{s_2}^{v_3} - \varepsilon_{s_1}^{v_1}\varepsilon_{s_2}^{v_3} + \varepsilon_{s_1}^{v_1}\varepsilon_{s_2}^{v_2} - \varepsilon_{s_1}^{v_2}\varepsilon_{s_2}^{v_1}} \tag{17}$$

$$C_2^{s_1} = \frac{\varepsilon_{s_2}^{v_1} - \varepsilon_{s_2}^{v_3}}{\varepsilon_{s_1}^{v_2}\varepsilon_{s_2}^{v_3} - \varepsilon_{s_1}^{v_1}\varepsilon_{s_2}^{v_3} + \varepsilon_{s_1}^{v_1}\varepsilon_{s_2}^{v_2} - \varepsilon_{s_1}^{v_2}\varepsilon_{s_2}^{v_1}} \tag{18}$$

$$C_3^{s_1} = \frac{\varepsilon_{s_2}^{v_2} - \varepsilon_{s_2}^{v_1}}{\varepsilon_{s_1}^{v_2}\varepsilon_{s_2}^{v_3} - \varepsilon_{s_1}^{v_1}\varepsilon_{s_2}^{v_3} + \varepsilon_{s_1}^{v_1}\varepsilon_{s_2}^{v_2} - \varepsilon_{s_1}^{v_2}\varepsilon_{s_2}^{v_1}} \tag{19}$$

These expressions have the same denominator as those for the flux-control coefficients, Eqs. (14–16), but now each numerator term contains one fewer elasticities than the denominator terms, because the concentration s_1 that occurs as a superscript on the left-hand side of the equation does not occur in the numerator elasticities. As before, the modulated enzyme is missing from all products, and one additional enzyme is also missing from each product. Each numerator term occurs twice in the three expressions, with opposite signs; for example, the term $\varepsilon_{s_2}^{v_3}$ in Eq. (17) is matched by the term $-\varepsilon_{s_2}^{v_3}$ in Eq. (18). This matching of numerator terms ensures that the summation relationship, Eq. (13), is obeyed.

These relations between control coefficients and elasticities (and the corresponding ones for s_2) can be derived from a matrix equation that expresses all the summation relationships (cf. Sauro et al., 1987; Cascante et al., 1989a,b):

$$\begin{bmatrix} C_1^J & C_2^J & C_3^J \\ C_1^{s_1} & C_2^{s_1} & C_3^{s_1} \\ C_1^{s_2} & C_2^{s_2} & C_3^{s_2} \end{bmatrix} \cdot \begin{bmatrix} 1 & -\varepsilon_{s_2}^{v_1} & -\varepsilon_{s_2}^{v_1} \\ 1 & -\varepsilon_{s_1}^{v_2} & -\varepsilon_{s_2}^{v_2} \\ 1 & 0 & -\varepsilon_{s_2}^{v_3} \end{bmatrix} = \begin{bmatrix} 1 & 0 & 0 \\ 0 & 1 & 0 \\ 0 & 0 & 1 \end{bmatrix} \tag{20}$$

The first matrix contains all the control coefficients, those for the flux in the first row, with each other row referring to one concentration. The second matrix contains a unit vector as first column, with all the other entries occupied by elasticities, one row per enzyme and one column per metabolite. [The zero in the bottom row reflects the fact that the model (Figure 4) assumes no direct effect of S_1 on E_3, i.e. $\varepsilon_{s_1}^{v_3} = 0$]. The product of these matrices is a unit matrix, and thus the control coefficient matrix is just the inverse of the elasticity matrix.

G. Connectivity Relationships

Figure 5 shows schematically how the two summation relationships, Eqs. (12, 13), result from multiplying rows of the first matrix in Eq. (20) by columns of the second, to get the elements of the first column of the unit matrix on the right-hand side. However, it is evident from Figure 5 that the summation relationships are not the only equations that can be generated in this way. In addition, there are three

$$\begin{bmatrix} C_1^J & C_2^J & C_3^J \\ C_1^{s_1} & C_2^{s_1} & C_3^{s_1} \\ C_1^{s_2} & C_2^{s_2} & C_3^{s_2} \end{bmatrix} \cdot \begin{bmatrix} 1 & -\varepsilon_{s_1}^{v_1} & -\varepsilon_{s_2}^{v_1} \\ 1 & -\varepsilon_{s_1}^{v_2} & -\varepsilon_{s_2}^{v_2} \\ 1 & 0 & -\varepsilon_{s_2}^{v_3} \end{bmatrix} = \begin{bmatrix} 1 & 0 & 0 \\ 0 & 1 & 0 \\ 0 & 0 & 1 \end{bmatrix}$$

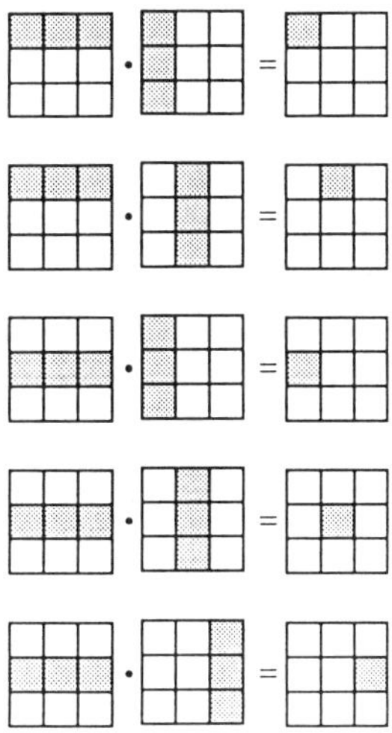

$\sum_{i=1}^{n} C_i^J = 1$ Flux control coefficient summation relationship

$\sum_{i=1}^{n} C_i^J \varepsilon_{s_j}^i = 0$ Flux connectivity relationship

$\sum_{i=1}^{n} C_i^{s_j} = 0$ Concentration control coefficient summation relationship

$\sum_{i=1}^{n} C_i^{s_j} \varepsilon_{s_j}^i = -1$ Concentration connectivity relationship (homogeneous)

$\sum_{i=1}^{n} C_i^{s_j} \varepsilon_{s_k}^i = 0$ Concentration connectivity relationship (heterogeneous)

Figure 5. Representation of summation and connectivity relationships as a matrix equation. The top line of the figure reproduces Eq. (20) from the text, and the other lines illustrates how each of the summation and connectivity relationships can be regarded as the result of multiplying a column of the control matrix by a row of the elasticity matrix to obtain one element of the unit matrix. In general, for a system of n enzymes there are n^2 such relationships, which fall into five classes, expressed by Eqs. (12, 13, 21, 24).

kinds of *connectivity relationships*, which express sums of products of control coefficients and elasticities. The first of these was discovered by Kacser and Burns (1973) and concerns flux control coefficients:

$$\sum_{i=1}^{n} C_i^J \varepsilon_{s_j}^i = 0 \tag{21}$$

Metabolites that are not involved in feedback or feedforward effects normally have only two nonzero elasticities, and this relationship then assumes rather a simple form. For example, for S_1 in Figure 4, we have

$$C_1^J \varepsilon_{s_1}^1 = -C_2^J \varepsilon_{s_1}^2 \tag{22}$$

which shows that the ratio of the flux control coefficients of two consecutive enzymes is equal to minus the reciprocal of the ratio of elasticities of the connecting metabolite (hence the name "connectivity relationship"):

$$\frac{C_1^J}{C_2^J} = -\frac{\varepsilon_{s_1}^2}{\varepsilon_{s_1}^1} \tag{23}$$

This relationship allows one to "walk" along a pathway relating control coefficients in pairs, and as control coefficients are in principle much more difficult to measure directly than elasticities, this is an important advantage.

The concentration connectivity relationships are represented by the bottom two lines of Figure 5, and were discovered by Westerhoff and Chen (1984):

$$\sum_{i=1}^{n} C_i^{s_j} \varepsilon_{s_k}^{v_i} = -\delta_{jk} = \begin{cases} -1 \text{ if } j = k \\ 0 \text{ if } j \neq k \end{cases} \tag{24}$$

H. Response Coefficients: The Partitioned Response

Earlier in this article I quoted the view of Kacser and Burns (1973) that any effect on the activity of an enzyme can be treated as if it were due to a change in the concentration of the enzyme. However, given the amount of misunderstanding that this has caused, it is not sufficient just to quote Kacser and Burns; one must also explain why they are correct. The justification lies in the treatment of external effectors on enzymes.

As an analog of a control coefficient that expresses the dependence of a system variable such as flux on an internal parameter such as enzyme activity, one can define a *response coefficient* R_z^J to express the dependence of a system variable on an external parameter, such as the concentration z of an external effector Z:

$$R_z^J = \frac{\partial \ln J}{\partial \ln z} \tag{25}$$

An external effector such as Z can only produce a systemic effect by acting on one or more enzymes in the system. Thus it must have at least one non-zero elasticity $\varepsilon_z^{v_i}$, defined in exactly the same way as any other elasticity:

$$\varepsilon_z^{v_i} = \frac{\partial \ln v_i}{\partial \ln z} \tag{26}$$

Some workers use a presuperscript κ to distinguish external elasticities (which they call "kappa elasticities") from internal elasticities, i.e. ${}^{\kappa}\varepsilon_z^{v_i}$. However, this seems an unnecessary source of confusion, suggesting a difference where none exists.

Kacser and Burns (1973) proved that the response coefficient is the product of this elasticity and the control coefficient of the enzyme in question, i.e.

$$R_z^J = C_i^J \varepsilon_z^{v_i} \tag{27}$$

This important relationship is called the *partitioned response*. It explains the relationship between the different definitions of a control coefficient represented by Eqs. (9) and (10), as it shows that these are apparently equivalent only because Eq. (7) has been assumed to be true; there is an implicit elasticity of unity connecting the response coefficient defined by Eq. (10) with the control coefficient defined by Eq. (9). In general, any response coefficient can be written as the product of a control coefficient and an elasticity.

IV. CONTROL-PATTERN ANALYSIS

The approach embodied in Eq. (20) can be generalized to encompass branched pathways, non-unit enzyme elasticities, enzyme–enzyme interactions, etc. However, even for pathways as short as the one shown in Figure 4 derivation of the individual expressions shown in Eqs. (14–19) is quite laborious, and for more complicated cases it becomes too tedious and error-prone to be practical (unless supported by computer programs capable of handling the algebra involved). The situation is analogous to that existing in steady-state enzyme kinetics in the 1950s, when the principles of deriving rate equations were well understood, but the actual procedures could only be applied to any but the most trivial mechanisms if one was willing to undertake very tedious algebra; they were also very frustrating, as much of the effort went into the derivation of terms that would subsequently cancel. In effect, deriving rate equations is a problem of symbolic matrix inversion, and steady-state enzyme kinetics was revolutionized when King and Altman (1956) showed how this could be done by means of geometrical rules.

If we look at the impact of these geometrical rules on enzyme kinetics today, we can see that their importance lies not so much in the actual derivation of equations—after all, most of the equations one is likely to want have been derived already—but in the insight that they give into the steady-state behavior of enzymes: once one has thorough experience in applying the rules of King and Altman, one can understand a lot about the kinetic behavior associated with a mechanism just by inspecting the mechanism (see Cornish-Bowden, 1979). As derivation of relationships between control coefficients and elasticities is also a matter of symbolic matrix inversion, it is natural to enquire whether any corresponding geometrical rules exist for simplifying it and for allowing the structure of the result to be discerned by inspection.

Hofmeyr (1989; 1990) has addressed this question, and has shown that such rules do exist. As the problem itself is considerably more complicated than that in steady-state enzyme kinetics, his rules are more complicated than those of King and Altman, and there is not space in this article to describe them in full. However,

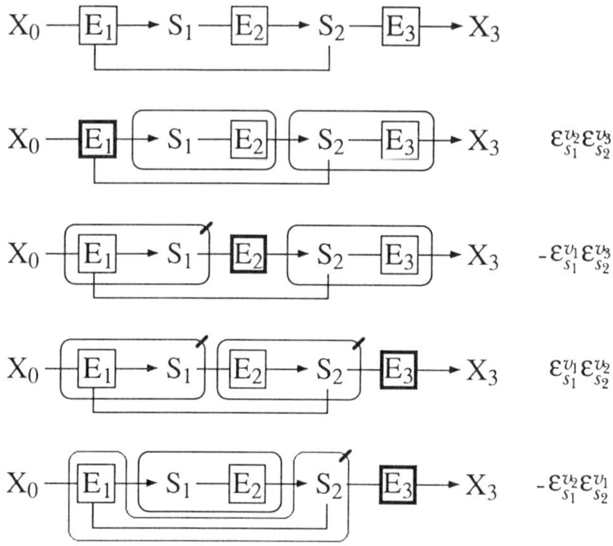

Figure 6. Control pattern analysis of fluxes. The top line of the figure shows how the pathway of Figure 4 must be redrawn before Hofmeyr's method of deducing control coefficients can be applied. The lower lines illustrate the method for each enzyme modulated (shown in an emphasized box) in turn. At the right of each pattern is listed the product of elasticities that it generates. The rules for drawing balloons, marking them and writing the elasticity products are detailed in the text.

they may be found in Hofmeyr (1989), with a somewhat simplified account in Hofmeyr (1990).

To apply them to the pathway of Figure 4, it is first necessary to represent the mechanism in a special way, as in the top line of Figure 6: each enzyme is shown in the box in the middle of the arrow representing the reaction that it catalyzes. External (X_0 and X_3) and internal (S_1 and S_2) metabolites must be represented by clearly distinguishable symbols: as this was already done in Figure 4, no change is required in Figure 6. All reactions are represented by unidirectional arrows: this does *not* imply that they are irreversible reactions, but only that the method requires a consistent direction of flux to be defined; it does not matter if the actual direction is the opposite of that shown. All nonzero elasticities are shown as lines connecting the metabolite to the enzyme that it acts on: in the case of substrate or product elasticities the reaction arrows serve to indicate the connections, but for the feedback effect of S_2 on E_1 a line must be added to the diagram.

To derive flux control coefficients, we now define each enzyme in turn as the modulated enzyme, representing this by emphasizing the box around it. The numerator for its flux control coefficient then consists of a sum of elasticity products, each of which products contains every internal metabolite and every enzyme except the modulated enzyme, with no enzyme or metabolite appearing more than once in the same product. For each modulated enzyme there are as many numerator terms as there are ways of drawing balloons around enzyme–metabolite pairs: for E_1 there is only one such way, shown on the second line of Figure 6, with one balloon containing S_1 and E_2 and a second containing S_2 and E_3; these lead to the elasticity product $\varepsilon_{s_1}^{v_2}\varepsilon_{s_2}^{v_3}$. In both balloons, the direction of the arrow is from metabolite to enzyme, and for these we make no special mark on the balloons.

The third line of Figure 6 shows the corresponding procedure with E_2 as modulated enzyme. There is again only one way of drawing the balloons, and only one product, $-\varepsilon_{s_1}^{v_1}\varepsilon_{s_2}^{v_3}$. The minus sign in this product is decided in the following way. The balloon containing E_1 and S_1 has an arrow direction from enzyme to metabolite, and when this is the case the balloon is marked. If the number of marked balloons is odd the elasticity product is prefixed by a minus sign.

With E_3 as modulated enzyme there are two different ways of drawing the balloons, shown on the bottom two lines of Figure 6. These lead in the same way as before to a sum of two elasticity products, $\varepsilon_{s_1}^{v_1}\varepsilon_{s_2}^{v_2} - \varepsilon_{s_1}^{v_2}\varepsilon_{s_2}^{v_1}$.

Comparing Figure 6 with Eqs. (14–16), the relationship of these equations to the geometrical rules becomes clear: each of the three enzyme modulations generates the numerator of one expressions, and in each expression the denominator is the same and is the sum of the three numerators.

Figure 7 illustrates the corresponding procedure for deriving concentration control coefficients. The rules are similar to those for flux control coefficients, with some additions. The reference metabolite, i.e. the metabolite to which the concentration control coefficient refers, must not appear in the numerator of its expression: to prevent it from appearing in a balloon a circle is drawn around it before the balloons are drawn. As this leaves fewer metabolites available than enzymes, one enzyme apart from the modulated enzyme is not included in a balloon: this is called the "excluded enzyme" and its box is shaded in Figure 7. No particular choice needs to be made of excluded enzyme: one simply takes all possible ways of drawing the balloons.

One additional rule is needed for deciding the sign of an elasticity product in the numerator of the expression for a concentration control coefficient. To understand this it is necessary to introduce the idea of a "concentration control path," which can be regarded as the route that must be traversed in the control pattern (one line of Figure 6 or Figure 7) to get from the modulated enzyme to the reference metabolite without passing through the excluded enzyme. In lines 1 and 2 of Figure 7 this is a one-step path from E_1 to S_1, in the same direction as the arrows; in lines 3 and 4 it is a one-step path from E_2 to S_1, against the direction of the arrows; in line 5 it passes through the balloon containing S_2 and E_1, and in line 6 it passes through the balloon containing S_2 and E_2. The circle containing the reference metabolite is now marked if the last step of this control path is against the direction

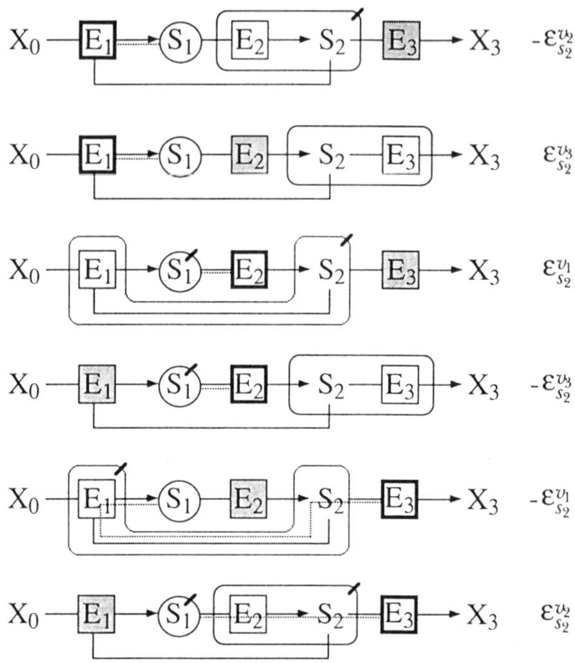

Figure 7. Control pattern analysis of concentrations. Details of how the patterns are drawn and elasticity products are given in the text. In this simple example each elasticity "product" consists of a single elasticity, but in more complex examples with *m* internal metabolites each elasticity product contains $m - 1$ elasticities.

of the arrow, as is the case in lines 3, 4 and 6. These marks are then counted in determining the signs of the elasticity products in the same way as before.

The use of unidirectional arrows in control pattern analysis might seem to introduce the risk of error in the case where the actual flux direction is opposite from that shown by the arrows. That this is not so may be seen by considering the effect of reversing the directions of all arrows: the signs of all elasticity products are then reversed, but as this happens in both numerators and denominators, there are no changes in the values of the final expressions. Thus the unidirectional arrows represent a device for ensuring consistency, not an assumption.

Alternative graphical approaches to the derivation of control coefficients have been developed by Sen (1990a; 1991a). These have the same objectives as Hofmeyr's method, and lead to the same results. For the ordinary biochemist they are somewhat more difficult to apply, and will not be described here, but readers familiar with the use of graph theory in the study of electrical circuits may find them preferable; indeed, in another paper Sen (1990b) has explored the analogy between metabolic and electrical circuits in some detail. Sen (1991b) has also described a graphical method that is applicable to the flux-orientated theory of Crabtree and Newsholme (1987a,b).

The obvious next step forward from graphical methods is to develop a computer program to do the derivations automatically, and this has now been done by Schulz (1991), who was also one of the first to develop a program for deriving steady-state rate equations, one of several listed by Cornish-Bowden (1977). However, all of these rate-equation programs are far less widely known and used than the graphical method of King and Altman (1956). One may expect the same to follow for programs for deriving expressions for control coefficients, because a program cannot offer the same insight into the structure of an equation that a good graphical method can, and, as argued above, much of the value of the method of King and Altman does not lie in the actual derivation of equations.

V. CONTROL AND REGULATION

Although the major ideas of metabolic control analysis date from 1973–1974, and have their roots in the work of Higgins (1965) a decade earlier, they have been absorbed into the mainstream of thought about metabolic regulation rather slowly. Indeed, it took almost ten years before any group other than those of the originators of the subject entered the field and tried to apply its ideas to new experimental systems. Before the work of Tager's group on respiration (Groen et al., 1982a) and gluconeogenesis (Groen et al., 1982b; 1983), therefore, the status of metabolic control analysis was the same as that of biochemical systems theory and flux-orientated theory today—essentially confined to its inventors.

This slow acceptance of metabolic control analysis by the biochemical community, still by no means complete, might seem to be of historical interest only; however, it needs to be understood by anyone seeking to advance the subject in the future, because advances will serve little purpose if they have no effect on experimental research in metabolic regulation. One obstacle to acceptance has probably been the one noted by Atkinson (1990):

> Being human, experimentalists are likely to be put off by . . . an intellectual value system in which it seems self-evident that a few hours of mathematical doodling by a theorist can invalidate any amount of work and thought by experimentalists—not so much because of the claim that they are wrong as by the implication that their work is so trivial that it would not be worth a few minutes of a theorist's time to consider whether it might perhaps have some validity.

One can argue that metabolic control analysis is not as disdainful of experimental work as Atkinson implies here, but that is irrelevant; what matters is that it is often perceived to be disdainful. Despite the fact that the last section of Kacser and Burns (1973) discusses the importance of feedback inhibition for effective control, many readers of this and other papers have come away with the impression that metabolic control analysis has little use for the classical ideas of metabolic regulation, e.g., feedback inhibition by end products (Yates and Pardee, 1956; Umbarger, 1956), allosteric and cooperative interactions (Monod et al., 1963; 1965; Koshland et al., 1966), and so on. Recognizing the enormous volume of productive research that was stimulated by these ideas during the 1960s and early 1970s, it is natural to conclude that a theory that has little use for them is probably of little value.

Part of the confusion has resulted from a lack of agreed definitions for certain crucial terms. "Control" is now widely accepted to have the meaning attributed to it by Kacser and Burns (1973), but "regulation" continues to give difficulties. For some, regulation is little different from control, for example Sauro (1990) took it to mean "some sort of of response of metabolism to a change in an external influence;" for others it is quite different, having to do with the properties of regulatory enzymes in isolation. Hofmeyr and Cornish-Bowden (1991) consider that its use in biochemistry ought to be brought as close as possible to its use in everyday life. When we say that a domestic refrigerator is well regulated, for example, we mean that it is capable of maintaining a predetermined internal temperature constant in the face of large variations in heat flux that result from opening the door or variations in the external temperature. Metabolism is in almost exact analogy to this if one considers a well regulated system to be one in which concentrations of internal metabolites (the "temperature") are maintained steady in the face of variations in metabolic flux. In economic terms, we usually regard a well regulated economy as one in which the rate at which goods are produced is determined largely by demand. Again, there are obvious metabolic analogies, and we should expect a well regulated organism to be one in which the supply of precursors for protein synthesis is determined by the need for protein synthesis, and not solely by the supply of food.

Another important term that seems to mean one thing but actually means something else is "end-product." It seems obvious that the end-product in a metabolic system ought to be the sink into which the flux flows. But let us examine how the term is actually used in the literature on metabolic regulation, such as the classic chapter by Stadtman (1970) in the 3rd edition of *The Enzymes*. Stadtman always uses "end product" (often even "ultimate end product") to refer to a metabolite such as threonine that is not excreted but is explicitly recognized as the starting point for other pathways. In virtually the entire experimental literature on metabolic regulation an end-product is understood in this way; it never means a genuine end-product of metabolism such as water or carbon dioxide.

It follows, therefore, that we cannot hope to understand the role of end-product inhibition in metabolic regulation unless we draw pathways as components of systems that explicitly recognize that there are steps after the release of "end-product." Thus in their discussion of feedback inhibition Kacser and Burns (1973) included a step after the formation of end-product, though they did not explain the reason for doing so. Likewise, in the example of Figures 4–6 of this article, feedback inhibition is assumed to be exerted by S_2, the last internal metabolite, not by X_3, the sink into which S_2 flows.

The metabolic utilization of an end-product is normally irreversible: threonine is not formed by step-by-step reversal of protein synthesis, glucose is not formed by reversal of glycolysis, etc. Thus it is often appropriate to analyze systems in which the final step is treated as irreversible, recognizing that it represents the total set of reactions that consume the end product rather than just one enzyme. The opinion of Mendes et al. (1992) that "having an absolutely irreversible final step seems rather unreal" apparently misunderstands this point.

For integrating the classical regulatory concepts into metabolic control analysis (Hofmeyr and Cornish-Bowden, 1991), one can represent a pathway with feedback inhibition as a two-step pathway, consisting of a supply block, consisting of all the reactions that lead to the end product, and a demand block, consisting of the reactions that consume it (Figure 8). If the supply block were the entire pathway,

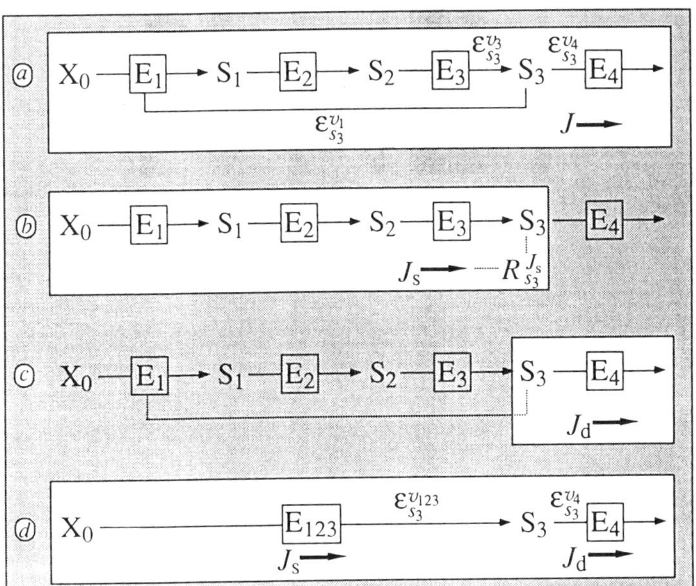

Figure 8. Partition of a biosynthetic pathway into supply and demand blocks. (**a**) The complete pathway, in which S_3 is the end product of a pathway from X_0 and the starting material for other pathways represented schematically by the single enzyme E_4. S_3 interacts with the flux J by virtue of its elasticities with E_3 (as product), E_4 (as substrate) and E_1 (as feedback inhibitor). (**b**) The supply block. In this subsystem s_3 is a constant external parameter and any influence that it has on the supply flux J_s is via a response coefficient $R^{J_s}_{s_3}$. (**c**) The demand block, with demand flux J_d. (**d**) When the full system is regarded as a two-step pathway, the first step being the whole of the supply block, the effects of S_3 on the original two enzymes E_1 and E_3 can be represented by a single elasticity $\varepsilon^{v_{123}}_{s_3}$, which is conceptually the same as the response coefficient $R^{J_s}_{s_3}$ for the supply block considered as a system. The two fluxes J_s and J_d are identical in the steady state, but for purposes of analyzing the block elasticities they can be treated as local rates.

the end product would be an external parameter and any effect that it had on the flux would have to be treated in terms of a response coefficient $R^{J_s}_{s_3}$, but this response coefficient is conceptually the same as the block elasticity $\varepsilon^{v_{123}}_{s_3}$ that defines its effect on the supply flux considered as the local rate.

It follows from this kind of discussion that the boundaries of a system and the distinctions between internal and external parameters or between local and systemic properties cannot be regarded as absolute. To understand how an end product such as S_3 in Figure 8 can fulfill its regulatory role, it is not sufficient to regard it solely as an internal metabolite; we must also study subsystems where it becomes an external parameter, so that we can ask questions like "if the supply block (Figure 8b) were the complete system, what effect would s_3 have on the supply flux?".

Using this type of analysis, one can study how to achieve effective regulation of a pathway such as that of Figure 8 by demand, by which we mean not only that the flux responds sensitively to changes in demand, but also that the concentration of end-product changes little when the flux changes. It turns out (see Figure 2e of Hofmeyr and Cornish-Bowden, 1991), that the main essential is that the supply elasticity, $\varepsilon^{v_{123}}_{s_3}$, in the complete system, which is the same as the response coefficient $R^{J_s}_{s_3}$ in the supply block considered in isolation, must be as large as possible in absolute magnitude. (Being an inhibitory elasticity it is negative, so "as large as possible" means "less than –2"). The demand elasticity $\varepsilon^{v_4}_{s_3}$ is less important, as effective regulation can be achieved over a broad range of values.

The results of a study of the importance of cooperativity in effective regulation (Hofmeyr and Cornish-Bowden, 1991) proved to be surprising, as they appeared at first to suggest that it was much less important than had been thought since the 1960s. However, it must be emphasized that this is an illusion: cooperativity is certainly as necessary for effective regulation as has been thought, but its role is somewhat different from what one might naively imagine.

Changing the degree of cooperativity in the feedback inhibition of E_1 by S_3 in Figure 8 in the range of Hill coefficients from one (no cooperativity) to four (approximately the maximum cooperativity observed in any single effector-enzyme interaction) has almost no effect on the control of flux by demand: the curves showing flux as a function of demand (expressed by the limiting rate V_4 of the demand block) show near-proportionality between flux and demand over a 25-fold range (Figure 4a of Hofmeyr and Cornish-Bowden, 1991), regardless of the Hill coefficient. This must surprise anyone who thinks that cooperativity is essential for flux regulation. However, as has been emphasized already, flux regulation is only part of regulation, and it is of little use without concentration regulation: we should not be very happy with a refrigerator that tolerated a wide range of heat fluxes but had no control over the internal temperature! When the concentration of end-product is considered as well as the flux, the effect of the Hill coefficient becomes very large: over the same 25-fold range of demand considered above, a Hill coefficient of four causes s_3 to be restricted to less than a threefold range, whereas with a Hill coefficient of one it varies more than 10-fold in a demand range of only about threefold.

In summary, cooperativity of feedback interactions is indeed essential for effective regulation, but it is not sufficient to say that it allows effective regulation of flux by demand; one must say that it allows effective regulation of flux by demand while maintaining homeostasis.

VI. MECHANISMS OF REGULATION

To a considerable extent, metabolic control analysis takes the properties of individual enzymes as it finds them, regarding a mechanistic explanation of these properties as outside its domain. Only a brief mention of them is needed here, therefore: the essential characteristics that distinguish the enzymes usually regarded as important in regulation from others are cooperative and allosteric interactions with substrates, products and effectors. Accounts of these properties may be found in numerous articles and books published around 1970. More recent accounts have been given by Ricard and Cornish-Bowden (1987), for a general view, and Cornish-Bowden and Cárdenas (1987) for discussion of how cooperativity can arise in monomeric enzymes, i.e. without the need for subunit interactions.

When we progress to regulatory mechanisms involving multiple enzymes, however, there are two kinds of mechanism that demand fuller discussion: these are channeling of intermediates between enzymes and interconvertible enzyme cascades involving covalent modification.

A. Metabolite Channeling

Channeling involves the idea that the metabolite shared by two consecutive enzymes in a pathway may be directly transferred from one to the other, without being released into free solution, or at least without achieving equilibrium with the metabolite in free solution. There are various versions of this idea (see Scheme 1 of Ovádi, 1991), but the essentials are shown in Figure 9a, which is based on the mechanism suggested by Gutfreund (1965). This can be regarded as a combination of a perfect channel (Figure 9b) with an ordinary free diffusion mechanism (Figure 9c).

Although there is at least one case, that of the multifunctional enzyme tryptophan synthase, for which the evidence for channeling (of indole) is overwhelming and generally accepted (see Yanofsky, 1989), channeling in general remains highly controversial, at least for enzymes forming "dynamic" complexes, i.e. complexes that exist only transiently during the transfer of metabolite.

A recent issue of the *Journal of Theoretical Biology* was devoted to a review of channeling by Ovádi (1991) followed by commentaries by numerous other authors. These covered an entire spectrum of views, from "metabolite channeling must be

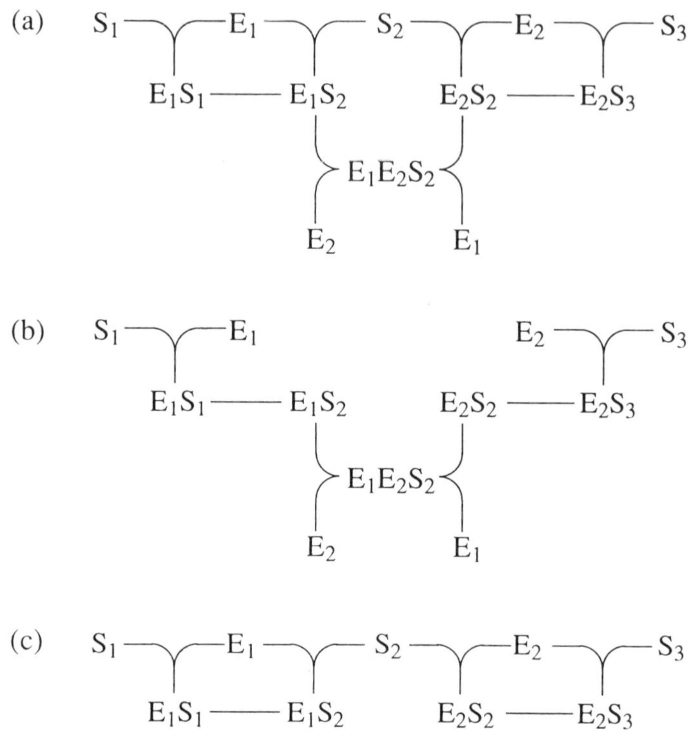

Figure 9. The Gutfreund channeling mechanism. (**a**) The complete mechanism, in which a metabolite S_1 is converted by the action of two enzymes E_1 and E_2 into another metabolite S_3, either via a free intermediate metabolite S_2 or via a ternary complex $E_1E_2S_2$ involving both enzymes. The mechanism can be regarded as a superimposition of a perfect channeling mechanism (**b**), in which no free S_2 is released, and a free diffusion mechanism (**c**), in which there is no channeling.

considered in any global account of metabolism. Dr Ovádi has provided a noble attempt to make that case for those (few?) who still doubt its importance" (Clegg, 1991) to "no convincing evidence is available for metabolite channelling between enzymes forming dynamic complexes" (Pettersson, 1991), statements that appeared on facing pages.

Most of this controversy concerns whether channeling actually occurs in most of the systems where it is proposed. This is certainly important in relation to metabolic control, because if it does not occur it cannot have any relevance to control. However, this is only part of the question, because it does not necessarily follow that channeling has any significant metabolic consequences even if it does occur. This point has been much less discussed, possibly because the advantages of channeling have been perceived to be obvious. Even authors who argue strongly against the existence of channeling nonetheless consider that if it were demonstrated it would constitute the discovery of the century (Gutfreund and Chock, 1991), surely an exaggerated view of its importance.

The idea that the advantages of channeling are obvious may result from confusion between the properties of a *perfect* channel (Figure 9b), for which there are certainly some metabolic consequences from the fact that no intermediate at all is released into free solution, and thus none is available to participate in competing reactions, with those of the sort of dynamic channeling mechanism that is the subject of the controversy (Figure 9a).

It might seem intuitively obvious that even in a mechanism such as that of Figure 9a increasing the rates of the channeling steps at the expense of the free-diffusion steps must decrease the steady-state concentration of free intermediate, but this is an illusion. Although this may well decrease the rate at which S_2 is released from E_2S_2, it will also decrease the rate at which S_2 is taken up by E_3, and the net result may be in either direction. This has been studied by computer simulation with the program MetaModel (Cornish-Bowden and Hofmeyr, 1991) of a model in which the mechanism of Figure 9a was incorporated into a pathway (Cornish-Bowden, 1991). This showed that when conditions were chosen to cause significant accumulation of free intermediate, varying the proportion of the total flux that passed through the channel (while keeping this total flux constant) had no effect at all on the free concentration of intermediate. Mendes et al. (1992) subsequently confirmed these results with their program GEPASI (Mendes, 1990), but also argued that they were based on too restricted a range of conditions. In extending them, they found that in some circumstances channeling could apparently decrease the concentration of intermediate at net flux especially if it was already very small in the absence of channeling. The important word here is "apparently," however, because all of the claimed counter-examples proved on careful examination to be artifacts of small but non-negligible changes in net flux, or of alterations of the catalytic activities of consecutive enzymes in opposite directions (Cornish-Bowden and Cárdenas, 1993). Simulation is also now unnecessary for reaching this conclusion, as it can be shown by algebraic analysis that channeling (even by "static" complexes) cannot affect metabolite concentrations at constant net flux (Heinrich and Schuster, 1991; Cornish-Bowden and Cárdenas, 1993).

B. Interconvertible Enzyme Cascades

As discussed above, cooperativity of interactions with individual enzymes is a very important way in which feedback inhibition can be more effective as a regulatory mechanism. However, it has a very serious drawback that prevents it from providing a universal way of increasing elasticities: the degree of cooperativity is severely limited by the need for site-site interactions, whether in space, as in the classical models of cooperativity (Monod et al., 1965; Koshland et al., 1966) or in time, as in models of cooperativity in monomers (Cornish-Bowden and

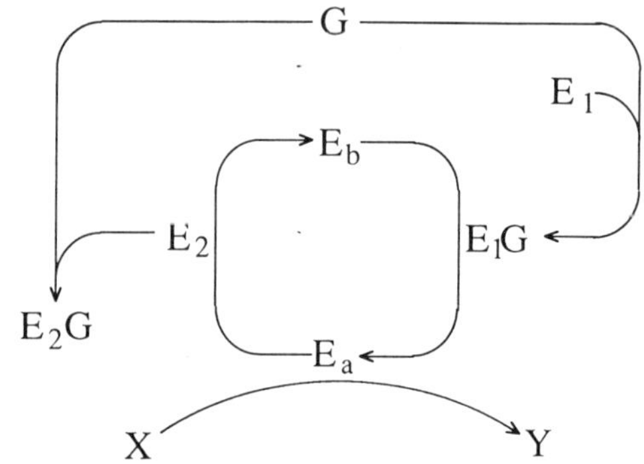

Figure 10. Interconvertible enzyme cascade. An enzyme E_a catalyzing a step from X to Y in a metabolic pathway exists in both active (E_a) and inactive (E_b) forms, which are interconvertible by two irreversible reactions catalyzed by different enzymes E_1G and E_2, of which E_1G requires activation by an effector G, whereas E_2 is inhibited by the same effector.

Cárdenas, 1987). In practice, enzymes showing Hill coefficients greater than four are virtually unknown; as the elasticity for an interaction cannot exceed the corresponding Hill coefficient, this means that individual enzyme-metabolite interactions do not result in elasticities greater than four. This is too small for a device intended to operate as a switch: it implies the need for a three-fold change in metabolite concentration to bring about a change from 10 to 90% of full activity.

Much higher effective elasticities become possible when one comes to consider multienzyme systems of the kind illustrated in Figure 10, which are called "cascades" by Stadtman and co-workers (Stadtman, 1970; Stadtman and Chock, 1977; Chock and Stadtman, 1977), though it is arguable that this term implies a system in which there are at least two linked cycles.

Interconvertible enzymes systems have been known for more than a quarter of a century, and many examples are known, including many involving protein kinases and phosphatases (Krebs and Beavo, 1979). Glutamine synthetase from *E. coli*, which is inactivated by adenylylation and reactivated by deadenylylation (Chock et al., 1980; 1990) has been studied in detail, and has served as the basis for extensive theoretical work (Chock and Stadtman, 1977; Stadtman and Chock, 1977; 1978).

In the context of this chapter, the essential point is that interconvertible enzyme systems can generate very high sensitivity to signals, much higher than is possible for single enzymes (Goldbeter and Koshland, 1981; 1984). It is tempting to suppose that this high sensitivity is inherent in the structure of the cycle, but in reality if the kinetic parameters of the cycle reactions are assigned arbitrary values, the typical result is a system that generates *less* sensitivity than a single noncooperative enzyme. Very high sensitivity results only if several conditions are satisfied

(Cárdenas and Cornish-Bowden, 1989): the interactions of the effector with the modifier enzymes should be predominantly catalytic rather than specific (uncompetitive rather than competitive in the terminology of inhibition); the inactivation reaction should be inhibited at much lower concentrations of effector than are needed to activate the activating reaction; both modifier enzymes should operate close to saturation, a condition especially emphasized by Goldbeter and Koshland (1981; 1982; 1984) under the name "zero-order ultrasensitivity." If all these conditions are satisfied the sensitivity possible with the mechanism of Figure 10 is enormously greater than is possible for a single enzyme; even with severe constraints allowed for the kinetic parameters of the modifier enzymes one can easily obtain the equivalent of Hill coefficient of 30, or even of 800 if one relaxes the constraints, while still staying within the range of behavior commonly observed (Cárdenas and Cornish-Bowden, 1989).

Small and Fell (1990) have examined these conditions in the framework of metabolic control analysis. They find that they are equivalent to minimizing the elasticities of the modification reactions with respect to their reactants (the active and inactive forms of the target enzyme) while maximizing the elasticities with respect to the effector. They also noted that the second condition could be partitioned into separate requirements that the effector concentration be large compared with the catalytic (uncompetitive) inhibition constant of the inactivating enzyme and small compared with the corresponding activation constant of the activating enzyme. This separation of effects becomes especially important if one considers the possibility that different effectors might act on the two reactions. Szedlacsek et al. (1992) considered this type of model for determining the magnitudes of response coefficients in interconvertible enzyme systems. They found that even if the effector acted only as inhibitor, a highly constrained system could generate a response coefficient of about nine for the effect of this inhibitor on the fraction of target enzyme in the active state. To be translated into a response of the metabolic flux this value must, of course, be multiplied by the control coefficient of the target enzyme [cf. Eq. (27)], but provided this is appreciable, it implies a very high level of responsiveness.

The first two of the conditions given above imply that experimenters should take special care to note what might appear to be insignificant kinetic properties of modifier enzymes. Even if the uncompetitive component of the inhibition of a modifier enzyme is an order of magnitude weaker than the competitive component it may still be essential to the effective working of the system. Likewise, if one observes that the phosphatase in a cycle is activated only at supposedly nonphysiological concentrations of an effector ten times higher than those effective for inhibiting the kinase, this does not mean that the effector is irrelevant to the action of the phosphatase; it means that the whole system is well designed for generating very high sensitivity.

As these two conditions are exceptions to the generalization made earlier that effects that are difficult to observe are unlikely to be important in metabolic control, how can one argue that they are indeed important? The explanation is that a system that is designed (or has evolved) to produce very high sensitivity must inevitably be capable of recognizing and responding to much smaller effects than would produce measurable effects in systems that were not so designed.

As interconvertible enzyme systems can generate so much more sensitivity than individual cooperative enzymes one may wonder why they are not universally used in metabolic regulation. However, unlike individual cooperative enzymes interconvertible enzyme systems consume energy, because both modification reactions are assumed to be irreversible; this is possible only if they involve different cosubstrates, for example the activation might be phosphorylation by ATP whereas the inactivation might be hydrolysis. Shacter et al. (1984) estimated that this energy consumption was trivial for two cases that they studied (pyruvate kinase and glycogen phosphorylase), but later work (Meinke et al., 1986; Goldbeter and Koshland, 1987) showed that they had used a literature value for the ATP regeneration rate in liver that was grossly in error, and that in reality the energy cost of covalent modification cycles is by no means negligible.

The interconvertible enzyme cascade shown in Figure 10 is an example of a *hierarchical control structure*, because in effect it contains two distinct metabolic networks that do not exchange matter between one another, though the interconversion of E_a and E_b influences the conversion of X to Y. More complex hierarchies are of great importance in biochemistry. For example, metabolism of DNA is separate from but influences metabolism of RNA, which is separate from but influences metabolism of enzymes, which is separate from but influences the metabolism of everything else. At first sight this type of hierarchy might be too complicated to be within the scope of metabolic control analysis. However, Westerhoff et al. (1990) have shown that control analysis is by no means hopeless and that one can introduce new response relationships between effects in different levels of the hierarchy. Kahn and Westerhoff (1991) have formalized this type of analysis, showing that systems containing two or more hierarchical levels can be decomposed into distinct *modules*, such that there are many nonzero elasticities within each module but comparatively few nonzero elasticities between modules.

VII. COMPUTER SIMULATION OF METABOLISM

In principle, metabolic control analysis is independent of computer modeling: it is not necessary to be able to simulate metabolism in the computer in order to verify its theorems. Nonetheless, most of us are not so confident of our algebra that we can accept everything without ever wanting to check it with computer models. Moreover, we may sometimes want to explore properties that are beyond the reach of current theory. Finally, for the teacher, a good program can provide the equivalent of an electronic rat: it is hardly possible to set up a "wet" experiment to compare the properties of an organism in which phosphofructokinase has its usual array of regulatory properties with one in which the enzyme lacks some of these properties, but a computer experiment may be quite easy to design.

As a digression, we may note that although it remains impractical to set up exactly the "wet" experiment suggested in the previous paragraph, *E. coli* does allow us to make an approximation to it, with results that should give pause to anyone who takes the classical view of metabolic regulation as the entire truth. The wild-type strain of *E. coli* contains about 10 percent of a variant phosphofructokinase that lacks some of the usual regulatory properties. Mutants that lack the "normal" form of the enzyme and have the normal amount of the variant form grow only slowly on glucose, as the classical view of the regulatory importance of phosphofructokinase would lead us to expect. However, other mutants that have high levels of the minor form as the only phosphofructokinase present grow well on glucose (Daldal et al., 1982). This observation is less easy to reconcile with the classical picture.

Computer modeling of metabolism was pioneered by Garfinkel and Hess (1964), during what we may regard as the heroic period of metabolic control analysis: at that time as much as 30 min of computer time were needed to simulate 75 ms of glycolysis, high-level programming languages hardly existed and programming was a matter of searching for ways of avoiding arbitrary restrictions such as the inability of the computer to handle numbers greater than one. The programs of that period were written by experts for experts to use, and would be regarded as extremely unfriendly by today's standards. However, in recent years, several programs for simulating metabolism on personal computers have been made available, and are listed in Table 1.

As far as I am aware, no one has attempted an unbiased comparison between these programs from a user's point of view, and it would be as unwise to compare them on the basis of their authors' descriptions as it would be to compare commer-

Table 1. Computer Programs for Metabolic Modeling[a]

Program	*Hardware*	*Reference*
CONTROL	IBM-PC	Letellier et al. (1991)
ESSYNS	IBM-PC	Irvine and Savageau (1990)
GEPASI	IBM-PC, Vax	Mendes (1990)
MetaModel	IBM-PC	Cornish-Bowden and Hofmeyr (1991)
SCAMP	IBM-PC, Atari	Sauro and Fell (1991)

Note: [a]Programs currently available for analysis of metabolic systems on small computers are listed. More detail (price, conditions of use, etc.) may be found in the references given, and also (apart from SCAMP) in Appendix B of Cornish-Bowden and Cárdenas (1990). The function of CONTROL is to convert elasticity matrices into matrices of control coefficients by means of the method of Reder (1988). The other programs are for simulating the kinetic behavior of pathways: ESSYNS uses the power–law equations of biochemical systems theory (Savageau, 1976); the others use kinetic equations defined by the user.

cial products solely on the basis of advertisements. Accordingly, Table 1 provides no more than a list, with information about where more detail may be found.

VIII. APPLICATIONS OF METABOLIC CONTROL ANALYSIS

Not many years ago one could realistically have hoped to discuss all published applications of metabolic control analysis in an article of this kind. However, the number of such applications has increased rapidly in recent years, and one now has to be more selective; anyway it hardly seemed that the needs of readers would be best served by covering all of the same ground as Fell (1992) does in his excellent and current review. He discusses nearly all of the current literature in some detail, but gives less attention to some topics that are discussed here, such as control pattern analysis, regulatory mechanisms, and the relationship between metabolic control analysis and classical ideas of regulation. Other sources of information are Porteous (1990), who includes a list of experimental applications (his Table 2) that was essentially complete at the time of writing, and several chapters in Cornish-Bowden and Cárdenas (1990), which contains almost a complete guide to metabolic control analysis as it was in 1989. In contrast to Fell (1992), therefore, the discussion of experimental examples will here be very general.

In the early development of metabolic control analysis, experiments were carried out primarily for establishing methods of measuring control coefficients, such as genetic manipulation (Flint et al., 1981; Middleton and Kacser, 1983) or "enzyme titration," i.e. addition of excess enzyme so as to "shorten" a pathway to more manageable proportions (Torres et al., 1986). It is a sign of the maturity of metabolic control analysis that experiments of this kind have now very largely been supplanted by ones motivated by a desire to address the central problems of metabolic regulation. In mitochondrial respiration, for example, the pioneering work of Groen et al. (1982a) has been followed and extended by other groups (e.g., Mazat et al., 1986; Brand et al., 1988; Padovan et al., 1989). Stitt and co-workers in particular have done extensive work in the study of the control of sucrose synthesis and other aspects of photosynthesis (e.g., Kruckeberg et al., 1989; Neuhaus and Stitt, 1990); other investigations of the control of photosynthesis include those of Woodrow (1986), Pettersson and Ryde-Pettersson (1988) and Giersch et al. (1990). In amino acid metabolism, there have been studies of aromatic amino acid metabolism by Salter et al. (1986a,b), and of serine biosynthesis by Fell and Snell (1988).

This list is far from complete, but it serves to illustrate how metabolic control analysis has ceased to be the preoccupation of a few people primarily interested in theory but has become an essential tool for understanding metabolic regulation. One may expect this trend to continue, now that nuclear magnetic resonance, already being used for metabolic control analysis in yeast (Brindle, 1988; Davies and Brindle, 1992), allows measurement of metabolic concentrations and fluxes in living organisms; this and other noninvasive techniques will leave less and less justification for making large and unjustified extrapolations from kinetic properties observed for isolated enzymes.

I close with a quotation from Kacser (1983). The revolution has taken longer to establish than he probably hoped in 1983, but it is now upon us:

> The lack of progress in certain areas of biochemistry (and genetics) has, in my opinion, been due to the addiction of its practitioners to false concepts and, like all addicts, they are not very effective workers. The pushers in this trade are, of course, the writers-of-books and the givers-of-lectures. They corrupt the young who are then hooked. All is not bleak, however. There *is* a cure. In the last few years a number of strong minded people have undergone the cold turkey treatment. This consists of reading the relevant papers until one has understood them. Once free of the addiction, unlike an ex-alcoholic, one is cured for life.

ACKNOWLEDGMENTS

Most of the work on this chapter was done during tenure of the Cátedra "Hermann Niemeyer" at the Laboratorio de Bioquímica, Facultad de Ciencias, Universidad de Chile, and I thank Dr. Tito Ureta and the other members of the laboratory for the use of facilities there. I also thank Drs. M. L. Cárdenas, D. A. Fell and J.-H. S. Hofmeyr for helpful criticisms of the first draft.

REFERENCES

Acerenza, L., Sauro, H. M., & Kacser, H. (1989). Control analysis of time-dependent metabolic systems. J. Theor. Biol. 137, 423–444.

Alberts, B., Bray, D., Lewis, J., Raff, M., Roberts, K., & Watson, J. D. (1983). Molecular Biology of the Cell, Garland, New York.

Atkinson, D. E. (1990). What should a theory of metabolic control offer to the experimenter? In: Control of Metabolic Processes (Cornish-Bowden, A. & Cárdenas, M. L., eds.), pp. 3–11, Plenum Press, New York.

Brand, M. D., Hafner, R. P., & Brown, G. C. (1988). Control of respiration in non-phosphorylating mitochondria is shared between the proton leak and the respiratory chain. Biochem. J. 255, 535–539.

Brindle, K. M. (1988). ^{31}P NMR magnetization–transfer measurements of flux between inorganic phosphate and adenosine 5′-triphosphate in yeast cells genetically modified to overproduce phosphoglycerate kinase. Biochemistry 27, 6187–6196.

Burns, J. A., Cornish-Bowden, A., Groen, A. K., Heinrich, R., Kacser, H., Porteous, J. W., Rapoport, S. M., Rapoport, T. A., Stucki, J. W., Tager, J. M., Wanders, R. J. A., & Westerhoff, H. V. (1985). Control analysis of metabolic systems. Trends Biochem. Sci. 10, 16.

Cárdenas, M. L. & Cornish-Bowden, A. (1989). Characteristics necessary for an interconvertible enzyme cascade to generate a highly sensitive response to an effector. Biochem. J. 257, 339–345.

Cascante, M., Franco, R., & Canela, E. I. (1989a). Use of implicit methods from general sensitivity theory to develop a systematic approach to metabolic control. I. Unbranched pathways. Math. Biosci. 94, 271–288.

Cascante, M., Franco, R., & Canela, E. I. (1989b). Use of implicit methods from general sensitivity theory to develop a systematic approach to metabolic control. II. Complex systems. Math. Biosci. 94, 289–309.

Chock, P. B. & Stadtman, E. R. (1977). Superiority of interconvertible enzyme cascades in metabolic regulation. Analysis of monocyclic systems. Proc. Natl. Acad. Sci. U.S.A. 74, 2766–2770.

Chock, P. B., Rhee, S. G., & Stadtman, E. R. (1980). Interconvertible enzyme cascades in cellular regulation. Ann. Rev. Biochem. 49, 813–843.

Chock, P. B., Rhee, S. G., & Stadtman, E. R. (1990). Metabolic control by the cyclic cascade mechanism: a study of *E. coli* glutamine synthetase. In: Control of Metabolic Processes (Cornish-Bowden, A. & Cárdenas, M. L., eds.), pp. 183–194, Plenum Press, New York.

Clegg, J. S. (1991). The physiological significance of metabolite channeling: an idea whose time has come. J. Theor. Biol. 152, 63–64.

Cornish-Bowden, A. (1977). An automatic method for deriving steady-state rate equations. Biochem. J. 165, 55–59.

Cornish-Bowden, A. (1995). (2nd ed.) Fundamentals of Enzyme Kinetics, Portland Press, London.

Cornish-Bowden, A. (1989). Metabolic control theory and biochemical systems analysis: different objectives, different assumptions, different results. J. Theor. Biol. 136, 365–377.

Cornish-Bowden (1991). Failure of channelling to maintain low concentrations of metabolic intermediates. Eur. J. Biochem. 195, 103–108.

Cornish-Bowden, A. & Cárdenas, M. L. (1987). Cooperativity in monomeric enzymes. J. Theor. Biol. 124, 1–23.

Cornish-Bowden, A. & Cárdenas, M. L. (1990). Control of Metabolic Processes, Plenum Press, New York.

Cornish-Bowden, A. & Cárdenas, M. L. (1993). Channelling can affect concentrations of metabolic intermediates at constant net flux: artefact or reality? Eur. J. Biochem. 213, 87–92.

Cornish-Bowden, A. & Hofmeyr, J.-H. S. (1991). MetaModel: a program for modelling and control analysis of metabolic pathways on the IBM PC and compatibles. Comput. Appl. Biosci. 7, 89–93.

Crabtree, B. & Newsholme, E. A. (1987a). The derivation and interpretation of control coefficients. Biochem. J. 247, 113–120.

Crabtree, B. & Newsholme, E. A. (1987b). A systematic approach to describing and analysing metabolic control systems. Trends Biochem. Sci. 12, 4–12.

Daldal, F., Babul, J., Guixé, V., & Fraenkel, D. G. (1982). An alteration in phosphofructokinase 2 of *Escherichia coli* which impairs gluconeogenic growth and improves growth on sugars. Eur. J. Biochem. 126, 373–379.

Davies, S. E. C. & Brindle, K. M. (1992). Effects of overexpression of phosphofructokinase on glycolysis in the yeast *Saccharomyces cerevisiae*. Biochemistry 31, 4729–4735.

Fell, D. A. (1992). Metabolic control analysis: a survey of its theoretical and experimental development. Biochem. J. 286, 313–330.

Fell, D. A. & Snell, K. (1988). Control analysis of mammalian serine biosynthesis. Feedback inhibition on the final step. Biochem. J. 256, 97–101.

Flint, H. J., Tateson, R. W., Barthelmess, I. B., Porteous, D. J., Donachie, W. D., & Kacser, H. (1981). Control of the arginine flux in the arginine pathway of *Neurospora crassa*. Modulations of enzyme activity and concentration. Biochem. J. 200, 231–246.

Garfinkel, D. & Hess, B. (1964). Metabolic control mechanisms. VII. A detailed computer model of the glycolytic pathway in ascites cells. J. Biol. Chem. 239, 971–983.

Giersch, C. (1988a). Control analysis of metabolic networks. 1. Homogeneous functions and the summation theorems for control coefficients. Eur. J. Biochem. 174, 509–513.

Giersch, C. (1988b). Control analysis of metabolic networks. 2. Total differentials and general formulation of the connectivity relations. Eur. J. Biochem. 174, 515–519.

Giersch, C., Lämmel, D., & Farquhar, G. (1990). Control analysis of photosynthetic CO_2 fixation. Photosynth. Res. 24, 151–165.

Goldbeter, A. & Koshland, D. E., Jr. (1981). An amplified sensitivity arising from covalent modification in biological systems. Proc. Natl. Acad. Sci. U.S.A. 78, 6840–6844.

Goldbeter, A. & Koshland, D. E., Jr. (1982). Sensitivity amplification in biochemical systems. Q. Rev. Biophys. 15, 555–591.

Goldbeter, A. & Koshland, D. E., Jr. (1984). Ultrasensitivity in biochemical systems controlled by covalent modification. Interplay between zero-order and multistep effects. J. Biol. Chem. 259, 14441–14447.

Goldbeter, A. & Koshland, D. E., Jr. (1987). Energy expenditure in control of biochemical systems by covalent modification. J. Biol. Chem. 262, 4460–4471.

Groen, A. K. & Westerhoff, H. V. (1990). Modern control theories: a consumers' test. In: Control of Metabolic Processes (Cornish-Bowden, A. & Cárdenas, M. L., eds.), pp. 101–118, Plenum Press, New York.

Groen, A. K., Wanders, R. J. A., Westerhoff, H. V., van der Meer, R., & Tager, J. M. (1982a). Quantification of the contribution of various steps to the control of mitochondrial respiration. J. Biol. Chem. 257, 2754–2757.

Groen, A. K., van Roermund, C. W. T., Vervoorn, R. C., & Tager, J. M. (1982b). Control of gluconeogenesis in rat liver cells. Flux control coefficients of the enzymes in the gluconeogenic pathway in the absence and presence of glucagon. Biochem. J. 237, 379–389.

Groen, A. K., Vervoorn, R. C., van der Meer, R., & Tager, J. M. (1983). Control of gluconeogenesis in rat liver cells. I. Kinetics of the individual enzymes and the effect of glucagon. J. Biol. Chem. 258, 14346–14353.

Gutfreund, H. (1965). An introduction to the study of enzymes, pp. 302–306, Blackwell, Oxford.

Gutfreund, H. & Chock, P. B. (1991). Substrate channeling among glycolytic enzymes: Fact or fiction. J. Theor. Biol. 152, 117–121.

Heinrich, R. & Rapoport, T. A. (1974). A linear steady-state treatment of enzymatic chains. General properties, control and effector strength. Eur. J. Biochem. 42, 89–95.

Heinrich, R. & Rapoport, T. A. (1975). Mathematical analysis of multienzyme systems. II. Steady-state and transient control. BioSystems 7, 130–136.

Heinrich, R. & Reder, C. (1991). Metabolic control analysis of relaxation processes. J. Theor. Biol. 151, 343–350.

Heinrich, R., Rapoport, S. M., & Rapoport, T. A. (1977). Metabolic regulation and mathematical models. Prog. Biophys. Mol. Biol. 32, 1–82.

Heinrich, R. & Schuster, S. (1991). Is metabolic channelling the complicated solution to the easy problem of reducing transient times? J. Theor. Biol. 152, 57–61.

Higgins, J. (1965). Dynamics and control in cellular reactions. In: Control of Energy Metabolism (Chance, B., Estabrook, R. W., & Williamson, J. R., eds.), pp. 13–46, Academic Press, New York.

Hofmeyr, J.-H. S. (1989). Control-pattern analysis of metabolic pathways. Flux and concentration control in linear pathways. Eur. J. Biochem. 186, 343–354.

Hofmeyr, J.-H. S. (1990). Control-pattern analysis of metabolic systems. In: Control of Metabolic Processes (Cornish-Bowden, A. & Cárdenas, M. L., eds.), pp. 239–248, Plenum Press, New York.

Hofmeyr, J.-H. S. & Cornish-Bowden, A. (1991). Quantitative assessment of regulation in metabolic systems. Eur. J. Biochem. 200, 223–236.

International Union of Biochemistry (1982). Symbolism and terminology in enzyme kinetics. Eur. J. Biochem. 128, 281–291.

International Union of Pure and Applied Chemistry (1981). Symbolism and terminology in chemical kinetics. Pure Appl. Chem. 53, 753–771.

Irvine, D. H. & Savageau, M. A. (1990). Efficient solution of ordinary nonlinear differential equations expressed in S-system canonical form. SIAM J. Num. Anal. 27, 704–735.

Kacser, H. (1983). Biochemists' alcohol problem: a case of addition to the wrong concepts? Trends Biochem. Sci. 8, 310–311.

Kacser, H. & Burns, J. A. (1973). The control of flux. Symp. Soc. Exp. Biol. 27, 65–104.

Kahn, D. & Westerhoff, H. V. (1991). Control theory of regulatory cascades. J. Theor. Biol. 153, 255–285.

King, E. L. & Altman, C. (1956). A schematic method for deriving the rate laws for enzyme-catalyzed reactions. J. Phys. Chem. 60, 1375–1378.

Kohn, M. C. & Chiang, E. (1982). Metabolic network sensitivity analysis. J. Theor. Biol. 98, 109–126.

Koshland, D. E., Jr., Némethy, G., & Filmer, D. (1966). Comparison of experimental binding data and theoretical models in proteins containing subunits. Biochemistry 5, 365–385.

Krebs, E. G. & Beavo, J. A. (1979). Phosphorylation-dephosphorylation of enzymes. Ann. Rev. Biochem. 48, 923–959.

Kruckeberg, A. L., Neuhaus, H. E., Feil, R., Gottlieb, L. D., & Stitt, M. (1989). Decreased-activity mutants of phosphoglucose isomerase in the cytosol and chloroplast of *Clarkia xantiana*. Impact of mass-action ratios and fluxes to sucrose and starch, and estimation of flux control coefficients and elasticity coefficients. Biochem. J. 261, 457–467.

Letellier, T., Reder, C., & Mazat, J.-P. (1991). CONTROL: software for the analysis of the control of metabolic networks. Comput. Appl. Biosci. 7, 383–390.

Mazat, J.-P., Jean-Bart, E., Rigoulet, M., & Guérin, B. (1986). Control of oxidative phosphorylation in yeast mitochondria. Role of the phosphate carrier. Biochim. Biophys. Acta 849, 7–15.

Meinke, M. H., Bishop, J. S., & Edstrom, R. D. (1986). Zero-order ultrasensitivity in the regulation of glycogen phosphorylase. Proc. Natl. Acad. Sci. U.S.A. 83, 2865–2868.

Meléndez-Hevia, E., Torres, N. V., Sicilia, J., & Kacser, H. (1990). Control analysis of transition times in metabolic systems. Biochem. J. 265, 195–202.

Mendes, P. (1990). GEPASI. In: Control of Metabolic Processes (Cornish-Bowden, A. & Cárdenas, M. L., eds.), p. 433, Plenum Press, New York.

Mendes, P., Kell, D. B., & Westerhoff, H. V. (1992). Channelling can decrease pool size. Eur. J. Biochem. 204, 257–266.

Middleton, R. J. & Kacser, H. (1983). Enzyme variation, metabolic flux and fitness: alcohol dehydrogenase in *Drosophila melanogaster.* Genetics 105, 633–650.

Monod, J., Changeux, J.-P., & Jacob, F. (1963). Allosteric proteins and cellular control systems. J. Mol. Biol. 6, 306–329.

Monod, J., Wyman, J., & Changeux, J.-P. (1965). On the nature of allosteric transitions: a plausible model. J. Mol. Biol. 12, 88–118.

Neuhaus, H. E. & Stitt, M. (1990). Decreased-activity mutants of phosphoglucose isomerase in the cytosol and chloroplast of *Clarkia xantiana*. II. Study of the mechanisms which regulate photosynthate partitioning. Planta 182, 445–454.

Ovádi, J. (1991). Physiological significance of metabolic channelling. J. Theor. Biol. 152, 1–22.

Padovan, A. C., Dry, I. B., & Wiskich, J. T. (1989). An analysis of the control of phosphorylation-coupled respiration in isolated plant mitochondria. Plant Physiol. 90, 928–933.

Pettersson, G. (1991). No convincing evidence is available for metabolite channelling between enzymes forming dynamic complexes. J. Theor. Biol. 152, 65–69.

Pettersson, G. & Ryde-Pettersson, U. (1988). A mathematical model of the Calvin photosynthesis cycle. Eur. J. Biochem. 175, 661–672.

Porteous, J. W. (1990). Control analysis: a theory that works. In: Control of Metabolic Processes (Cornish-Bowden, A. & Cárdenas, M. L., eds.), pp. 51–67, Plenum Press, New York.

Reder, C. (1988). Metabolic control theory: a structural approach. J. Theor. Biol. 135, 175–201.

Ricard, J. & Cornish-Bowden, A. (1987). Co-operative and allosteric enzymes: 20 years on. Eur. J. Biochem. 166, 255–272.

Salter, M., Knowles, R. G., & Pogson, C. I. (1986a). Transport of the aromatic amino acids into isolated rat liver cells. Properties of uptake by two distinct systems. Biochem. J. 233, 499–506.

Salter, M., Knowles, R. G., & Pogson, C. I. (1986b). Quantification of the importance of individual steps in the control of aromatic amino acid metabolism. Biochem. J. 234, 635–647.

Sauro, H. M. (1990). Regulatory responses and control analysis: assessment of the relative importance of internal effectors. In: Control of Metabolic Processes (Cornish-Bowden, A. & Cárdenas, M. L., eds.), pp. 225–230, Plenum Press, New York.

Sauro, H. M. & Fell, D. A. (1991). SCAMP: a metabolic simulator and control analysis program. Mathl. Comp. Modeling 15, 15–28.

Sauro, H. M., Small, J. R., & Fell, D. A. (1987). Metabolic control and its analysis. Extensions to the theory and matrix analysis. Eur. J. Biochem. 165, 215–221.

Savageau, M. A. (1969a). Biochemical systems analysis. 1. Some mathematical properties of the rate laws for the component enzymatic reactions. J. Theor. Biol. 25, 365–369.

Savageau, M. A. (1969b). Biochemical systems analysis. 2. The steady-state solutions for an n-pool system using a power-law approximation. J. Theor. Biol. 25, 370–379.

Savageau, M. A. (1970). Biochemical systems analysis. 3. Dynamic solutions using a power-law approximation. J. Theor. Biol. 26, 215–226.

Savageau, M. A. (1976). Biochemical Systems Analysis: A Study of Function and Design in Molecular Biology, Addison-Wesley, Reading, Massachusetts.

Savageau, M. A. (1992). Dominance according to metabolic control analysis; major achievement or house of cards? J. Theor. Biol. 154, 131–136.

Schulz, A. R. (1991). Algorithms for the derivation of flux and concentration control coefficients. Biochem. J. 278, 299–304.

Schuster, S. and Heinrich, R. (1992). The definitions of metabolic control analysis revisited. BioSystems 27, 1–15.

Sen, A. K. (1990a). Metabolic control analysis. An application of signal flow graphs. Biochem. J. 269, 141–147.

Sen, A. K. (1990b). Application of electrical analogues for control analysis of simple metabolic pathways. Biochem. J. 272, 65–70.

Sen, A. K. (1991a). Quantitative analysis of metabolic regulation. A graph-theoretic approach using spanning trees. Biochem. J. 275, 253–258.

Sen, A. K. (1991b). Calculation of control coefficients of metabolic pathways. A flux-oriented graph-theoretic approach. Biochem. J. 279, 55–65.

Shacter, E., Chock, P. B., & Stadtman, E. R. (1984). Energy consumption in a cyclic phosphorylation/dephosphorylation cascade. J. Biol. Chem. 259, 12260–12264; correction (1985): J. Biol. Chem. 260, 6501.

Small, J. R. & Fell, D. A. (1990). Covalent modification and metabolic control analysis. Modifications to the theory and their application to metabolic systems containing covalently modifiable enzymes. Eur. J. Biochem. 191, 405–411.

Srere, P. A. (1987). Complexes of sequential metabolic enzymes. Ann. Rev. Biochem. 56, 89–124.

Stadtman, E. R. (1970). Mechanisms of enzyme regulation in metabolism. In: The Enzymes (3rd ed., Boyer, P. D., ed.), vol. 1, pp. 397–459.

Stadtman, E. R. & Chock, P. B. (1977). Superiority of interconvertible enzyme cascades in metabolic regulation. Analysis of multicyclic systems. Proc. Natl. Acad. Sci. U.S.A. 74, 2761–2766.

Stadtman, E. R. & Chock, P. B. (1978). Interconvertible enzyme cascades in metabolic regulation. Curr. Topics Cell. Reg. 13, 53–95.

Szedlacsek, S. E., Cárdenas, M. L., & Cornish-Bowden, A. (1992). Response coefficients of interconvertible enzyme cascades towards effectors that act on one or both modifier enzymes. Eur. J. Biochem. 204, 807–813.

Torres, N. V., Mateo, F., Meléndez-Hevia, E., & Kacser, H. (1986). Kinetics of metabolic pathways. A system *in vitro* to study the control of flux. Biochem. J. 234, 169–174.

Umbarger, H. E. (1956). Evidence for a negative-feedback mechanism in the biosynthesis of isoleucine. Science 123, 848.

Westerhoff, H. V. & Chen, Y.-D. (1984). How do enzyme activities control metabolite concentrations? An additional theorem in the theory of metabolic control. Eur. J. Biochem. 142, 425–430.

Westerhoff, H. V., Groen, A. K., & Wanders, R. J. A. (1984). Modern theories of metabolic control and their applications. Biosci. Rep. 4, 1–22.

Westerhoff, H. V., Koster, J. G., van Workum, M., & Rudd, K. E. (1990). On the control of gene expression. In: Control of Metabolic Processes (Cornish-Bowden, A. & Cárdenas, M. L., eds.), pp. 399–412, Plenum Press, New York.

Woodrow, I. E. (1986). Control of the rate of photosynthetic carbon dioxide fixation. Biochim. Biophys. Acta 851, 181–192.

Wright, B. E. & Albe, K. R. (1990). A new method for estimating enzyme activity and control coefficients *in vivo*. In: Control of Metabolic Processes (Cornish-Bowden, A. & Cárdenas, M. L., eds.), pp. 317–328, Plenum Press, New York.

Yanofsky, C. (1989). Tryptophan synthetase of *E. coli*: a multifunctional multicomponent enzyme. Biochim. Biophys. Acta 1000, 133–137.

Yates, R. A. & Pardee, A. B. (1956). Control of pyrimidine biosynthesis in *Escherichia coli* by a feed-back mechanism. J. Biol. Chem. 221, 757–770.

EXPERIMENTAL APPROACHES TO STUDYING ENZYMES *IN VIVO*: THE APPLICATION OF NUCLEAR MAGNETIC RESONANCE METHODS TO GENETICALLY MANIPULATED ORGANISMS

Simon-Peter Williams, Alexandra M. Fulton, and Kevin M. Brindle

Advances in Molecular and Cell Biology
Volume 11, pages 65–91.

ISBN: 1-55938-844-7

I. INTRODUCTION

Many of today's most interesting problems in enzymology are also those most intractable to study with conventional techniques. There is a great debate about how the intracellular environment influences the behavior of enzymes, and especially how groups of enzymes work together as pathways in the intact cell. For example, the proposed benefits of one enzyme's product being directly transferred to another enzyme as its substrate ("substrate channeling") include rate enhancement and reduction in the free metabolite pool size. However even the theoretical advantages of such interactions remain deeply controversial, while experimental studies of enzyme–enzyme interactions have been criticized for employing markedly non-physiological conditions or for relying on inadequate theories of enzyme kinetics for their interpretation (see other chapters in this volume and, for example, Chock and Gutfreund, 1988; Walsh et al., 1989; Vas and Batke, 1990; Aragon and Sols, 1991; Brooks and Storey, 1991a, b; Cornish-Bowden, 1991; Ovádi, 1991; Wu et al., 1991; Mendes et al., 1992). The paper by Ovádi is the lead article in a special issue of the *Journal of Theoretical Biology*, which brings together many authors to critically discuss the evidence for the existence and significance of enzyme-enzyme interactions in the cell.

The properties of enzymes peculiar to the intracellular environment, particularly their high concentrations relative to their substrates and the very high total protein concentration, have been thought to render invalid many experimental studies *in vitro* as models of the situation *in vivo*. Models of fluxes through metabolic pathways, incorporating theories such as metabolic control analysis (Kacser and Burns, 1973; Heinrich and Rapoport, 1974; Kacser and Porteous, 1987), are also heavily influenced by assumptions made about the interactions between the components. High protein concentrations, for example, have been shown to have a large influence on the operation, and hence the appropriate mathematical description of moiety-conserved pathways (Kholodenko et al., 1992). Unfortunately, experimental validation of the available models has been rather limited. Aflalo and Segel have made some interesting theoretical and experimental studies of ATP distribution within cells (Aflalo and Segel 1992) based on different source/sink models, but the authors themselves discuss some difficulties with the experiments.

It is obvious that enzymes *in vitro* demonstrate a great many more properties than are physiologically relevant, and, conversely, that some physiologically important properties are not apparent *in vitro*. In view of the present confusion, which can seem like a re-emergence of the "vital force" controversies of previous centuries, there is a clear need to develop noninvasive experimental techniques which allow a specific enzyme to be studied in the intact cell. Some of these techniques will include nuclear magnetic resonance (NMR), which has been used in noninvasive studies of the biochemistry of tissues and animal organs for some twenty years (Moon and Richards, 1973; Hoult et al., 1974).

NMR can certainly now be regarded as a mature technique for the investigation of metabolism, particularly in studies of cellular energetics, pH, ion levels and transport, and metabolite fluxes and concentration profiles. The pathways of ATP generation have been extensively investigated by ^{31}P NMR in muscle, originally because muscle presented opportunities to work with a large, relatively homogenous excitable tissue mass containing readily detectable quantities of the bioenergetically important phosphorus metabolites. The scope of NMR in the biological sciences has been well reviewed (Iles et al., 1982; Avison et al., 1986; Radda, 1986; Brindle and Campbell, 1987; Radda, 1992).

Studies of the abnormal, especially human patients and animal models, have greatly advanced our understanding of biochemistry. NMR has been able to make some unique contributions to the understanding of certain disease states, helping to explain the underlying biochemistry of the abnormality. For example Radda (1992), cites the case of a patient with post-viral fatigue syndrome ("ME") who showed, in a ^{31}P NMR study, an abnormal pattern of phosphocreatine depletion and acidosis in skeletal muscle accompanying the onset of exercise. These observations suggested a loss of coordination between glycolytic and oxidative metabolism; it would have been difficult to gather the same data in a clinical setting using conventional invasive techniques. Chance and his colleagues studied a phosphofructokinase-deficient patient and were able to gain useful insights into the roles of ADP and inorganic phosphate in controlling mitochondrial activity, showing that the ATP and P_i concentrations were nearly constant in the post-exercise recovery period and that the increased ADP concentration must be largely responsible for the stimulation of oxidative phosphorylation (Chance et al., 1982).

Even when a final diagnosis is made by other techniques, NMR can be invaluable in showing an investigator where to look. For example, Radda (1992) also cites the example of a patient with a variety of symptoms that identified an unspecified myopathy. NMR examinations showed that utilization of PCr and P_i during exercise were faster, post-exercise recovery slower, and the sugar phosphates accumulated to higher levels than normal. However, pH changes were normal. These observations suggested a defect in the phosphate-dependent malate–aspartate shuttle, leading to slow phosphate uptake by the mitochondria and reduced oxidative phosphorylation. Armed with this hypothesis based on NMR measurements, a biopsy and "wet" biochemistry confirmed the nature of the defect.

The burgeoning repertoire of molecular genetic techniques now makes it possible to tailor-make "abnormal" organisms for study (Goodridge, 1990). We are no longer limited to studies of simple lesions; it is possible to alter the activity of a particular enzyme either up or down, and to express engineered proteins with altered catalytic, regulatory, or structural properties. Advances in our understanding of enzyme function allow rational selection of these mutations (Bone and Agard, 1991; Hedstrom et al., 1991), for example modulation of some property of the protein such as substrate specificity or regulatory response (Khouri et al., 1991). These proteins can be expressed in cells with genetic backgrounds tailored to facilitate the interpretation of experimental results, for example by removing a wild-type activity prior to expressing a mutant enzyme. For example, Kispal and co-workers have used mutant yeasts, site-directed citrate synthase mutants, and ^{13}C NMR to demonstrate a functionally important structural role for the mitochondrial citrate synthase (Kispal et al., 1989). This work is described more fully in the chapter by Srere et al. (this volume). The relative ease with which genetic manipulations can now be executed means that metabolic studies will be designed more and more around such systems, extending the already wide range of *in vitro* mutagenesis studies to the intact cell.

Although proteins tend to be thought of as catalytic or structural elements, they can, of course, act as indicators of their ligand concentrations if they have an appropriate reacting chromophore or similar reporting group. The color changes accompanying the oxygenation of hemoglobin have long been known, and can now be measured quite reliably near the skin surface to determine the blood oxygen concentration (Sevick et al., 1991). The introduction of fluorescent or NMR-detectable groups into a particular protein can make it possible to observe changes in its ligand concentrations in a similar manner, allowing the protein to serve as a "probe" or indicator group just as small, dye-like molecules have been used, for example, as probes of the intracellular magnesium ion concentrations (London, 1991). Recent advances in fluorescent and NMR methodologies make it possible to observe such labeled proteins *in situ* (Brindle et al., 1989; Adams et al., 1991; Knight et al., 1991; Williams et al., 1993), opening up a new range of investigations into the microenvironments of proteins in the intact cell.

The combination of genetically manipulated experimental models and the non-invasive techniques of NMR offer some promise in advancing our understanding of enzymology *in vivo*. Already, this combination of relatively new methodologies is bringing new insights into the metabolism of intact cells. The following sections illustrate some of the more important uses of NMR in metabolic investigations, particularly highlighting those instances where NMR has been used to study genetically manipulated organisms. NMR has had many uses in measuring metabolite concentrations and fluxes, and is also now beginning to provide information about macromolecules in the cell.

II. MEASUREMENT OF METABOLITE CONCENTRATIONS IN THE INTACT CELL

A. Direct Measurement of Metabolites

Appropriately calibrated, NMR is one of the most powerful tools for the noninvasive measurement of intracellular metabolite concentrations. It has been particularly useful in measuring the pH (Moon and Richards, 1973; Roberts et al., 1981; Seo et al., 1983) and the levels of the major phosphorylated metabolites such as

ATP, inorganic phosphate, and phosphocreatine. Following the changes in NMR-observable metabolite levels has become a routine part of many metabolic studies, and the technique has been applied in many different systems. For reviews of NMR in the study of metabolism, see Iles et al., 1982; Avison et al., 1986; and Cohen et al., 1989. Since the artifactual generation of phosphate during cell extraction procedures makes it difficult to determine the intracellular inorganic phosphate concentration from cell extracts (Seraydarian et al., 1961), the noninvasive measurement of phosphate by ^{31}P NMR has been particularly valuable. NMR reveals the presence of unexpected compounds rather than simply those specifically sought by the experimenter, and is nondestructive.

Unfortunately, the inherent insensitivity of NMR has largely confined it to studies of metabolites present in at least tens of micromolar concentration. Recent work has focused on trying to extend the range of metabolites that can be detected by

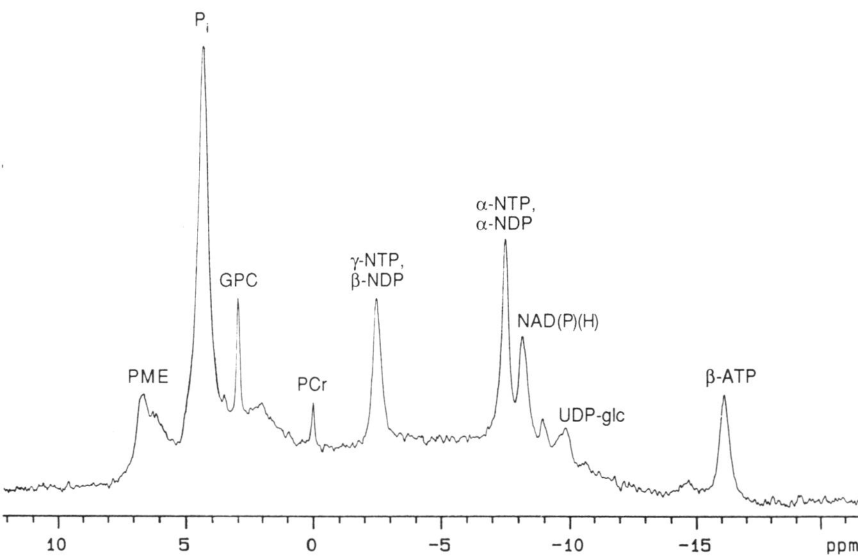

Figure 1. ^{31}P NMR spectrum of mammalian cells (CHO-K1) growing on macroporous beads within a hollow-fiber cartridge, in which the cells can continue growing to reach tissue densities. The spectrum was acquired over several hours. The peaks can be identified as follows: PME: phosphomonoesters (including sugar phosphates); P_i: inorganic phosphate, mostly that in the perfusion medium; GPC: the phosphodiester glycerophosphocholine; PCr: phosphocreatine; γ-NTP,β-NDP: γ-phosphates of nucleoside triphosphates and β-phosphates of nucleoside diphosphates (principally γ-ATP); α-NTP, α-NDP: α-phosphates of NDP and NTP (principally α-ATP); NAD(P)(H): $NAD(P)^+$ and their reduced forms; UDP-glc: other diphosphodiesters, especially UDP-glucose. The spectrum was kindly provided by Michelle E. Jackson of this laboratory.

NMR, and through spatial localization, the tissues in which they can be specifically detected (Koretsky and Williams, 1992).

B. Indirect Measurements of Metabolites

Use of Equilibria

The intracellular free ADP concentration is an important parameter in regulating cellular energy metabolism (Atkinson, 1977). Although ADP and ATP are readily distinguished by ^{31}P NMR in cell-free extracts, the deterioration in spectral resolution that accompanies recording spectra from cells and animal organs means that it is hardly ever possible to distinguish ADP from ATP reliably in the intact cell. Unfortunately, therefore, it is not possible to obtain information about the free ADP concentration from simple ^{31}P NMR spectra of cells. There is in any case some evidence that ADP and possibly ATP may not be wholly visible in normal NMR spectra, perhaps because they are sequestered in the mitochondrion or bound to macromolecules where their resonances relax too quickly to be detected in the NMR experiment (Veech et al., 1979; Iles et al., 1982; Stubbs et al., 1984; Murphy et al., 1988; Ikai et al., 1991; Masson and Quistorff, 1992). However, with information that can be obtained from simple ^{31}P NMR spectra of cells or organs (the pH, ATP, and PCr concentrations), and the total creatine concentration, the equilibrium constant for the creatine kinase reaction allows the calculation of the free ADP concentration provided that all of the components are in fact present in the same compartment and that the enzyme is catalyzing a reaction that is near-to-equilibrium. The creatine kinase equilibrium has been used to make free ADP measurements in a number of systems (Lawson and Veech, 1979; Matthews et al., 1982; Shoubridge et al., 1984; Brindle et al., 1990; Brosnan et al., 1990), but this has, of course, only been possible in tissues where creatine kinase is expressed. One of the most exciting applications of molecular genetics in this field has been to facilitate the expression of creatine kinase, and hence allow measurements of free ADP, in tissues such as yeast (Brindle et al., 1990) and the livers of transgenic mice (Brosnan et al., 1990) which do not normally express creatine kinase. The applications in transgenic mice are discussed fully in the chapter by Koretsky (this volume).

Use of Indicator or "Probe" Molecules

Fluorescent or NMR-detectable indicators can be used to measure pH and cation concentrations (London, 1991). Probes or indicators interact with some target molecule in a way that changes the spectroscopic properties of the probe, and reveals the concentration of the target molecule. Of course, most of the probes of intracellular ion concentrations or pH, for example, are detected optically and are based on relatively small synthetic organic molecules. Some of these fluorescent probes have been systematically modified to make them NMR-detectable, for example the fluorinated divalent cation probes (Kirschenlohr et al., 1988).

To avoid the inevitable difficulties of building small-molecule probes to specifically detect a metabolite even as simple as, say, glucose, use has recently begun to be made of proteins as probes. Since virtually all metabolites bind to an enzyme involved in their synthesis or degradation, nature has effectively provided us with a repertoire of ready-made metabolite-binding domains from which to construct protein probes of the intracellular environment. As with the small-molecule probes, these too have been constructed both for detection optically and with NMR.

Cytochromes and other pigmented proteins have been studied in the intact cell by optical methods for many years (Keilin, 1925). Millikan measured the changes in oxygen tension during muscle contraction by examining cat muscle globins *in vivo* with a combined microscope and colorimeter over 50 years ago (Millikan, 1937). Recent improvements in optical instrumentation and theory now allow the oxygenation state of the blood in human brain to be determined by optical means (Sevick et al., 1991). NMR studies of specific proteins have also been possible in circumstances where the protein observed was naturally present at high concentration and had some serendipitous property that allowed unambiguous resonance assignment. For example, in the human erythrocyte ^1H resonances from surface histidines of hemoglobin have been observed, and in muscle hyperfine- and ring-current-shifted ^1H resonances from myoglobin have been detected. The chemical shifts of these resonances titrate with pH (Brown et al., 1977) and pO_2 (Kreutzer et al., 1992) respectively, and thus could be used to estimate these parameters in the intact cell.

The techniques of molecular genetics have made it possible to express protein probes in any tissue of interest, and even to target them to subcellular compartments. Several recent examples use the naturally luminescent calcium-binding protein aequorin, from a luminescent jellyfish, as a probe of intracellular free calcium concentrations. The gene for aequorin has been cloned and expressed in various cell types, resulting in cells which contain apoaequorin. Soaking the cells in a solution of the luminescence cofactor converts the apoaequorin to native, bioluminescent, aequorin on a time scale of several hours. Measurements of the luminescence then allow changes in the calcium concentration to be directly visualized. Calcium ion transients have been monitored in cells of tobacco seedlings under conditions of environmental and infective stress using aequorin (Knight et al., 1991). Engineering aequorin to make it responsive to ligands other than calcium has been suggested as a general way of investigating intracellular metabolite concentrations (Campbell, 1990). Adding a leader sequence peptide directing the apoaequorin to the mitochondrion has enabled Rizzuto et al. to measure intramitochondrial calcium transients (Rizzuto et al., 1992). The complementary approach, to derivatize a ligand-binding protein to give it the desired optical properties, has been illustrated with cAMP-dependent protein kinase and, more recently, with calmodulin. Both these proteins have been fluorescently tagged by chemical derivatization in vitro, microinjected and studied in single cells. Fluorescence ratio imaging of labeled calmodulin allowed investigations of its activation within the cell (Hahn et al., 1992) while resonance energy transfer studies of the kinase allowed the measurement of cAMP levels within single cells (Adams et al., 1991).

The approach of modifying a ligand-binding protein to elicit a detectable optical or NMR response requires some form of specific protein labeling. This may be simpler with NMR detection than optical detection, because the label for NMR detection can be as simple a single atom, which should be easier to introduce into a protein than a relatively complex fluorophore. Although fluorescence detection is very sensitive, it and other optical methods are limited to situations where the light used can penetrate the tissue examined. This would seem to confine the optical techniques to studies of single cells or cell suspensions and the surface layers of organs, whereas with NMR it is possible, in principle, to monitor labeled proteins deep within an animal organ, for example. However, advances in low-light imaging and image processing are allowing optical techniques to operate at ever lower chromophore concentrations and in smaller sample volumes. See Chance (1991) for a recent review of the development and state of optical spectroscopic methods. Several of the studies cited above examine fluorophores at subcellular resolution. This is likely to be increasingly important as the heterogeneity of cell populations and intracellular locations becomes more important in our understanding of how cells and tissues function. The reader is referred to the chapter by Pagliaro (this volume).

NMR techniques are also improving, both by the application of higher magnetic field strengths and with novel labeling strategies to create probes *in situ*. We have developed, in the yeast *Saccharomyces cerevisiae*, a practical method to achieve the specific labeling required. The system uses ^{19}F-labeled amino acids which can be monitored *in vivo* using ^{19}F NMR. Fluorine-19, the 100 percent naturally-abundant isotope, is an excellent label in this context. It is, like ^{1}H, ^{31}P and ^{13}C, a spin-1/2 nucleus which gives rise to relatively simple NMR spectra (Harris, 1986). Unlike these other commonly studied nuclei, it also combines high receptivity to detection by NMR with a wide chemical shift dispersion, and it is absent from most living tissue, avoiding interference from background signals (Gerig, 1989). Fluorine labels have been simply and successfully introduced into many different proteins using biosynthetic incorporation of fluoroamino acids (Sykes and Weiner, 1980), and the purified proteins have then been studied *in vitro* using ^{19}F NMR. We have extended such studies to the intact cell, using an inducible expression system in yeast to achieve the specific labeling of a single protein species.

A single protein can be selectively labeled in the intact cell using an inducible expression vector containing the protein coding sequence. When induced in stationary phase cells, the target protein is the predominant protein synthesized in the cell and a labeled amino acid added to the medium at this point is incorporated essentially into a single protein. A protein labeled with 5-fluorotryptophan, for example, can then be observed in the intact cell using ^{19}F NMR. In principle, the

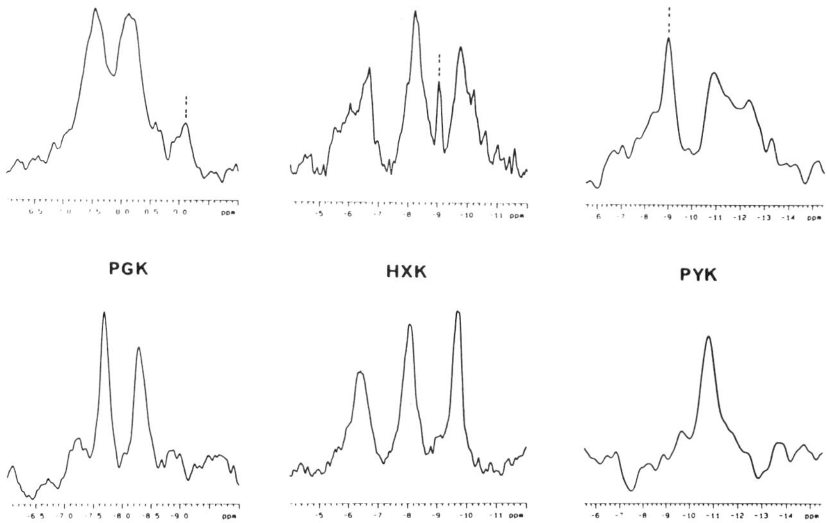

Figure 2. ^{19}F NMR spectra of 5-fluorotryptophan-labeled 3-phosphoglycerate kinase (PGK), hexokinase (HXK), and pyruvate kinase (PYK) in the intact yeast cell (top) and *in vitro* (bottom). 5-fluorotryptophan was biosynthetically incorporated into the specified protein using an induction protocol as described in the text. The spectra were obtained from 4 g of cells immobilized in agarose-gel threads and perifused aerobically with a glucose-containing buffer at 30 °C. *In vivo* (top series), some unincorporated fluorotryptophan is visible at around 9 ppm, indicated by dashed lines over the relevant peaks. The isolated enzymes were studied in buffers of physiological ionic strength (0.15 M) and pH (7.2). The relative viscosity of the hexokinase and pyruvate kinase solutions was increased to 2 with glycerol, while the PGK solution was in simple aqueous solution. The PGK solution contained approximately 3 mM MgATP, 50 μM MgADP; the HXK solution contained approximately 5 mM glucose; the PYK solution contained approximately 5 mM fructose-1,6-bisphosphate. The chemical shifts of the resonances were responsive to the presence of these metabolites, which are substrates (PGK, HXK) or allosteric effectors (PYK) of these enzymes. More detailed characterization of the enzymes *in vitro* allows the deduction of properties *in vivo*, in terms of apparent ligand concentrations and solution viscosities.

method we have developed may be applied to other organisms, including higher animals, but its application in these systems will depend on the development of appropriate methods for rapidly inducing the synthesis of a particular protein species.

We have fluorine-labeled the two tryptophan residues in yeast 3-phosphoglycerate kinase (PGK) by biosynthetic incorporation of 5-fluorotryptophan (Brindle et al., 1989), and shown that the ^{19}F NMR chemical shifts of these tryptophan resonances are sensitive primarily to the MgADP concentration. Calibrated *in vitro*

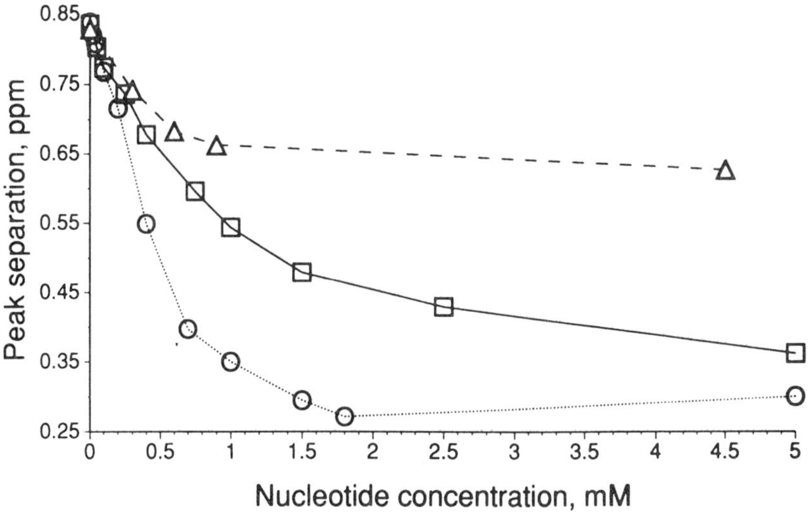

Figure 3. Titration of the ^{19}F NMR chemical shift separation (Δppm) between the two resonances of 5-fluorotryptophan-labeled 3-phosphoglycerate kinase with magnesium nucleotides. AMP(△), ADP(○), ATP(□). Solutions of similar PGK concentration (0.6 mM), pH (7.2) and ionic strength (0.15 M) to those measured in intact yeast cells (see Figure 2) were studied in the presence of different nucleotide concentrations. The simple interpretation of these titration curves is compromised by the complexity of the binding processes to PGK, which has two non-independent nucleotide-binding sites (Williams et al., 1993), but they illustrate the general principle of using ^{19}F NMR titrations of fluorine-labeled proteins to measure ligand concentrations.

(as is necessary with any probe) and studied *in vivo*, the labeled protein can be used to measure the intracellular free MgADP concentration in the same way that hemoglobin and myoglobin were used as probes of pH and pO_2 respectively (Brown et al., 1977; Kreutzer et al., 1992). We have been able to place an upper limit of approximately 50 μM on the free cytoplasmic ADP concentration in stationary phase anaerobic glucose-fed yeast cells, a value that is much lower than the total extractable ADP concentration of ~400 μM (Williams et al., 1993). Earlier calculations of the free ADP level in yeast tentatively arrived at a similar value, based on measurements of glycolytic metabolite concentrations and assumptions about the equilibrium constants of several coupled reactions (Brindle, 1988).

III. THE REGULATION OF METABOLISM

The introduction of metabolic control analysis (MCA) has promoted a more rigorous and quantitative approach to the study of metabolic control (Kacser and Burns, 1973; Heinrich and Rapoport, 1974) . The theory provides a quantitative description of the operation of a pathway, parameterized to show how it can be expected to respond to perturbations in enzyme activities, effector or metabolite concentrations. The theory has been extended since its first introduction to cover many physiologically relevant phenomena such as regulatory cascades, branched pathways, multifunctional enzymes, substrate cycles, and high protein concentrations. The principles and application of MCA have recently been reviewed by Fell (1992) and are described in some detail in the chapter by Cornish-Bowden (this volume). Enzymes in the pathway are characterized by flux control coefficients which describe how sensitive the flux is to small changes in the activity of the enzyme. They are defined as the ratio of the fractional change in flux to the fractional change in enzyme activity. The greater the flux control coefficient, the greater the effect on the flux that a given change in enzyme activity will have. The origin of the change in activity does not matter; it could be a change in the quantity of an enzyme or a change in its kinetic properties (perhaps the result of phosphorylation or a change in the concentration of an allosteric effector). The influence that a change in metabolite concentration will have on the flux is defined as the elasticity of the enzyme for that metabolite.

Metabolic control analysis not only shows how to relate the flux and intermediary metabolite levels in a pathway to the kinetic properties of the enzymes involved but it also offers a simple practical way to determine the relevant coefficients. For example the flux control coefficient of an enzyme can be determined by measuring pathway flux at different levels of enzyme activity. Changes in enzyme levels can be readily achieved *in vivo* using molecular genetic techniques. These make it possible to elevate or to reduce the activity of any chosen enzyme in the cell, in a way quite specific for a single isoform of a chosen enzyme and without the specificity, availability, and permeability problems associated with drugs. MCA has been extended to deal with the large stepwise changes in enzyme activity that typically result from such experiments; see, for example, (Niederberger et al., 1992). The characterization of such manipulations by NMR in the intact cell should bring further understanding of the control of cellular metabolism.

Several *in vivo* NMR studies of genetically induced metabolic perturbations have been undertaken, particularly in the major pathways of fuel utilization and energy production. Information can be gained from measuring the fluxes in these systems and, using isotope labels, the fate of the original metabolite. The use of NMR-detectable isotope labels (especially ^{13}C and ^{15}N) complements and extends the long established techniques of radiochemistry in metabolism, because NMR offers more information and, since it uses stable isotopes, can readily be applied even in human studies (see, for example, Gruetter et al., 1992). Not only does NMR enable the investigator to know which compounds the label is in, but also which position(s) in the molecule the label occupies, without the need for stepwise chemical degradation. Srere and co-workers have made good use of ^{13}C isotopomer analysis in their investigations of the structural organization of the Krebs cycle in genetically modified yeast (Kispal et al., 1989). The reader is referred to the chapter by Srere et al. (this volume) for further details.

Studies of nitrogen metabolism have been relatively rare because of the difficulty in working with the radioisotope, ^{13}N, which has a half-life of just ten minutes, although some studies have been performed with ^{13}N and ^{15}N using mass spectrometry for isotope detection. NMR allows the isotope labels to be studied noninvasively without the separation and derivatization treatments required by mass spectrometry studies. The use of an inverse detection technique to improve the ordinarily very poor sensitivity of ^{15}N NMR detection (Brindle et al., 1984) also allows the fractional labeling to be determined in the intact tissue, and has made the study of ^{15}N-labeled metabolites much more practical. Brindle and co-workers have made a study of cellular nitrogen metabolism using ^{15}N-labeled compounds and cell lines in which the glutamine synthase (GS) genes have been amplified (Street et al., 1993). The rate of ^{15}N assimilation from ammonia into the amide of glutamine was measured and used to show that GS activity in the intact cell was negligible in normal cells, and low even in cells with GS activities tens of times higher than normal. The authors were able to refute the suggestion that a significant "futile cycle" operates between GS and glutaminase, thought by some to be an important component of the cell's "maintenance energy" requirements.

Other studies have also concentrated on cellular energy metabolism, for example the roles of the glycolytic and oxidative contributions to ATP synthesis. In glycolysis, noninvasive ^{31}P NMR magnetization transfer techniques (Forsén and Hoffman, 1963; Campbell et al., 1977; Brindle and Campbell, 1987) were used to measure the exchange between P_i and ATP in the near-equilibrium reactions catalyzed by glyceraldehyde-3-phosphate dehydrogenase (GAPDH) and phosphoglycerate kinase (PGK) (Brindle and Krikler, 1985; Brindle and Radda, 1987; Campbell-Burk et al., 1987; Brindle, 1988). The reactions are as follows:

$$\text{GA3-P} + \text{NAD}^+ + \text{P}_i \leftrightarrow \text{1,3-BPG} + \text{NADH} + \text{H}^+ \quad (1)$$

(GAPDH)

$$\text{1,3-BPG} + \text{Mg.ADP} \leftrightarrow \text{Mg.ATP} + \text{3-PG} \quad (2)$$

(PGK)

Brindle undertook a study of yeasts in which multicopy plasmids were used to elevate the activity of PGK in the cytoplasm (Brindle, 1988). Conventional biochemical assays of glucose consumption and ethanol production were used to estimate the net glycolytic flux. The net glycolytic and unidirectional P_i to ATP fluxes were measured in anaerobic cells at levels of PGK activity ranging from wild-type to twenty times higher than wild-type, and at levels of GAPDH activity reduced by iodoacetate treatment to approximately 25% of wild-type levels. Changes in the enzyme activities resulted in relatively small changes in the net flux,

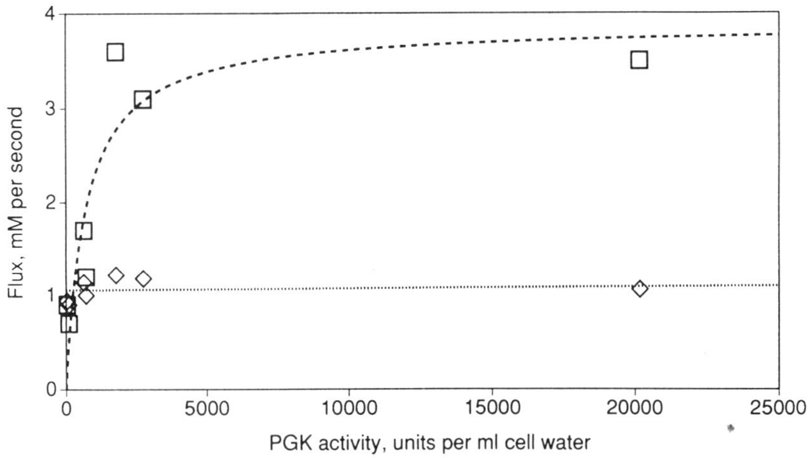

Figure 4. Effect of genetically manipulating the activity of the glycolytic enzyme 3-phosphoglycerate kinase (PGK) in yeast on the net glycolytic flux (□) and on the γ-phosphate of ATP to inorganic phosphate exchange flux (◇). The net glycolytic flux was estimated from the average rate of glucose consumption and ethanol production from anaerobically perifused agarose-gel-entrapped cells. The ATP↔P_i exchange was measured by a ^{31}P NMR magnetization transfer method (see text). At wild-type and higher levels of PGK, the exchange flux is greater than the net flux. As the PGK activity is lowered, the difference decreases until the ATP↔P_i exchange flux is equal to the net glycolytic flux, when the reaction at PGK must be proceeding essentially unidirectionally.

showing that both enzymes had near-zero flux control coefficients for glycolysis. However, the P_i to ATP exchange flux increased in approximate proportion to the PGK activity, showing that for that flux, PGK had a flux control coefficient of almost one in wild-type cells.

Further experiments using a PGK-deficient strain with low-copy-number centromeric plasmids to introduce low levels of PGK activity (Williams, 1992) showed that reducing the PGK activity to approximately 2% of wild-type levels caused a drop in glycolytic rate of only 25%. With these low concentrations of PGK, the P_i to ATP exchange flux was equal to the net glycolytic flux, indicating that the PGK reaction was now operating far from equilibrium.

Thus, ^{31}P NMR magnetization transfer measurements of $P_i \rightarrow$ ATP flux in wild-type cells reflects primarily the activity of PGK and can be used to investigate the kinetic properties of the enzyme in the intact cell. The activity expressed by the enzyme in the cell was shown to be similar to that expressed by the isolated enzyme *in vitro* under conditions designed to resemble those *in vivo* in terms of pH, substrate concentrations etc. (Brindle, 1988). Thus, there are no unusual kinetic properties conferred upon this enzyme by the intracellular environment in the yeast cell.

The exchange reaction catalyzed by GAPDH and PGK also contributes to the $P_i \rightarrow$ ATP flux determined by ^{31}P NMR magnetization transfer measurements in other tissues such as heart muscle and must be allowed for before the flux due to oxidative phosphorylation can be calculated (Brindle, 1988; see also Doumen and La Noue, this volume). Preliminary measurements of the P:O ratio for oxidative phosphorylation in yeast strains expressing low levels of PGK showed that it was approximately three (Williams, 1992), which is similar to previous estimates in which the glycolytic contribution to the $P_i \rightarrow$ ATP flux had been allowed for (Brindle, 1988). The similarity between this P:O ratio and that determined in isolated mitochondria catalyzing net ATP synthesis suggests the ATP synthase is operating unidirectionally *in vivo*, although it has been shown to be reversible *in vitro* (La Noue et al., 1986; see Doumen and La Noue, this volume).

One of the most complex patterns of enzymatic regulation in glycolysis is shown by phosphofructo–kinase, PFK1. It has a wide range of allosteric effectors *in vitro* and catalyzes a reaction that is far from equilibrium *in vivo*. Its effector concentrations change significantly under different metabolic conditions, and these observations have led to the suggestion that PFK1 is a key point of regulation in the glycolytic pathway. PFK1 may be largely responsible for the Pasteur effect (the sparing of glucose consumption under aerobic relative to anaerobic conditions) (Boscá and Corredor, 1984; Ramaiah, 1984; Porteous, 1985). To investigate the effect of increasing the PFK1 concentration in the cell, Davies and Brindle used a multicopy plasmid expressing both the α and β subunits of PFK1 to increase the enzyme's activity fivefold in the yeast *Saccharomyces cerevisiae*. ^{31}P NMR measurements on the intact cells showed that there were no significant differences between control and PFK1 overproducing cells in terms of their intracellular pH, nor in the concentrations of sugar phosphates, P_i, NTP or NAD(H). At wild-type levels of PFK1, the cells showed a significant Pasteur effect, the glucose consumption rate being reduced by some 20% upon aerobiosis. However, cells with five times the normal level of PFK1 showed the same glucose consumption rate, equal to the rate in anaerobic control cells, under both aerobic and anaerobic conditions, i.e. the Pasteur effect was abolished by overexpressing PFK1 (Davies and Brindle, 1992). The total ATP turnover remained the same, but the balance between glycolytic and oxidative phosphorylation was changed. A similar increase in the aerobic glycolytic flux was observed in mammalian cells overexpressing PFK1 by Elson and colleagues, in work which is discussed briefly later.

Davies and Brindle further showed, in cell extracts, that there were no significant differences in the levels of some other glycolytic and mitochondrial enzymes (hexokinase, pyruvate kinase, citrate synthase, cytochrome c oxidase) or metabolites and effectors (cAMP, AMP, ADP, ATP, citrate, fructose-1,6-bisphosphate). The only enzyme or effector concentrations that were significantly altered appeared to be those of phosphofructo-2-kinase, PFK2, and its product fructose-2,6-bisphosphate, a potent allosteric effector of PFK1 (van Schaftingen, 1987). The changes in fructose-2,6-bisphosphate were shown to quantitatively account for the gly-

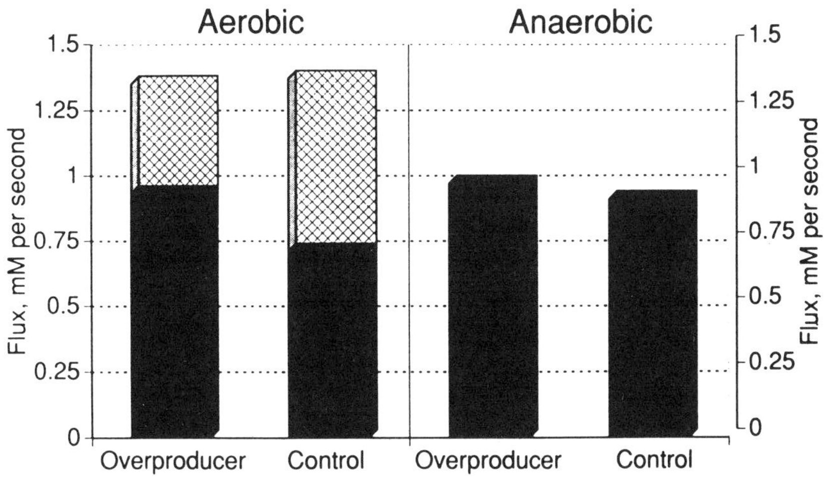

Figure 5. ATP turnover flux under aerobic and anaerobic conditions in yeast cells expressing 6-phosphofructo-1-kinase (PFK1) at 100 percent or 500 percent of wild-type levels (Davies and Brindle, 1992). The dark-hatched boxes show the glycolytic contribution to ATP turnover; the light-hatched boxes show the contribution from oxidative phosphorylation. Yeast cells were immobilized in agarose gel threads and perifused aerobically or anaerobically. ATP turnover was calculated from glucose and oxygen consumption rates assuming there are 2 mol ATP produced per mol glucose consumed and a P:O ratio in oxidative phosphorylation of 3. Note that the overproduction of PFK1 has very little effect on total ATP turnover but significantly shifts the balance between glycolytic and oxidative contributions during aerobiosis, and that aerobiosis has a glucose-sparing effect in control cells but not in the PFK overproducer, i.e. the Pasteur effect in these cells is abolished.

colytic rates observed, reducing the specific activity of the overexpressed PFK1 such that the net activity was unaltered. Other studies have also suggested a reciprocal relationship between the PFK1 level and the concentration of fructose-2,6-bisphosphate (Clifton and Fraenkel, 1983), in a way which may be correlated with the activity of PFK2 (Francois et al., 1984; Francois et al., 1987). The origin of the change in PFK2 activity was not elucidated, but the constancy of the cAMP levels implied it was not through the cAMP-dependent protein kinase which activates PFK2. Davies and Brindle draw attention to the difficulties of applying metabolic control theory to systems which have adapted to the change in enzyme concentration, but conclude that the flux control coefficient for PFK1 in glycolysis in resting cells is low. Importantly, the control and PFK1 overproducing cells showed the same total ATP turnover, but the balance between glycolytic and mitochondrial ATP synthesis was shifted in the overproducer, which was driven to a more glycolytic state. This demonstrated that there must be a tight coupling between the regulation of glycolysis and oxidative phosphorylation, through mechanisms which were unknown, but were probably changes in the concentration of one or more metabolites. ADP has been suggested (Racker, 1974), but changes in the free ADP may be masked in measurements of the total extractable ADP, which remained constant. It would certainly be interesting to repeat these experiments in a system where the free ADP level could also be monitored.

IV. MODELS OF HUMAN DISEASES

Although there is no scientific basis for considering human disease models apart from metabolic investigation, there is a convenient distinction in the motives for study and the experimenter's expectations. At least three models of human diseases have been constructed by molecular genetics for characterization noninvasively with NMR. One of the models is based on a transgenic mouse, a system which looks increasingly like the "tool of choice" for studying the effects of metabolic manipulations in higher organisms (Koretsky, 1992).

Cohen and co-workers have transfected mammalian cells with the cloned gene for the P-glycoprotein (Kaplan et al., 1991) thought to be responsible for the multiple drug resistance (MDR) phenotype that is of great clinical importance in the chemotherapeutic treatment of cancer. Their study aimed to identify changes in ^{31}P NMR spectra associated with MDR, and employed a variety of transfected and drug-selected cell lines. They showed that ATP and phosphocreatine both increase significantly in resistant cells, which they suggest is compatible with previously reported increases in the glucose utilization of resistant cells. Glycerophosphocholine and glycerophosphoethanolamine were also significantly altered, but the direction of these and other changes did not directly correlate with the expression of the P-glycoprotein, suggesting it may have a relatively complex role in altering the metabolism of resistant cells.

Elson and co-workers have examined the role of PFK1 overexpression in Down's syndrome, trisomy 21 (Elson et al., 1992). The liver isoform of PFK1 maps to the region of chromosome 21 that is triplicated in Down's syndrome, and could conceivably be responsible for the increased glycolytic rate seen in erythrocytes and fibroblasts from Down's syndrome patients. The gene for the liver isoform of human PFK1 was cloned and transfected into PC12 cells, derived from a rat adrenal tumor. Conventional characterization of the PFK1 present in the transfected PC12 cells showed it had similar properties to that isolated from Down's syndrome fibroblasts. The transfected cells had PFK1 activities and mRNA levels ~50% higher than control cells, similar to Down's syndrome cells. ^{13}C NMR was used to measure the glycolytic rate in control and transfected cells perifused on microcarrier beads, by monitoring the appearance of the C_3-peak of lactate. These observations showed that the glycolytic rate was enhanced about 40% in high-PFK1 cells. Since the fractional change in flux through glycolysis was almost directly proportional to the fractional change in PFK1 activity, PFK1 might be described by some as the rate-limiting enzyme in glycolysis under the conditions used. In the more quantitative terminology of metabolic control theory (Kacser and Burns, 1973; Heinrich and Rapoport, 1974), the flux control coefficient of PFK1 for glycolysis under the conditions used was found to be near to one. ^{31}P NMR was used to show that there were no significant changes in the concentrations of the major soluble phosphate metabolites, particularly ATP and phosphocreatine. By implication, the ADP and AMP levels were also constant, through the action of the near-equilibrium reactions catalyzed by creatine kinase and adenylate kinase, interconverting PCr and ADP with ATP, and ADP with AMP and ATP, respectively. The authors propose to extend their model to the transgenic mouse for further characterization.

Kauppinen et al. have studied the involvement of polyamine synthesis via ornithine decarboxylase (ODC) in the brain pathologies associated with ischemia and epilepsy. They established a transgenic mouse line expressing ornithine decarboxylase at approximately 70 times the normal level of activity, resulting in steady-state levels of putrescine, the ODC product, rising from virtually 0 to 60 μM, a level comparable to that found in post-ischemic and neurotoxin-damaged tissue (Kauppinen et al., 1992). Levels of the higher polyamines, such as spermidine and spermine, remained unaffected. ^{31}P NMR allowed the noninvasive measurement of the pH, energy status and free magnesium ion levels in the brains of the transgenic mice. The free magnesium ion concentration was shown to drop by approximately 40% in the ODC-overexpressing mice, which is important because of the role magnesium plays in calcium transport and neuronal activity. There was no change in pH or energy status, nor in other indicators of brain function such as the choline/N-acetyl aspartate or total creatine/N-acetyl aspartate ratios determined noninvasively with ^{1}H NMR.

V. INTRACELLULAR RHEOLOGY

The studies discussed so far have been concerned with metabolite concentrations and enzyme kinetics, since these are important parameters to measure in considering the operation of enzymes in the intact cell. However, NMR and molecular genetics also have a contribution to make in the understanding of the physical surroundings of enzymes and metabolites, which are increasingly thought to be very relevant to their operation (see chapters by Srere et al. and Pagliaro, this volume).

The nature of the cytoplasm can clearly have a great impact on the reactions occurring in the cell (Fulton, 1982). The properties of water in the cell are poorly understood. Since the activity coefficients of water and metabolites in the cytoplasm may be rather different from those in simple solutions, even the equilibrium positions of the simplest reactions may be very different *in vivo* and *in vitro* (Minton, 1990). An enzyme may behave very differently if its molecular motions are altered

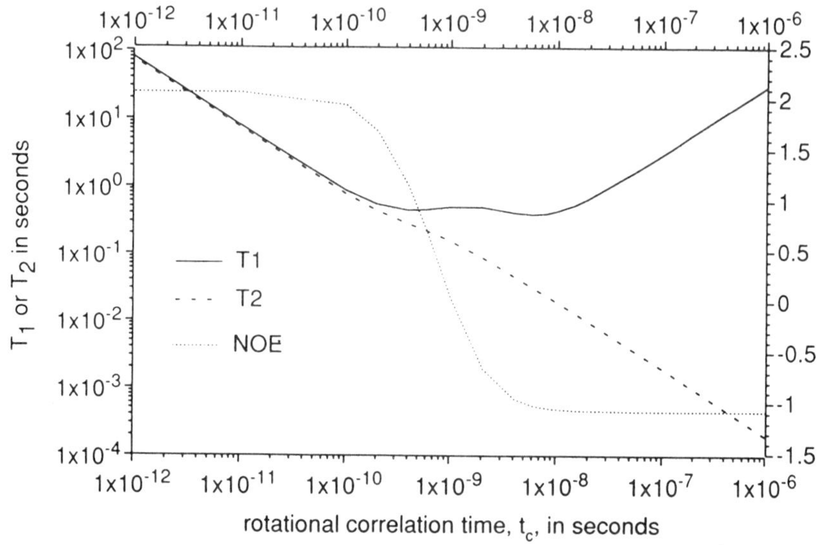

Figure 6. NMR relaxation parameters calculated for the fluorine atom of 5-fluorotryptophan at 9.4 T (^{19}F frequency 376 MHz) and a range of rotational correlation times (Williams, 1992). The calculations include relaxation by chemical shift anisotropy and by dipolar relaxation both to the adjacent indole protons and the protein proton matrix (Post et al., 1984). The rotational correlation time of a globular protein in aqueous solution typically ranges from 10^{-8} to 10^{-6} s (approximately 10 to 1000 kDa). In this range, the T_1 and T_2 relaxation time constants change linearly and oppositely with correlation time, giving a good measure of the correlation time. The magnitude of the nuclear Overhauser effect (NOE), on the other hand, remains virtually constant and is relatively uninformative for protein-size molecules. Provided the relaxation behavior *in vivo* can be modeled *in vitro* to rule out segmental motion artifacts, measurements can be made of the rotational correlation time (and hence viscosity or molecular size) *in vivo.*

by immobilization on a surface or suspension in a viscous medium because its internal conformational motions are slowed, reducing the turnover number significantly (Ghosh and McCammon, 1987; Demchenko et al., 1989). These issues raise questions about the validity of extrapolating biochemical information obtained from experiments *in vitro* to the intact cell (Clegg, 1984).

A variety of different experimental approaches will be needed to characterize a system as complicated as the cytoplasm. There is electron microscopic evidence for some sort of three-dimensional matrix within the cell, termed the microtrabecular lattice (MTL) (Porter, 1984). This is thought to have a dynamic mesh size averaging 700 to 1000 Å (Gershon et al., 1985). In simple solutions, the viscosity of the medium, rotational correlation time and translational diffusion coefficient are all interdependent, but in a crowded solution permeated by a three-dimensional dynamic mesh each must be considered separately. Perhaps the most important effect of the cytomatrix is binding of the proteins to it, preventing translational diffusion of some proteins but allowing others free passage. Large molecules in solution, but trapped within the matrix, would have rotational correlation times near to those in free solution, possibly reduced by frictional interactions with the bounding matrix. However, they would have much smaller translational diffusion coefficients. Very large particles would be enmeshed and move very little at all (Luby-Phelps et al., 1987; Luby-Phelps et al., 1988).

In both optical and NMR spectroscopy, the relaxation properties of the excited states that give rise to the spectra can be quantitatively related by classical physics to the underlying molecular motion (Cantor and Schimmel, 1980). For near-spherical molecules with rigidly attached reporting groups, the molecular motion can be adequately characterized by a single rotational correlation time, τ_c. For markedly asymmetrical molecules or those with significant internal motions, several correlation time parameters are needed and the analyses are more complex, although little different in principle. Translational diffusion can be measured directly with fluorescence techniques such as fluorescence recovery after photobleaching (FRAP), in which a defined region of a sample is "bleached" by illumination and then the recovery of fluorescence intensity is related to the diffusion of fluorophore into the bleached region (Wojcieszyn et al., 1981; Luby-Phelps et al., 1986; Luby-Phelps et al., 1987; Luby-Phelps et al., 1988). The reader is referred to the contribution by Pagliaro (this volume). Signal attenuation following diffusion through a magnetic field gradient can be used to obtain similar direct measurements of translational diffusion in pulsed-field-gradient spin-echo (PFGSE) NMR experiments (Stejskal and Tanner, 1965). This has been done with phosphocreatine in muscle (Moonen et al., 1990), and with hemoglobin in the erythrocyte (Kuchel and Chapman, 1991; see also Berthon and Kuchel, this volume).

Applied to labeled proteins, noninvasive relaxation measurements can give valuable information about the physical nature of the intracellular environment. For example, the apparent size (aggregation state or intermolecular associations, for example) of the labeled molecules *in vivo* can be calculated, or conversely the apparent viscosity determined if the size of the hydrated molecule is assumed, or even known from crystallographic studies and molecular modeling. Relating the apparent viscosity from rotational correlation times with the apparent motion from diffusion measurements may reveal more about the nature of enzymes within the cytoplasm.

The retardant effect of the cytomatrix has been investigated with FRAP studies of a series of fluorescently tagged proteins under different conditions of tonicity and cytoskeletal structure. Jacobson and Wojcieszyn (1984) showed that the diffusion coefficients of the proteins were almost independent of size in the cell, suggesting that strong protein–protein interactions with the cytomatrix limited their diffusion rather than a simple mechanical hindrance. Luby-Phelps and co-workers used fluorescently tagged dextrans and Ficoll™s of different sizes in FRAP studies similar to those of Jacobson and Wojcieszyn (Luby-Phelps et al., 1986; Luby-Phelps et al., 1987). These neutral probe molecules were chosen instead of proteins because they were expected to show fewer specific interactions with the cytomatrix and to have a constant surface chemistry over a wide range of molecular weights. These studies have shown that the translational diffusion coefficients in the cytoplasm are close to those measured in water for small particles, but for very large particles a mechanical "sieving" effect appeared to restrict the diffusion. The molecular radius in this case was approximately 180 Å, which is large compared to a glycolytic enzyme. For example, the radius of the yeast pyruvate kinase tetramer is only about 42 Å.

The conformational flexibility of the relatively mobile dextran molecules and the dynamic nature of the cytomatrix make it difficult to produce anything other than "average" figures for the mesh size in the cell. Similar difficulties exist in measuring the viscosity, i.e. isolating the effects of true solvent viscosity from specific interactions of probe and cytomatrix. Cytoplasmic viscosity has been estimated using a variety of techniques including NMR, Electron Spin Resonance, several fluorescence techniques, and even direct microscopic observation of moving particles within the cell (see Jacobson and Wojcieszyn, 1984; Fushimi and Verkman, 1991 and references therein). The various fluorescence approaches to studying tagged macromolecules in the cell, particularly actin, tubulin and other cytoskeletal proteins, have been well reviewed (Luby-Phelps et al., 1988).

NMR detection of proteins allows the extension of this type of experiment into opaque tissues and animal organs where fluorescence techniques cannot, at present, be applied. For example NMR studies of a labeled protein were used to measure the intracellular viscosity in mouse erythrocytes almost twenty years ago, using spin-lattice relaxation time measurements of [2-^{13}C]histidine-labeled hemoglobin (London et al., 1975). That study suggested that the intracellular viscosity was relatively low, about twice that of water, even at the high protein concentration within the erythrocyte. Studies of the magnetic field dependence of the linewidth of the proximal histidyl Nδ proton resonance in myoglobin in bovine heart muscle revealed a similarly low viscosity of 2.5 cP (Livingston et al., 1983). These studies were possible as hemoglobin is the predominant protein in the erythrocyte and myoglobin is both abundant in muscle and possesses unique resonances which are shifted out from under the envelope of ^{1}H resonances in the tissue. Clearly the study of other proteins will require some form of protein-specific labeling.

Using our yeast-based system for labeling a single protein in the intact cell with a fluoroamino acid for ^{19}F NMR detection (Brindle et al., 1989), we have made some preliminary measurements of intracellular viscosity by measuring the relaxation parameters of fluorotryptophan-labeled phosphoglycerate kinase (45-kDa monomer) and hexokinase (102-kDa dimer). Our data, based on measurements of the linewidths and the longitudinal relaxation times (T_1) of the labeled proteins *in vivo*, indicate that the viscosity of the cytoplasm in which these proteins tumble is

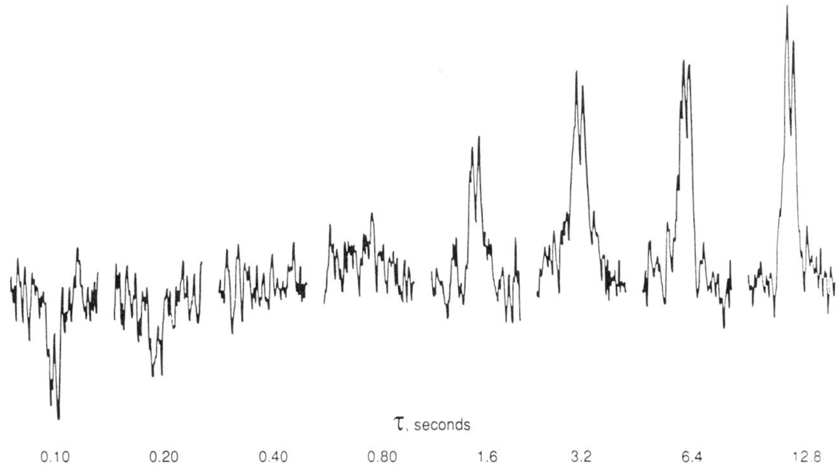

Figure 7. ^{19}F NMR fast inversion-recovery experiment to measure the longitudinal relaxation time constant, T_1, for fluorine-labeled 3-phosphoglycerate kinase in the yeast cell. The cells contained approximately 200 μM labeled PGK in their cell water. A series of 8 interleaved spectra were collected in a total time of approximately 12 h from 4 g of aerobically perifused agarose-gel-immobilized yeast cells. The T_1 can be determined from the signal intensities which depend on upon the T_1 and the variable-delay parameter, τ, in the experiment. The spin–spin relaxation time constant (T_2) can be estimated from the linewidths in these spectra.

very low, less than twice that of water. Studies of the labeled proteins *in vitro* showed that the NMR-derived correlation times at different viscosities and temperatures were in excellent agreement with those predicted from the classical physics used to interpret the data obtained *in vivo*. We plan to extend these measurements using other proteins spanning a range of molecular weights. It should also be possible to use NMR to measure the translational diffusion of these molecules. Berthon and Kuchel describe elsewhere in this volume the application of pulsed field gradient spin echo techniques (PFGSE) to measure the diffusion of proteins in cells. These techniques could be applied to the fluorinated proteins described here. However, since the transverse relaxation times of the fluorine resonances in these labeled proteins are relatively short, these experiments are technically demanding and will require large and rapidly switchable magnetic field gradients.

Molecular genetics should allow similar measurements to be made in specific subcellular locations, such as the nucleus or mitochondrion, by targeting the fluorine-labeled protein to the appropriate compartment. The low values for cytoplasmic viscosity from NMR data in yeast are in good agreement with values determined recently with picosecond polarization decay fluorescence techniques in Swiss 3T3 fibroblasts (Fushimi and Verkman, 1991).

Review of the more recent literature suggests that a consensus is forming about the intracellular viscosity of the bulk cellular water; that it is not much greater than free water. This result, although useful, leaves the various phenomena of limited diffusion and macromolecular organization within the cell to be wholly explained on the basis of specific interactions between proteins — an even greater challenge for the experimentalist to tackle in the intact cell.

VI. CONCLUSIONS

Our view of the intracellular environment is becoming increasingly complex, with a dynamic meshwork of structural proteins and associated enzymes permeating cytoplasmic spaces of water-like viscosity. This structure can have a profound influence on the operation of enzymes in the cell. Our understanding of enzymology in the intact cell has grown rapidly in the last twenty years as there have been improvements in the theory of enzyme and pathway kinetics, in the engineering of spectroscopic equipment, and especially in the availability and application of molecular genetic manipulations. We can numerically model the behavior of many complex biological systems, be it the details of catalysis in a single enzyme or the flux in a metabolic pathway, but it seems that our experimental approaches to testing the models by studying the cell interior and its biochemistry are somewhat inadequate, although improving rapidly.

We now have some more experimental tools to characterize the microenvironment of enzymes in the intact cell and to measure the free metabolite concentrations to which they are exposed. NMR spectroscopy and optical techniques would appear to offer complementary approaches which will allow studies ranging from measurements of subcellular localization in microinjected cells through to whole-organ intact-animal studies. The techniques of molecular genetics will undoubtedly be of great value in creating the right experimental models to propel our understanding into the next century.

REFERENCES

Adams, S. R., Harootunian, A. T., Buechler, Y. J., Taylor, S. S., & Tsien, R. Y. (1991). Fluorescence ratio imaging of cyclic AMP in single cells. Nature 349, 694–696.

Aflalo, C. & Segel, L. A. (1992). Local probes and heterogenous catalysis: A case study of a mitochondria-luciferase-hexokinase coupled system. J. Theor. Biol. 158, 67–108.

Aragon, J. J. & Sols, A. (1991). Regulation of enzyme activity in the cell: effect of enzyme concentration. FASEB J. 5, 2945–2950.

Atkinson, D. E. (1977). Cellular Energy Metabolism and its Regulation. Academic Press, Inc., New York.

Avison, M. J., Hetherington, H. P., & Shulman, R. G. (1986). Applications of NMR studies to tissue metabolism. Annu. Rev. Biophys. Chem. 15, 377–402.

Bone, R. & Agard, D. A. (1991). Mutational remodelling of enzyme specificity. Methods Enzymol. 202, 643–670.

Boscá, L. & Corredor, C. (1984). Is phosphofructokinase the rate-limiting step of glycolysis? Trends Biochem. Sci. 9, 372–373.

Brindle, K. M. (1988). ^{31}P NMR magnetization-transfer measurements of flux between inorganic phosphate and adenosine 5′-triphosphate in yeast cells genetically modified to overproduce phosphoglycerate kinase. Biochemistry 27, 6187–6196.

Brindle, K. M. & Krikler, S. (1985). ^{31}P-NMR saturation transfer measurements of phosphate consumption in *Saccharomyces cerevisiae*. Biochim. Biophys. Acta 847, 285–292.

Brindle, K. M. & Campbell, I. D. (1987). NMR studies of kinetics in cells and tissues. Q. Rev. Biophys. 19, 159–182.

Brindle, K. M. & Radda, G. K. (1987). ^{31}P NMR saturation transfer measurements of exchange between P_i and ATP in the reactions catalyzed by glyceraldehyde-3-phosphate dehydrogenase and phosphoglycerate kinase in vitro. Biochim. Biophys. Acta 928, 47–55.

Brindle, K. M., Porteous, R., & Campbell, I. D. (1984). ^{1}H NMR measurements of enzyme-catalyzed ^{15}N-label exchange. J. Magn. Reson. 56, 543–547.

Brindle, K. M., Williams, S.-P. & Boulton, M. (1989). ^{19}F NMR detection of a fluorine-labeled enzyme in vivo. FEBS Lett. 255, 121–124.

Brindle, K., Braddock, P., & Fulton, S. (1990). ^{31}P NMR measurements of the ADP concentration in yeast cells genetically modified to express creatine kinase. Biochemistry 29, 3295–3302.

Brooks, S. P. J. & Storey, K. B. (1991a). A quantitative evaluation of the effect of enzyme complexes on the glycolytic rate in vivo: mathematical modeling of the glycolytic complex. J. Theor. Biol. 149, 361–375.

Brooks, S. P. J. & Storey, K. B. (1991b). Where is the glycolytic complex? A critical evaluation of present data from muscle tissue. FEBS Lett. 278, 135–138.

Brosnan, M. J., Chen, L., Van Dyke, T. A., & Koretsky, A. P. (1990). Free ADP levels in transgenic mouse liver expressing creatine kinase. J. Biol. Chem. 265, 20849–20855.

Brown, F. F., Campbell, I. D., Kuchel, P. W., & Rabenstein, D. L. (1977). Human erythrocyte metabolism studied by ^{1}H spin echo NMR. FEBS Lett. 82, 12–16.

Campbell, A. K. (1990). Bioluminescent proteins; rainbow proteins. European Patent Office PCT/GB90/01131.

Campbell, I. D., Dobson, C. M., Ratcliffe, R. G., & Williams, R. J. P. (1977). Fourier transform NMR pulse methods for the measurement of slow exchange rates. J. Magn. Reson. 29, 397–405.

Campbell-Burk, S. L., Jones, K. A., & Shulman, R. G. (1987). ^{31}P NMR saturation-transfer measurements in *Saccharomyces cerevisiae*: Characterization of phosphate exchange reactions by iodoacetate and antimycin A inhibition. Biochemistry 26, 7483–7492.

Cantor, C. R. & Schimmel, P. R. (1980). Biophysical chemistry. W.H. Freeman, San Francisco.

Chance, B. (1991). Optical method. Ann. Rev. Biophys. Biophys. Chem. 20, 1–28.

Chance, B., Eleff, S., Bank, W., Leigh, J. S., & Warnell, R. (1982). ^{31}P NMR studies of control of mitochondrial function in phosphofructokinase-deficient human skeletal muscle. Proc. Natl. Acad. Sci. USA 79, 7714–7718.

Chock, P. B. & Gutfreund, H. (1988). Reexamination of the kinetics of the transfer of NADH between its complexes with glycerol-3-phosphate dehydrogenase and with lactate dehydrogenase. Proc. Natl. Acad. Sci. USA 85, 8870–8874.

Clegg, J. S. (1984). Intracellular water and the cytomatrix: some methods of study and current views. J. Cell Biol. 99, 167s–171s.

Clifton, D. & Fraenkel, D. G. (1983). Fructose-2,6-bisphosphate and fructose-6-phosphate-2-kinase in *Saccharomyces cerevisiae* in relation to metabolic state in wild type and fructose-6-phosphate-1-kinase mutant strains. J. Biol. Chem. 258, 9245–9249.

Cohen, J. S., Lyon, R. C., & Daly, P. F. (1989). Monitoring intracellular metabolism by nuclear magnetic resonance. Methods Enzymol. 177, 435–452.

Cornish-Bowden, A. (1991). Failure of channelling to maintain low concentrations of metabolic intermediates. Eur. J. Biochem. 195, 103–108.

Davies, S. E. C. & Brindle, K. M. (1992). Effects of overexpression of phosphofructokinase on glycolysis in the yeast *Saccharomyces cerevisiae*. Biochemistry 31, 4729–4735.

Demchenko, A. P., Rusyn, O. I., & Saburova, E. A. (1989). Kinetics of lactate dehydrogenase reaction in high-viscosity media. Biochim. Biophys. Acta 998, 196–203.

Elson, A., Bernstein, Y., Degani, H., Levanon, D., Ben-Hur, H., & Groner, Y. (1992). Gene dosage and Down's syndrome: metabolic and enzymatic changes in PC12 cells overexpressing transfected human liver-type phosphofructokinase. Somat. Cell. Mol. Genet. 18, 143–161.

Fell, D. A. (1992). Metabolic control analysis—a survey of its theoretical and experimental development. Biochem. J. 286, 313–330.

Forsén, S. & Hoffman, R. A. (1963). Study of moderately rapid chemical exchange reactions by means of nuclear magnetic double resonance. J. Chem. Phys. 39, 2982–2901.

Francois, J., van Schaftingen, E., & Hers, H.-G. (1984). The mechanism by which glucose increases fructose-2,6-bisphosphate concentration in *Saccharomyces cerevisiae*. A cyclic-AMP-dependent activation of phosphofructokinase-2. Eur. J. Biochem. 145, 187–193.

Francois, J., Eraso, P., & Gancedo, C. (1987). Changes in the concentration of cAMP, fructose-2,6-bisphosphate and related metabolites and enzymes in *Saccharomyces cerevisiae* during growth on glucose. Eur. J. Biochem. 164, 369–373.

Fulton, A. B. (1982). How crowded is the cytoplasm? Cell 30, 345–347.

Fushimi, K. & Verkman, A. S. (1991). Low viscosity in the aqueous domain of cell cytoplasm measured by picosecond polarization microfluorimetry. J. Cell Biol. 112, 719–725.

Gerig, J. T. (1989). Fluorine nuclear magnetic resonance of fluorinated ligands. Methods Enzymol. 177, 3–23.

Gershon, N. L., Porter, K. R., & Trus, B. L. (1985). The cytomatrix—its volume and surface area and the diffusion of molecules through it. Proc. Natl. Acad. Sci. USA 82, 5030–5034.

Ghosh, I. & McCammon, J. A. (1987). Solvent viscosity effects on the rate of side-chain rotational isomerization in a protein molecule. J. Phys. Chem. 91, 4878–4881.

Goodridge, A. G. (1990). The new metabolism: molecular genetics in the analysis of metabolic regulation. FASEB J. 4, 3099–3110.

Gruetter, R., Novotny, E. J., Boularare, S. D., Rothman, D. L., Mason, G. F., Shulman, G. I., Shulman, R. G., & Tamborlane, W. V. (1992). Direct measurement of brain glucose concentration in humans by C-13 NMR spectroscopy. Proc. Natl. Acad. Sci. USA 89, 1109–1112.

Hahn, K., DeBiasio, R., & Taylor, D. L. (1992). Patterns of elevated free calcium and calmodulin activation in living cells. Nature 359, 736–738.

Harris, R. K. (1986). Nuclear Magnetic Resonance Spectroscopy. A physicochemical view. Longman, Harlow, Essex.

Hedstrom, L., Graf, L., & Stewart, C.-B. (1991). Modulation of enzyme specificity by site-directed mutagenesis. Methods Enzymol. 202, 671–687.

Heinrich, R. & Rapoport, T. A. (1974). A linear steady-state treatment of enzymatic chains: general properties, control and effector strength. Eur. J. Biochem. 42, 89.

Hoult, D. I., Busby, S. J. W., Gadian, D. G., Radda, G. K., Richards, R. E., & Seeley, P. J. (1974). Observation of tissue metabolites by ^{31}P nuclear magnetic resonance. Nature 252, 285.

Ikai, I., Okuda, M., Doliba, N., & Chance, B. (1991). Visibility of ADP and ATP in freeze-trapped tissue from perfused rat liver during normoxia and ischemia using 31-P cryo-NMR. Biochim. Biophys. Acta 1074, 289–293.

Iles, R. A., Stevens, A. N., & Griffiths, J. R. (1982). NMR studies of metabolites in living tissue. Prog. Nucl. Magn. Reson. Spectrosc. 15, 49–200.

Jacobson, K. & Wojcieszyn, J. (1984). The translational mobility of substances within the cytoplasmic matrix. Proc. Natl. Acad. Sci. USA 81, 6747–6751.

Kacser, H. & Burns, J. A. (1973). The control of flux. Symp. Soc. Exp. Biol. 27, 65–104.

Kacser, H. & Porteous, J. W. (1987). Control of metabolism: what do we have to measure. Trends Biochem. Sci. January, 5–14.

Kaplan, O., Jaroszewski, J. W., Clarke, R., Fairchild, C. R., Schoenlein, P., Goldenberg, S., Gottesman, M. M., & Cohen, J. S. (1991). The multidrug resistance phenotype: ^{31}P nuclear magnetic resonance characterization and 2-deoxyglucose toxicity. Cancer Res. 51, 1638–1644.

Kauppinen, R. A., Halmekyto, M., Alhonen, L., & Janne, J. (1992). Nuclear magnetic resonance study on energy metabolism, intracellular pH, and free Mg^{2+} concentration in the brain of transgenic mice overexpressing human ornithine decarboxylase gene. J. Neurochem. 58, 831–836.

Keilin, D. (1925). On cytochrome, a respiratory pigment, common to animals, yeast, and higher plants. Proc. Roy. Soc. London, series B 98, 312–320.

Kholodenko, B. N., Lyubarev, A. E., & Kurganov, B. I. (1992). Control of the metabolic flux in a system with high enzyme concentrations and moiety-conserved cycles. The sum of the flux control coefficients can drop significantly below unity. Eur. J. Biochem. 210, 147–153.

Khouri, H. E., Vernet, T., Menard, R., Parlati, F., Laflamme, P., Tessier, D. C., Gour-Salin, B., Thomas, D. Y., & Storer, A. C. (1991). Engineering of papain: selective alteration of substrate specificity by site-directed mutagenesis. Biochemistry 30, 8929–8936.

Kirschenlohr, H. L., Metcalfe, J. C., Morris, P. G., & Rodrigo, G. C. (1988). Ca^{2+} transient, Mg^{2+}, and pH measurements in the cardiac cycle by ^{19}F NMR. Proc. Natl. Acad. Sci. USA 85, 9017–9021.

Kispal, G., Evans, C. T., Malloy, C., & Srere, P. A. (1989). Metabolic studies on citrate synthase mutants of yeast—a change in phenotype following transformation with an inactive enzyme. J. Biol. Chem. 264, 11204–11210.

Knight, M. R., Campbell, A. K., Smith, S. M., & Trewevas, A. J. (1991). Transgenic plant aequorin reports the effects of touch and cold-shock and elicitors on cytoplasmic calcium. Nature 352, 524–526.

Koretsky, A. P. (1992). Investigation of cell physiology in the animal using transgenic technology. Am. J. Physiol. 262, C261–C275.

Koretsky, A. P. & Williams, D. S. (1992). Application of localized *in vivo* NMR to whole organ physiology in the animal. Annu. Rev. Physiol. 54, 799–826.

Kreutzer, U., Wang, D. S., & Jue, T. (1992). Observing the ^{1}H NMR signal of the myoglobin Val-E11 in myocardium: An index of cellular oxygenation. Proc. Natl. Acad. Sci. USA 89, 4731–4733.

Kuchel, P. W. & Chapman, B. E. (1991). Translational diffusion of hemoglobin in erythrocytes and hemolysates. J. Magn. Reson. 94, 574–580.

LaNoue, K., Jeffries, F. M. H., & Radda, G. K. (1986). Kinetic control of mitochondrial ATP synthesis. Biochemistry 25, 7667–7678.

Lawson, J. W. R. & Veech, R. L. (1979). Effects of pH and free Mg^{2+} on the K_{eq} of the creatine kinase reaction and other phosphate hydrolyses and phosphate transfer reactions. J. Biol. Chem. 254, 6528–6537.

Livingston, D. J., La Mar, G. N., & Brown, W. D. (1983). Myoglobin diffusion in bovine heart muscle. Science 220, 71–73.

London, R. (1991). Methods for measurement of intracellular magnesium: NMR and fluorescence. Annu. Rev. Physiol. 53, 241–258.

London, R. E., Gregg, C. T., & Matwiyoff, N. A. (1975). NMR of rotational mobility of a mouse hemoglobin labeled with [2-^{13}C]histidine. Science (Washington, D.C.) 188, 266–268.

Luby-Phelps, K., Taylor, D. L., & Lanni, F. (1986). Probing the structure of cytoplasm. J. Cell. Biol. 102, 2015–2022.

Luby-Phelps, K., Castle, P. E., Taylor, D. L., & Lanni, F. (1987). Hindered diffusion of inert tracer particles in the cytoplasm of mouse 3T3 cells. Proc. Natl. Acad. Sci. USA 84, 4910–4913.

Luby-Phelps, K., Lanni, F., & Taylor, D. L. (1988). The submicroscopic properties of cytoplasm as a determinant of cellular function. Ann. Rev. Biophys. Biophys. Chem. 17, 369–396.

Masson, S. & and Quistorff, B. (1992). The ^{31}P NMR visibility of ATP in perfused rat liver remains about 90% unaffected by changes of metabolic state. Biochemistry 31, 7488–7493.

Matthews, P. M., Bland, J. L., Gadian, D. G., & Radda, G. K. (1982). A ^{31}P NMR saturation transfer study of the regulation of creatine kinase in the rat heart. Biochim. Biophys. Acta 721, 312–320.

Mendes, P., Kell, D. B., & Westerhoff, H. V. (1992). Channelling can decrease pool size. Eur. J. Biochem. 204, 257–266.

Millikan, G. A. (1937). Experiments on muscle haemoglobin *in vivo*; the instantaneous measurement of muscle metabolism. Proc. Roy. Soc. London, series B 123, 218–241.

Minton, A. P. (1990). Holobiochemistry: The effect of local environment upon the equilibria and rates of biochemical reactions. Int. J. Biochem. 22, 1063–1067.

Moon, R. B. & Richards, J. H. (1973). Determination of intracellular pH by ^{31}P magnetic resonance. J. Biol. Chem. 248, 7276–7278.

Moonen, C. T. W., van Zijl, P. C. M., Le Bihan, D., & DesPres, D. (1990). *In vivo* NMR diffusion spectroscopy: ^{31}P Application to phosphorus metabolites in muscle. Mag. Reson. Med. 13, 467–477.

Murphy, E., Gabel, S. A., Funk, A., & London, R. E. (1988). NMR observability of ATP: Preferential depletion of cytosolic ATP during ischemia in perfused rat liver. Biochemistry 27, 526–528.

Niederberger, P., Prasad, R., Miozzari, G., & Kacser, H. (1992). A strategy for increasing an *in vivo* flux by genetic manipulations. The tryptophan system of yeast. Biochem. J. 287, 473–479.

Ovádi, J. (1991). Physiological significance of metabolite channeling. J. Theor. Biol. 152, 1–22.

Porteous, J. W. (1985). Control confused. Trends Biochem. Sci. 1, 14–15.

Porter, K. R. (1984). The cytomatrix—a short history of its study. J. Cell Biol. 99, 3s–12s.

Post, J. F., Cottam, P. F., Simplaceanu, V., & Ho, C. (1984). Fluorine-19 nuclear magnetic resonance study of 5-fluorotryptophan-labeled histidine-binding protein J of *Salmonella typhimurium*. J. Mol. Biol. 179, 729–743.

Racker, E. (1974). History of the Pasteur effect and its pathobiology. Mol. Cell. Biochem. 5, 17–23.

Radda, G. K. (1986). The use of NMR spectroscopy for the understanding of disease. Science (Washington, D.C.) 233, 640–645.

Radda, G. K. (1992). Control, bioenergetics, and adaptation in health and disease—noninvasive biochemistry from nuclear magnetic resonance. FASEB J. 6, 3032–3038.

Ramaiah, A. (1984). Pasteur effect and phosphofructokinase. Curr. Top. Cell. Regul. 8, 297–345.

Rizzuto, R., Simpson, A. W. M., Brini, M., & Pozzan, T. (1992). Rapid changes of mitochondrial Ca^{2+} revealed by specifically targeted recombinant aequorin. Nature 358, 325–327.

Roberts, J. K. M., Wade-Jardetzky, N., & Jardetzky, O. (1981). Intracellular pH measurements by ^{31}P NMR. Influence of factors other than pH on ^{31}P chemical shifts. Biochemistry 20, 5389–5394.

Seo, Y., Murakami, M., Watari, H., Imai, Y., Yoshizaki, K., Nishikawa, H., & Morimoto, T. (1983). Intracellular pH determination by a ^{31}P NMR technique. The second dissociation constant of phosphoric acid in a biological system. J. Biochem. 94, 729–734.

Seraydarian, K., Mommaerts, W.F.H.M., Wallner, A., & Guillory, R.J. (1961). An estimation of the true inorganic phosphate content of frog sartorius muscle. J. Biol. Chem. 236, 2071–2075.

Sevick, E. M., Chance, B., Leigh, J., Nioka, S., & Maris, M. (1991). Quantitation of time- and frequency-resolved optical spectra for the determination of tissue oxygenation. Anal. Biochem. 195, 330–351.

Shoubridge, E. A., Bland, J. L., & Radda, G. K. (1984). Regulation of creatine kinase during steady state isometric twitch contraction in rat skeletal muscle. Biochim. Biophys. Acta 805, 72–78.

Stejskal, E. O. & Tanner, J. E. (1965). Spin diffusion measurements: spin echoes in the presence of a time-dependent field gradient. J. Chem. Phys. 42, 288–292.

Street, J. C., Delort, A.-M., Braddock, P. S. H., & Brindle, K. M. (1993). A ^{1}H/^{15}N n.m.r. study of nitrogen metabolism in cultured mammalian cells. Biochem. J. 291, 485–492.

Stubbs, M., Freeman, D., & Ross, B. D. (1984). Formation of n.m.r.-invisible ADP during renal ischaemia in rats. Biochem. J. 224, 2421–2426.

Sykes, B. D. & Weiner, J. H. (1980). In: Magnetic Resonance in Biology (Cohen, J.S., ed.), pp. 171–196, Wiley, New York.

van Schaftingen, E. (1987). Fructose-2,6-bisphosphate. Adv. Enzymol. Relat. Areas Mol. Biol. 59, 315–395.

Vas, M. & Batke, J. (1990). Kinetic misinterpretation of a coupled enzyme reaction can lead to the assumption of an enzyme-enzyme interaction. The example of 3-phospho-D-glycerate kinase and glyceraldehyde-3-phosphate dehydrogenase couple. Eur. J. Biochem. 191, 679–683.

Veech, R. L., Lawson, J. W. R., Cornell, N. W., & Krebs, H. A. (1979). Cytosolic phosphorylation potential. J. Biol. Chem. 254, 6538–6547.

Walsh, J. L., Keith, T. J., & Knull, H. R. (1989). Glycolytic enzyme interactions with tubulin and microtubules. Biochim. Biophys. Acta 999, 64–70.

Williams, S.-P. (1992). Studies of enzyme kinetics and aspects of enzyme structure *in vivo* using NMR and molecular genetics. D. Phil. thesis, Oxford University, U.K.

Williams, S. P., Fulton, A. M., & Brindle, K. M. (1993). Estimation of the intracellular free ADP concentration by ^{19}F NMR studies of fluorine-labeled yeast phosphoglycerate kinase *in vivo*. Biochemistry 32, 4895–4902.

Wojcieszyn, J. W., Schlegel, R. A., Wu, E. S., & Jacobson, K. A. (1981). Diffusion of injected macromolecules within the intact cell. Proc. Natl. Acad. Sci. USA 78, 4407–4410.

Wu, X., Gutfreund, H., Lakatos, S., & Chock, P. B. (1991). Substrate channeling in glycolysis: A phantom phenomenon. Proc. Natl. Acad. Sci. USA 88, 497–501.

GLYCOLYSIS *IN VIVO*: FLUORESCENCE MICROSCOPY AS A TOOL FOR STUDYING ENZYME ORGANIZATION IN LIVING CELLS

Len Pagliaro

Advances in Molecular and Cell Biology
Volume 11, pages 93–123.

ISBN: 1-55938-844-7

ABSTRACT

Indirect evidence that glycolysis does not exist as a series of fully solvated, diffusion-limited reactions in cytoplasm has been accumulating for more than two decades, but many biologists remain unconvinced due to a lack of direct evidence. Models of alternative regimes of glycolytic enzyme organization *in vivo* may be significant; if verified experimentally, they could substantially modify our thinking about this ubiquitous metabolic pathway and the dual roles it may play in cytoplasmic organization and metabolism. In addition, evidence for such unconventional enzyme organization could have far reaching implications for a variety of metabolic and cellular diseases. The most direct approach to these questions is to study glycolysis *in vivo*. Toward this end, we are studying functional fluorescent analogs of the glycolytic enzymes aldolase and enolase *in vivo*, using fluorescence microscope imaging and fluorescence photobleaching to characterize enzyme distribution and mobility. In this chapter, I will outline our approach to glycolysis, and review both the tools we are currently using and some of the new experimental approaches we are developing. A series of conceptually simple questions about the organization and dynamics of enzymes in living cells is defined, and fluorescence microscope approaches to each of the questions are discussed. While this chapter will focus on glycolysis, the approach we have used for glycolytic enzymes could be used to study other enzymes *in vivo*.

I. INTRODUCTION

> As has often been remarked, the chemical conditions existing in the living cell are of necessity imperfectly known, because every attempt to examine them by precise methods kills the cell. In the main, therefore, our statements concerning these conditions are limited to inferences based on the chemical behavior of dead cells or their components. (Wilson, 1925).

In his uncannily prescient manner, E. B. Wilson identified a paradox that is no less relevant to biologists today than it was almost 70 years ago, when he published the third edition of his classic book, *The Cell in Inheritance and Development*. This is remarkable, given the enormous amount of biochemical research that has been done since 1925. We now know the sequence, metabolites, and intermediates of many metabolic pathways, and we know the ultrastructure of cells and their organelles in exquisite detail. In spite of this substantial body of knowledge, a number of important questions regarding the organization and regulation of cell metabolism remain unanswered. The answer to this paradox lies in the fact that it is still difficult to study specific metabolic pathways in living cells, but a growing array of biological tools and technologies are making new kinds of metabolic experiments *in vivo* possible (see Brindle, 1991, for a recent review).

The reductionist approach to biology and the use of biochemical and molecular biological tools have given us powerful methods for studying the logic and informatics of biological systems. As with any research tools, however, these tools have inherent experimental biases that are difficult to overcome. Thus, while molecular and biochemical approaches are the method of choice for determining the sequence of reactions in a metabolic pathway, they have limitations for studying the dynamics of pathways. New approaches must be used to complement the wealth of metabolic information we already have, but the use of new approaches mandates an important caveat: we must be careful to evaluate new techniques in the proper context. We should not apply older standards to new techniques if the purpose of the new techniques is to give us new kinds of information. Thus, for example, live cell experiments involving relatively small sample sizes cannot be held to the same statistical standards as biochemical or genetic experiments involving large numbers of cells or organisms. The corresponding benefit is that experiments on individual living cells allow us a unique opportunity to characterize an important, but largely unexplored, area of enzymology: enzyme dynamics in living cells.

In this chapter, I will discuss the current capabilities and future potential of fluorescence microscope imaging techniques for studying enzymology *in vivo.* I use the term 'fluorescence microscope imaging' in a broad sense here; a variety of associated techniques are at least as important as the actual imaging. The preparation and characterization of active, fluorescently labeled enzyme analogs, and the use of quantitative fluorescence redistribution after photobleaching (FRAP) are two associated techniques I will discuss at length. In addition, I will discuss several kinds of experiments that can be done *in vitro* to help generate specific predictions about enzyme behavior that can be tested *in vivo*. Finally, I will outline the potential for several evolving technologies that will give us the capability to perform elegant metabolic manipulations in living cells in real time.

II. WHY GLYCOLYSIS?

Every student of biology learns the central role glycolysis plays in metabolism. We are taught the substrates, intermediates and enzymes involved, the sequence of reactions, and the role glycolysis plays in providing precursors for the citric acid cycle and other pathways, as well as its role in providing a small percentage of cellular ATP requirements. We are also taught that "the ten reactions of glycolysis occur in the cytosol" (Stryer, 1988). While undeniable, the statement that glycolysis occurs in the cytosol is misleading, since it implies that the organizational scheme of glycolysis has been characterized in the same detail as, for example, the respiratory enzymes of the mitochondrion.

To better understand the misleading nature of the term 'cytosol,' we must remember that, while well integrated into the vernacular of most biologists, 'cytosol' remains a rather vague operational definition. First coined by Lardy (1965), cytosol refers to "that portion of the cell which is found in the supernatant fraction after centrifuging the homogenate at 105,000 × *g* for 1 hour." The ability to cleanly separate membrane-bound, aldehyde fixable organelles (such as mitochondria) by ultracentrifugation was an important experimental breakthrough, but drawing conclusions about the biochemical organization of the supernatant fraction of such a process is problematic at best. An enormous range of biochemical reactions occurs in the cytosol, and localizing a reaction to the cytosol is only marginally more precise than saying that it does not occur in the nucleus or mitochondrion. In addition, the increasing evidence for non-membrane-bound mechanisms of compartmentation in the cytoplasmic space should caution us against treating the cytosol as a single, homogeneous compartment. It is possible that precisely the kinds of weak interactions that do not survive ultracentrifugation could be the essential forces in establishing and maintaining organization in living cytoplasm.

Using techniques analogous to those used in the elucidation of mitochondrial structure and function 35 years ago, de Duve and colleagues were unable to isolate and characterize a 'glycolytic particle,' leading them to conclude that a glycolytic organelle did not exist (reviewed by de Duve, 1972). During the search for a glycolytic particle, however, several investigators found what they termed a 'contaminating protein' of about 45 kDa in sedimented preparations having high glycolytic activity. This 'contaminant' turned out to be actin, and several of the glycolytic enzymes were among the first actin-binding proteins identified, although they were not called that at the time.

Over the past 25 years, work from a number of laboratories has generated substantial evidence that some glycolytic enzymes bind to cytoskeletal components. These studies have used biochemical (Arnold and Pette, 1968; Stewart et al., 1980), histochemical (Arnold et al., 1969; Sigel and Pette, 1969), immunofluorescence (Dolken et al., 1975), and ultrastructural (Clarke and Morton, 1976; Morton et al., 1977) techniques to demonstrate specific interactions between aldolase and the actin cytoskeleton. Well developed, elegant models of unconventional organizational schemes for glycolysis have been in the literature for over 15 years (Ottaway and Mowbray, 1977; Welch, 1977). In addition, there is growing evidence that some glycolytic enzymes also bind to the tubulin cytoskeleton (Huitorel and Pantaloni, 1985; Walsh et al., 1989), suggesting that cytoskeleton-based organization of glycolytic enzymes is not limited to any particular cytoskeletal system, but may represent a more general phenomenon.

Two lines of evidence for interactions between some glycolytic enzymes and the actin cytoskeleton are shown in Figure 1. The binding of enzymes to an F-actin affinity column (solid bars) and the F-actin cross-linking activity of enzymes,

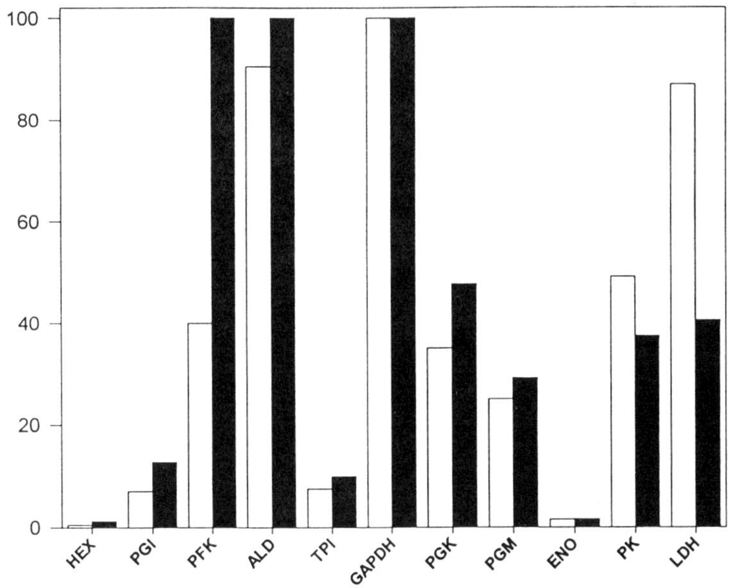

Figure 1. Comparison of the actin-binding and actin-gelling activities of the glycolytic enzymes. The binding of enzymes (in percent catalytic activity) from a crude muscle extract to an F-actin affinity column is shown by the solid bars (data replotted from Bronstein & Knull, 1981). No binding data are shown for hexokinase. The actin-gelling activities of the glycolytic enzymes assayed with falling-ball viscometry are shown by the open bars (Pagliaro, unpublished data). Each enzyme was incubated in 100 μl assay tubes at 1 μm concentration with 1 mg/ml F-actin for 30 min at 37°C in 40 mM KCl, 1mM $MgCl_2$, 0.5 mM dithiothreitol, and 10 mM imidazole at pH 6.8. After incubation, falling ball viscometry was performed in triplicate for each sample as described previously (Wang et al., 1994). Actin-gelling activity resulting in a gel with an apparent viscosity of greater than 2000 centipoise is defined as 100%; lower readings are scaled as a percentage of 2000 centipoise.

assayed by falling-ball viscometry (open bars) both show significant interactions for most glycolytic enzymes. F-actin affinity column binding of an enzyme requires monovalent binding with adequate affinity to prevent washoff; F-actin gelling activity requires cross-linking, hence at least divalent binding of enzyme to F-actin. Both techniques reveal that phosphofructokinase (PFK), aldolase, and glyceraldehyde-3-phosphate dehydrogenase (GAPDH) have very high actin-binding activity, making these the most desirable enzymes to study with respect to possible actin–enzyme interactions *in vivo*. These techniques detect relatively little direct triose phosphate isomerase (TPI) binding activity, but there is evidence that indirect, or 'piggyback', binding of TPI may occur (Stephan et al., 1986). Enolase and hexokinase are conspicuous due to their lack of activities; note that affinity column data for hexokinase are absent, and that other data suggest significant binding of hexokinase to the mitochondrial outer membrane (Kropp and Wilson, 1970; Wilson, 1980). Based in part on the data in Figure 1, we have studied aldolase and enolase; aldolase is a well studied enzyme with known actin-binding characteristics, and enolase is useful as a glycolytic enzyme with no known actin- or tubulin-binding activity.

From a large body of work, including that cited above, several models for organization of glycolytic enzymes around the cytoskeleton have been developed; models proposed by Masters (1984) and Clegg (1988) are representative of these. Masters proposed a three-way dynamic equilibrium which would involve not only the polymerization of F-actin from a pool of actin monomer, but also reversible 'plating' of glycolytic enzymes in a domain surrounding the polymerized actin

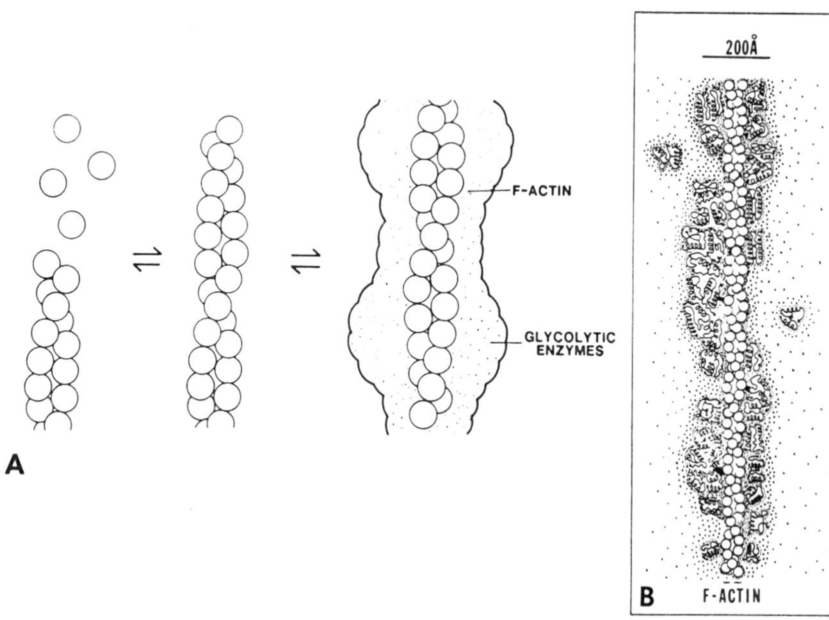

Figure 2. Two models for glycolytic enzyme organization. Masters proposed a three way dynamic equilibrium among G-actin, F-actin, and glycolytic enzymes (**A**). The enzyme-rich domain represented by the stipled area is maintained by numerous transient interactions between enzymes and F-actin. Clegg proposed sequential ordering of several enzymes along cytoskeletal elements; the resulting array would be a functional metabolic unit (**B**). The corollary organization of water, and a zone of 'vicinal' water play an important role in this model. (Figures reproduced from Masters, 1984 (**A**) and Clegg, 1988 (**B**), with permission.)

filaments (Figure 2A). The transient binding of enzymes to actin filaments proposed in this model would result in an enzyme-rich domain surrounding actin filaments. The model also suggests that an element of periodicity may also be present in the enzyme-rich domain; presumably such periodicity would arise from the actin template around which the enzymes were 'plated', rather than from the enzymes themselves.

Clegg's model is fundamentally similar, but it makes an additional important prediction about enzyme organization (Figure 2B). This model is based on the microtrabecular lattice as the structural scaffold for enzyme organization, and it predicts sequential ordering of several enzymes in a metabolic pathway. The implication is that the physical organization of enzymes in cytoplasm might actually correspond with the logical reaction sequence deduced from biochemical experiments. Sequential organization of this kind, even if dynamic, transient, and involving only small portions of a pathway at any one time, could generate a non-membrane-bound metabolic compartment, and could be a widespread phenomenon in cytoplasmic organization (Srere, 1987). Another important consequence of this model is that sequential binding of enzymes should be testable *in vivo* using some of the techniques described below.

Several important consequences could result from the binding of glycolytic enzymes to cytoskeletal elements *in vivo*. First, it is clear that kinetic parameters of enzymes can be significantly affected by binding to cytoskeletal components, suggesting unconventional mechanisms of metabolic regulation (Arnold and Pette, 1970; Clarke et al., 1974; Walsh et al., 1980). Second, modes of enzyme organization, such as functional duality of glycolytic enzymes, could be a result of the ability of glycolytic enzymes to bind to structural components as well as substrate (Clarke et al., 1985a,b). Third, direct mechanisms linking glycolytic metabolism and cell motility may exist, and could explain a variety of unpredicted findings which suggest that glycolysis may fuel some kinds of cell motility directly (Kamiya et al., 1957; Bricknell et al., 1981; Gibbins, 1982). Finally, there is growing evidence that glycolytic (and not bulk) ATP specifically fuels both Na^+/K^+ and Ca^{2+} ion channels, due to 'functional compartmentation' (Mercer and Dunham, 1981; Lynch and Paul, 1983; Weiss and Lamp, 1987; Paul et al., 1989; Hardin et al., 1992). Individually, these are intriguing examples of metabolic organization; together they are compelling evidence that we need to reevaluate the role enzymes play in cytoplasmic organization.

We have been studying functional fluorescent analogs of the glycolytic enzymes aldolase and enolase in living Swiss 3T3 cells. I will describe both the approach we have been using, and the potential application of some new techniques to the study of glycolysis *in vivo*. Figure 3 summarizes our initial findings, which will be discussed below.

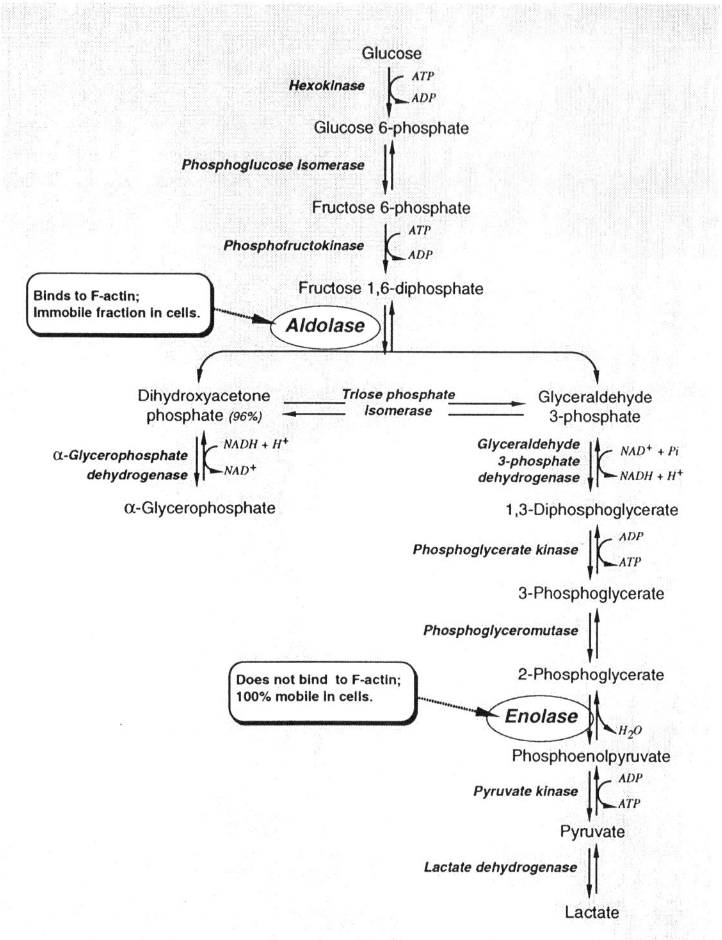

Figure 3. Summary of FRAP measurements of fluorescent analogs of aldolase and enolase, both *in vitro* and *in vivo*. Aldolase binds to F-actin *in vitro*, and exhibits an immobile fraction *in vivo*, while enolase does not bind to F-actin *in vitro*, and is 100 percent mobile in cells. These results are consistent with models in which the actin-binding activity of aldolase is responsible for its immobile fraction *in vivo*.

III. FLUORESCENCE MEASUREMENTS IN LIVING CELLS

Generating a 'wish list' of the kinds of measurements one would like to make in living cells is, of course, far easier than devising methods for actually making those measurements. However, such a list can help us put our current capabilities into perspective, and evaluate the impact that new technologies may have on specific enzymologic questions. The details of individual systems will vary, but most of what follows could be generalized to the study of other enzymes *in vivo*. The techniques outlined below require fluorescent labeling and loading of the enzyme in question into living cells, a technique known as fluorescent analog cytochemistry (Taylor et al., 1984).

A. How Much Enzyme?

Quantifying the amount of a specific enzyme in a cell can be done well with conventional biochemical techniques. Two parameters are important: the catalytic activity present and the total enzyme present; their ratio represents the specific activity. Methods that measure how much of a particular *activity* is present generally rely on direct or indirect (coupled) measurements of the appearance or disappearance of a substrate, metabolite or cofactor from a defined system (Bergmeyer et al., 1974). Methods that measure how much of a particular *protein* is present, thereby allowing us to calculate how many molecules there are, generally rely on immunologic recognition (independent of activity) of the enzyme in question, using a western blot or dot blot to quantify a specific protein. Of course, the shortcoming of these approaches is that they require total disruption of the cell (or cellular compartment) in which the enzyme is to be assayed.

Quantifying specific activity of an enzyme *in vivo* is difficult, if not impossible, but the specific activity, in the strict biochemical sense, may be less important than relative activity *in vivo*. If we consider the possibility that some enzymes may be spatially regulated *in vivo*, through mechanisms such as ambiquity (Wilson, 1978; 1988), then spatial differences in catalytic activity (evaluated by activity mapping) may be more metabolically relevant than net specific activity (for the whole cell). Some approaches to quantifying relative activity *in vivo* are outlined below (Section D); activity maps could be generated in time series to follow the course of a physiological experiment, and spatial differences in enzyme activity could then be visualized.

B. What Is the Enzyme Distribution?

The spatial distribution of specific enzymes in living cells can be mapped with fluorescence microscope ratio imaging if two conditions can be met. First, it must be possible to generate an active, fluorescently labeled analog of the enzyme to be studied. Activity refers not only to verifying the catalytic K_m and V_{max} of the labeled enzyme, but also to other binding characteristics of that enzyme. In the case of most glycolytic enzymes, for example, significant specific binding activity to structural proteins has been established (Arnold et al., 1971; Clarke and Masters, 1975; 1976; Walsh et al., 1989). Second, a neutral, inert reference molecule must be developed for use as a nonbinding intracellular control. We have used fluorescently labeled

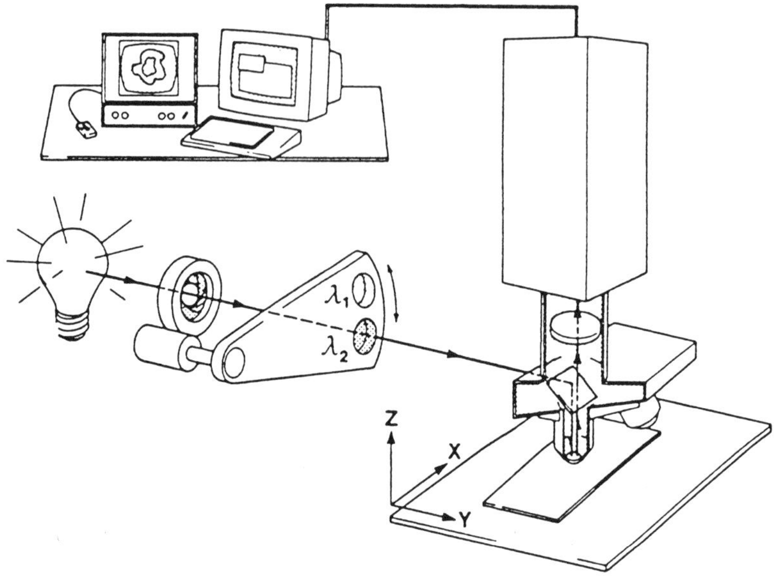

Figure 4. Schematic ratio imaging system. The enzyme of interest and an appropriate reference molecule are labeled with contrasting fluorophores and loaded into cells; images of the distribution of each species are then acquired at the appropriate wavelength by changing the fluorescence filter sets. Image pairs are acquired with a low-light level camera and digitized for processing and storage in the computer. (Figure reproduced from Bright, 1987, with permission.)

dextrans and ficolls as reference molecules to date, though it appears that enolase may serve well as a non-binding reference molecule for other glycolytic enzymes (see Section IV.B.).

The strategy of mapping enzyme distribution *in vivo* with ratio imaging is conceptually simple. Ratio imaging has been described at length elsewhere (Bright et al., 1989, and references therein), and is diagrammed schematically in Figure 4. For studies on enzymes, both the enzyme and a companion, neutral, inert reference molecule are microinjected into cells in tracer quantities. The enzyme and the reference molecule are labeled with contrasting fluorophores which allow their distributions to be visualized separately using the appropriate filter sets. For example, the enzyme to be studied may be labeled with rhodamine, and the dextran or ficoll reference molecule may be labeled with fluorescein. Pairs of fluorescence images of the enzyme and reference molecules are then acquired using a low light level camera, such as an intensified SIT (silicon-intensified target) or CCD (charge-coupled device) camera, and digitized for storage in a computer. The stored images can then be manipulated in a variety of ways, including subtraction of background noise, and division of one image by another to provide the distribution of the numerator image relative to the denominator image. Figure 5 summarizes a typical enzyme-mapping experiment using ratio imaging; the finding was that enolase enters the nucleus of 3T3 cells, but that a reference dextran did not. The mapping results indicate that a parameter other than hydrodynamic radius was responsible for preferential partitioning of enolase in the nucleus.

Ideally, both the enzyme and reference molecules can be mixed and co-injected into cells, but this may not always be possible (see Section IV.C.). Recent improvements in interference filter and probe technology allow for excellent discrimination of multiple fluorophores, but control experiments must be done to determine the

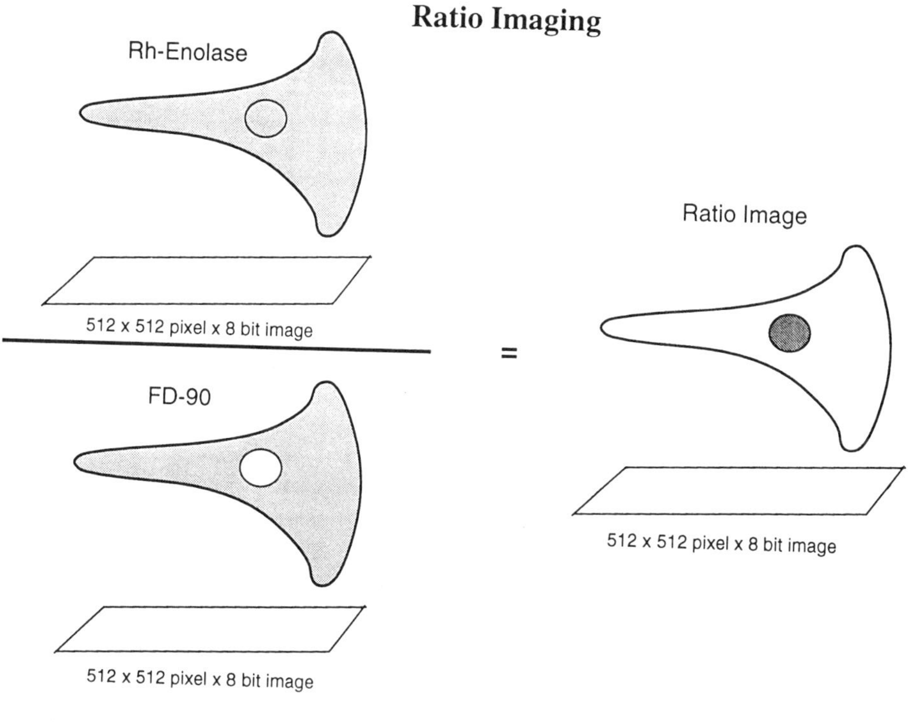

Figure 5. Summary of enzyme-mapping experiment using fluorescence ratio imaging. In this case, enolase labeled with rhodamine (Rh-enolase), and size-fractionated fluorescein-labeled dextran (FD-90) have been co-microinjected into the cytoplasmic space of a living Swiss 3T3 cell. An image of the Rh–enolase distribution is acquired with the rhodamine filter set, and an image of the FD-90 distribution is acquired with the fluorescein filter set. The digitized enolase image is then divided, pixel for pixel, by the digitized FD-90 image, normalizing intensity artifacts due to pathlength and accessible volume. The resulting ratio image (right) reveals that enolase is more concentrated in the nucleus than FD-90 is, indicating that there is specificity of transport of enolase into the nucleus.

amount of unwanted signal generated by contrasting fluorophores; e.g. how much signal is generated by fluorescein with the rhodamine filter set. We have found this 'signal crossover' to be very small, and uniformly << 5% with carefully selected filter sets, such as the 'quantitative' series of six-cavity interference filters available from Omega Optical (Brattleboro, Vermont).

The use of tracer quantities is essential for experiments from which metabolic conclusions are to be drawn. Clearly, a pool of labeled, exogenous enzyme that is introduced into cells could significantly affect the equilibria it is intended to detect. However, if the labeled enzyme is in small enough quantity, equilibration effects can be minimized. The opposing constraint is the need to load adequate labeled enzyme into cells to generate an acceptable signal-to-noise ratio with the image detector used. In our experiments with glycolytic enzymes, we have been able to image labeled enzyme when it is present at approximately 2–4% of the concentration of endogenous enzyme. This level of injected analog represents a reasonable compromise between the need to minimize metabolic interference, and the need to maximize signal.

Under the best of conditions, the use of fluorescence microscope ratio imaging to map the relative distribution of a fluorescent analog of an enzyme represents a delicate balance. On the one hand, it is desirable to minimize the dye-to-protein ratio of the labeled enzyme, and to inject the smallest possible amount of labeled enzyme into cells to minimize biological perturbation of the system under study. On the other hand, clean detection of the small signals involved presents a challenge for the best instrumentation available today. It is clear, however, that available technology is adequate to reveal biologically significant spatial differences in enzyme distribution using fluorescence ratio imaging.

C. Are the Enzyme Molecules Bound or Freely Diffusing?

A map of the relative distribution of enzyme molecules, while important, does not in itself give us information about the behavior of the molecules in different regions of the cell. In addition to location, we can also measure diffusion coefficient and bound fraction *in vivo*, using FRAP. Originally developed in the mid-1970s (Axelrod et al., 1976; Edidin et al., 1976), the use of quantitative spot FRAP measurements was limited for over a decade after its introduction, due to the initial use of FRAP to measure only two-dimensional diffusion of membrane molecules, and due to the significant instrumentation required. As the technique evolved, however, it became clear that it was more generally applicable to three-dimensional bulk diffusion measurements in solution and in cells (Wojcieszyn et al., 1981; Jacobson et al., 1982). Equally important, rapid improvements in the hardware and software needed to implement a semiautomated photobleaching instrument have made the programming and interfacing tasks involved much less formidable. The few commercial photobleaching instruments currently on the market have been limited by an inability to make quantitative diffusion coefficient measurements;

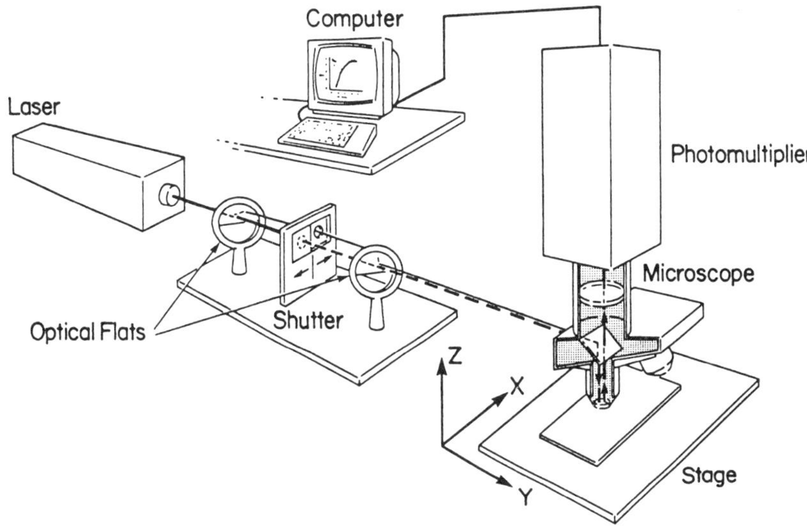

Figure 6. Schematic fluorescence redistribution after photobleaching (FRAP) system. A series of optical components and shutters allows the beam of an argon ion laser to be switched rapidly between a low intensity monitoring beam, and a high intensity bleaching beam. The bleaching beam is 5000- to 10,000-fold as intense as the monitoring beam. The Gaussian beam is focused onto a spot on the cell through a modified epifluorescence microscope, and the intensity of the fluorescence (but not an image) is recorded with a photomultiplier tube. The signal is digitized and stored in a computer for analysis and curve-fitting; the system of shutters is operated under computer control.

thus building an instrument in one's laboratory remains an attractive, and relatively economical, alternative.

The instrumentation for FRAP is outlined in Figure 6, and the design and construction of a quantitative photobleaching instrument has been described in detail by Wolf (1989). Our instrument has several modifications which optimize it for making FRAP measurements of fluorescent enzyme analogs *in vivo*. Perhaps the most important consideration is the need for the operator to visualize the position of the laser beam (via the spot of fluorescence emitted from the analog in the cell) while the full field is being illuminated by a conventional epifluorescence illuminator. This allows for the beam to be positioned precisely within a given cell by translating the cell under the beam with the microscope stage. The requirement for simultaneous full-field and spot imaging is satisfied by bouncing the full-field illumination off a mirror with a hole drilled in it; the laser beam is passed through the hole, allowing it to be superimposed upon the image of the cell. This modification allows for the beam to be easily and precisely positioned within a single cell, or from cell to cell, so that measurements may be made rapidly after a perfusion or other physiological stimulation.

FRAP measurements yield two kinds of data: immobile (or bound) fraction and diffusion coefficient (D) (see Figure 7). The immobile fraction represents labeled molecules which remained bound during the timecourse of the experiment, and it is expressed as a percentage. The diffusion coefficient is calculated from the half-time of fluorescence recovery, and is expressed in units of cm^2/sec.; measured values of diffusion in aqueous solution (D_{aq}) and cytoplasm (D_{cyto}) are distinguished. Cytoplasmic measurements cannot rigorously be expressed as true diffusion coefficient values, since cytoplasm is a complex, heterogeneous medium; thus D_{cyto} typically refers to an 'apparent' diffusion coefficient value.

The data analysis algorithms used for FRAP measurements were originally optimized for two-dimensional cases, primarily membrane mobility experiments (Yguerabide et al., 1982; van Zoelen et al., 1983), however FRAP measurements for three-dimensional diffusion are remarkably accurate, given two caveats. First, the specimen to be studied must be relatively thin and flat. This does not represent a significant problem with well-spread, adherent fibroblasts, which are often only one or two microns thick throughout most of their lamellae. The nuclear region of the cell is thicker, perhaps 10–12 microns thick, thus potentially creating an unacceptable experimental geometry; in practice, however, the use of a neutral reference molecule will reveal whether unexpected results in the nuclear region are artifactual. Solution FRAP measurements are made in 50 μm thick flat capillary tubes (Vitro Dynamics, Inc., Rockaway, NJ), which yield very accurate D_{aq} measurements of standard molecules.

Second, it must be appreciated that measurement of an immobile fraction will not be compromised in three-dimensional FRAP experiments. This is important, since transient enzyme binding may be the most important physiological parameter FRAP can detect in the experiments described. In addition, experience has shown that values of D obtained with FRAP, both in solution and in cytoplasm, have been in good agreement with those obtained by other methods. Relative changes in D during an experiment are significant, regardless of absolute values, and the use of a well-characterized reference molecule can provide a good internal control.

A final caution in the use of FRAP in living cells is the potential for heat and/or photodamage, since irradiation energy levels in cells during bleaching pulses can be as high as 20–30 kW/cm^2 (Pagliaro and Taylor, 1988). Photodamage has the potential to create artifactually high or low measurements of both D_{cyto} and immobile fraction, depending on a number of variables, including the fluorescent dye used, the wavelength to which the laser is tuned, and the species which is labeled in a given experiment. Bleaching irradiation may induce crosslinking, hence an artifactual bound fraction, of the labeled species; alternatively, the laser energy may destabilize existing cytoskeletal structure, thereby artifactually decreasing the

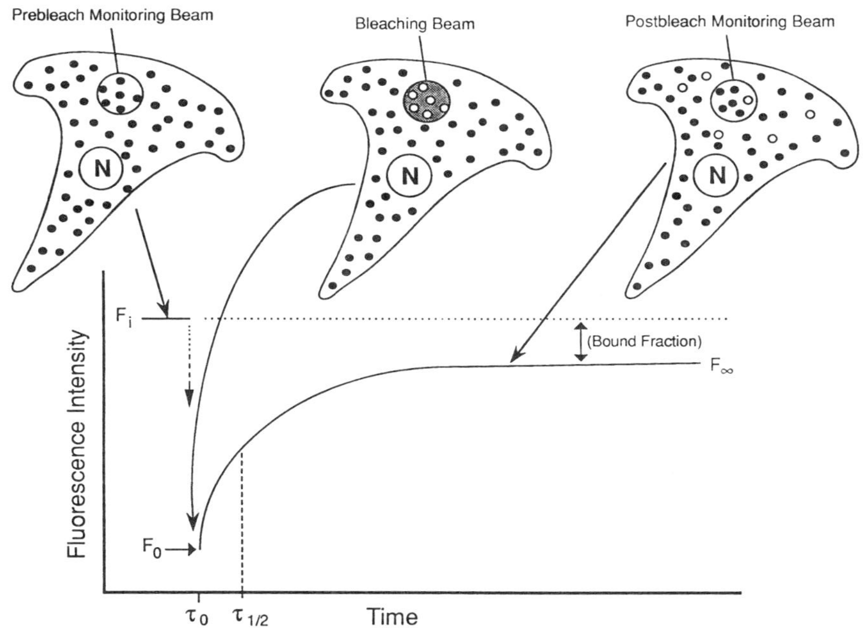

Figure 7. A typical FRAP experiment. The cell has been microinjected with a fluorescent analog of an enzyme molecule. The laser beam is focused to a 4–6 μm diameter spot on the cell; spread fibroblasts typically have dimensions of 30–40 μm, allowing for relatively high resolution mapping with multiple FRAP measurements. Prebleach fluorescence is measured with the low-intensity monitoring beam to establish an initial fluorescence level. The high intensity bleaching beam illuminates the spot briefly, typically for a 10–100 ms pulse, bleaching a fraction (typically 50–60%) of the labeled molecules in the spot. Monitoring and bleaching beams are precisely aligned to illuminate identical portions of the field. Return of fluorescence in the spot is then measured periodically with the monitoring beam as the bleached molecules redistribute over time. The half-time of recovery ($t_{1/2}$) is used to calculate diffusion coefficient, and the extent of recovery (F_∞) detects the presence of a bound fraction.

bound fraction of the labeled species. While there is evidence that both kinds of artifact may occur under extreme conditions (Simon et al., 1988), it is also clear that, under the bleaching conditions used *in vivo*, the V_{max} of labeled aldolase decreased by 30–40%, and the K_m was not significantly affected (Pagliaro and Taylor, 1988). Since this level of photodamage occurs only in the enzyme analog directly under the bleaching spot (roughly 2–4% of the pool of analog in the cell), the vast majority of the specimen is unaffected.

D. Are the Enzyme Molecules Active?

Perhaps the greatest technological challenge for those studying enzymology *in vivo* is to measure specific catalytic activity with spatial resolution in living cells. Activity measurements *in vivo* are important, since the environment is so unlike the solutions in which enzyme activity is normally assayed (Halling, 1989). This is a formidable task, however, due to the inherent difficulty of obtaining an absolute measure of enzyme (protein) *in vivo*, as outlined above in Section A. For many metabolic questions, however, a spatial map of the relative catalytic activity would be quite valuable, and may be experimentally feasible using fluorescence imaging approaches. A relative catalytic activity map would be a ratio image of catalytic activity/enzyme distribution, thus revealing enzyme regulation *in vivo*. A variety of microfluorometric techniques have been used to measure enzyme activity in single cells for over 30 years (reviewed in Dolbeare, 1981). Techniques used in these studies include early fluorogenic substrates, such as the fluorescein fluorochromasia reaction (Rotman and Papermaster, 1966), and measurements of endogenous NADH and NADPH fluorescence (reviewed in Kohen et al., 1983). Fluorogenic substrates are substrate analogs for enzymes which undergo a significant change (generally a large increase) in fluorescence emission after enzymatic modification. While early fluorogenic substrate work generally took advantage of endogenous dehydrogenase or esterase activity, an increasing array of fluorogenic substrates designed to interact with specific enzymes is available. For example, several fluorogenic substrates for β-galactosidase are commercially available and have been used to assay β-galactosidase activity in living cells (Jongkind et al., 1986).

Several possibilities for developing fluorogenic substrates for glycolytic enzymes exist. The dehydrogenase activity of GAPDH is one clear candidate. Another possibility would take advantage of the hexose cleaving activity of aldolase; the release of quenched fluorophores on fructose 1,6-bisphosphate could result in a large change in fluorescence output. Further work is needed to evaluate the best initial approach to developing a fluorogenic substrate for a glycolytic enzyme, but the prospects are good.

An image of fluorescent product distribution (from a fluorogenic substrate) does not, however, give us information about the relative activity of the enzyme molecules in question. In order to do this, substrate output relative to enzyme distribution must be mapped. A suitable fluorogenic substrate would make possible an experiment in which the enzyme activity was imaged and divided by an image of enzyme distribution, showing spatial differences in activity per unit enzyme. Since the enzyme distribution map generated with a fluorescent analog is relative, the resulting activity map will also be relative. If both the fluorescent analog of the enzyme and the fluorogenic substrate incorporate into the appropriate intracellular pools; however, the activity map, while relative, should be representative of endogenous activity.

An inherent limitation of a fluorogenic substrate approach involves the stability and accumulation of the fluorescent product *in vivo*, as well as the size of the fluorogenic substrate pool. A variety of coupling agents have been used for solution and whole-cell averaging fluorimetric assays (reviewed in Dolbeare, 1981). Coupling agents generally react with, and precipitate the fluorescent product of a fluorogenic substrate; in some cases, fluorescence is also quenched in the coupling reaction. While advantageous for more precise short-term kinetics, the use of conventional coupling reagents *in vivo* creates an increasing pool of insoluble product in cells over time. Without coupling agents, accurate kinetic measurements *in vivo* can be continued for 5–10 minutes, long enough to enable a number of worthwhile physiological experiments. The prospect for improved fluorogenic substrates is good, and coupled with other techniques, they will become increasingly useful tools for studying enzyme dynamics *in vivo*.

E. How Do Specific Molecules/Ions Affect Enzymes?

One approach to testing hypotheses about enzyme behavior and organization *in vivo* is to manipulate the concentration of a known effector (ion, intermediate, nucleotide, etc.) in a precise way. Perfusion chambers are routinely used to treat cells with effectors which are freely diffusible across the cell membrane, or which are known to bind to a cell surface receptor. Perfusion, however, is of limited usefulness for non-diffusible and short-lived effectors. An important potential set of tools for enzymology *in vivo* are photoactivatable, or 'caged' compounds.

Caged compounds have been in use in physiological experiments for over a decade (reviewed in Gurney and Lester, 1987), but the array of caged compounds available has diversified considerable in recent years, and many are commercially available (e.g. Molecular Probes, Eugene, Oregon). A number of questions about glycolytic metabolism are immediately approachable with caged compounds, while other questions will depend on the development of new probes. One of the first uses of caged compounds was development of caged nucleotide analogs to produce rapid jumps in intracellular ATP concentration upon photolysis (Kaplan et al., 1978; McCray et al., 1980). In such an experiment, caged ATP is introduced into the cell, and a pulse of photolysis illumination (typically at 340–350 nm for nitro–phenyl ester derivatives) releases the caging groups on a 10–100 ms timescale. A physiological parameter is monitored before and after the sudden ATP increase and any effects quantified.

Caged ATP could be used to investigate the role of nucleotides in glycolytic enzyme organization, perhaps of particular interest for the kinases in the pathway.

It is also clear that several of the glycolytic enzymes are calcium sensitive, either directly (Harris and Winzor, 1989), or via a complex with calmodulin (Orosz et al., 1988). Intracellular calcium can be manipulated with photoactivatable chelators that release Ca^{2+} upon photolysis (Adams et al., 1988), thus creating the possibility of using rapid increases in intracellular Ca^{2+} concentration as a physiological probe *in vivo*. The widespread effects of Ca^{2+} will mandate good controls and careful interpretation of such experiments *in vivo*, but such experiments are quite feasible.

Finally, caged analogs of specific substrates and intermediates could be developed. Such analogs would enable rapid experimental increases of a specific metabolite during an experiment, and could be used to investigate, for example, the role of substrates in cytoskeletal enzyme binding.

IV. SOME PRACTICAL DETAILS

A. Enzyme Labeling and Characterization

Techniques for producing fluorescent analogs of proteins have been dealt with extensively elsewhere (Taylor and Wang, 1980; Wang, 1989), but several points should be stressed for the preparation of enzyme analogs. A basic tenet of fluorescent analog cytochemistry is that, if tracer quantities of a protein analog are to accurately represent the endogenous protein pool they represent, then the analog must possess all the functionalities of the native molecule. Thus, a fluorescently labeled enzyme must also have demonstrated catalytic activity, in addition to the usual characterization of fluorescent analogs for parameters such as the dye-to-protein ratio, fluorescence spectrum, and electrophoretic mobility. In addition, the enzyme analog must have other binding activities (such as actin-binding activity) that have been documented in the native molecule.

Fluorescent labeling of an active tracer molecule represents a compromise. If the molecule is labeled heavily, it will be more easily detectable, but activity will inevitably be compromised at some point. On the other hand, optimizing activity with minimal labeling may result in an undetectable analog with excellent functionality. Clearly, neither extreme is acceptable, and a dye-to-protein ratio that gives both adequate signal–noise and good activity must be used. For aldolase and enolase, our working molar dye–protein is 1.5–1.7, calculated for tetramer (aldolase) or dimer (enolase), since gel-filtration chromatography demonstrated that the analogs retain native conformation (Pagliaro and Taylor, 1988; Pagliaro et al., 1989).

If a particular activity is compromised excessively during labeling, a valuable strategy is to protect the active site of the molecule with the appropriate ligand during labeling. For an enzyme, labeling may be performed in the presence of an excess of substrate, divalent cation, actin, etc. In general we have not found this precaution to be necessary with aldolase or enolase; neither catalytic activity nor actin-binding activity is compromised significantly until molar dye–protein significantly exceeds 2.0. Preliminary work in developing a fluorescent analog of PFK indicates that the Ca^{2+} sensitivity of this enzyme may require protection of the divalent cation binding site during labeling (L. Pagliaro, unpublished results).

B. Reference Molecules

Fluorescence ratio imaging and FRAP are proven techniques for mapping the distribution and measuring the diffusion coefficient of labeled molecules *in vivo*, but as with any other technique, experimental artifact is possible. One of the major sources of artifact in the kinds of experiments described here is the effect of cell geometry, which is inherently irregular. The resulting differences in volume accessible to a given probe, and pathlength variation across the cell, can result in large differences in total fluorescence, even though the probe concentration in a given compartment is uniform. Of course, the situation becomes even more complex if the probe of interest has a nonuniform intracellular distribution. In short, cells are not ideal cuvettes.

A reference molecule can be used to correct for much of the geometric artifact inherent in live cell studies. A reference molecule is an essential part of our fluorescence ratio imaging approach, but we have found them equally valuable for FRAP, particularly when multiple wavelengths are used. In addition, the possibility of fluorophore-specific effects can be ruled out with reference molecules. The most widely used reference molecules for tracer studies in cytoplasm are neutral, inert dextrans and ficolls that have been labeled with the appropriate fluorophore (Luby-Phelps, 1989). Fluorescently labeled dextrans can be purchased from a number of sources, but it is essential that they be size-fractionated before use. Dextrans are polydisperse polymers with a wide range of molecular weights; with the proper gel filtration chromatography, however, fractions can be obtained that have K_{av} values close to the enzyme they are a control for, and acceptably narrow elution profiles. Ficolls have the advantage that their hydrodynamic behavior more nearly resembles that of globular proteins, but fewer ficolls with suitable molecular weights are commercially available.

We have found that enolase does not bind to actin under a wide variety of conditions, and that it remains entirely mobile *in vivo* (Pagliaro et al., 1989). This makes enolase an ideal candidate for a reference molecule for other glycolytic enzymes; if enolase does not bind to cytoskeletal structures under conditions in which other glycolytic enzymes do, then nonspecific binding becomes a difficult explanation to justify.

C. Microinjection and Other Methods of Cell Loading

The preparation of a well characterized, active fluorescent analog of an enzyme is a significant research investment, and results in a relatively precious reagent. The kinds of live cell experiments outlined above require the ability to load tracer quantities of enzyme analogs into hundreds of cells rapidly and reproducibly. We have found that manual microinjection, for all its shortcomings, is the best technique for loading enzyme analogs into adherent fibroblasts. Although a number of mechanical methods of bulk-loading cells have been developed, including scrape loading and bead loading (McNeil et al., 1984; McNeil, 1989), and electroporation (Shigekawa and Dower, 1988), these methods require the use of relatively large quantities of reagent (hundreds of microliters of mM solution) in the culture vessel, and they tend to yield a broad distribution of the quantity of enzyme that is loaded into individual cells.

Microinjection, on the other hand, requires only tens of microliters of enzyme solution per experiment, and can be learned in only a few weeks. The technical requirements for loading cells by microinjection are generally less demanding than for electrophysiological applications, but a quality micromanipulator is an essential investment. We routinely microinject 20–40 adherent cells on a single coverslip, and prepare several coverslips for a given experiment, resulting in a population of 100–200 injected cells per experiment. The trauma of microinjection results in the loss of some cells, but an 80–90% recovery rate is a very attainable goal, giving an ample population of injected cells to work with.

A new approach for loading molecules into large numbers of cells takes advantage of endogenous folate receptors and constitutive receptor-mediated endocytosis of folate (Leamon and Low, 1991). The probe molecule of interest is covalently liked to folate; normal endocytosis of this vitamin then transports the molecule into the cell. The technique has been used with enzymes, and functional enzyme is delivered into the cytoplasmic space after the enzyme folate complex is endosomally processed (Leamon and Low, 1992). This approach enables bulk loading of large numbers of cells, while minimizing the quantity of probe molecule that must be used. A possible disadvantage of this technique is that, although folate receptors are widespread, receptor density may be too low on some cultured cell types to allow efficient probe loading. Similar biologically mediated strategies for cell loading are worth investigation for loading large numbers of cells using minimum quantities of probes.

V. FINDINGS TO DATE WITH ALDOLASE AND ENOLASE

A. Experiments *In Vitro*

We developed a rhodamine-labeled analog of aldolase (Rh–aldolase) with demonstrated catalytic and actin-binding activities, and native tetrameric conformation (Pagliaro and Taylor, 1988). We also produced a companion size-fractionated fluorescein-labeled dextran (FD-150) for use as a reference molecule. With these probes, we demonstrated binding of aldolase to F-actin *in vitro* directly, using FRAP measurements in reconstituted aqueous solutions, and confirming previous bio-

Table 1. *In Vitro* FRAP Data: Rhodamine–Aldolase, Fluorescein–Enolase, and F-Actin in Solution

	$D_{37,cyto} \times 10^{7a}$	*Percent Mobile*
1 µm Rh–Aldolase (26)[b]	4.72 ± 1.78[c]	101.7 ± 6.9
Rh–Aldolase + 1 mg/ml F-Actin (20)	1.04 ± 0.12	63.7 ± 10.4
Rh–Aldolase + FBP + F-Actin (26)	4.18 ± 1.57	98.6 ± 5.1
1 µm Fl–Enolase (20)	6.08 ± 0.43	99.1 ± 3.9
Fl–Enolase + 1 mg/ml F-Actin (11)	6.11 ± 0.41	105.0 ± 1.9
Fl–Enolase + F-Actin + Aldolase (17)	5.98 ± 0.18	107.1 ± 1.1

Notes: [a]Aqueous diffusion coefficient, in cm^2/sec.
[b]Number of measurements used in calculating the mean.
[c]Mean ± the sample standard deviation.

chemical data supporting binding of aldolase to actin. In aqueous solution, aldolase had a diffusion coefficient (D_{aq}) of 4.7×10^{-7} cm^2/sec, and it was 100% mobile, consistent with its known molecular weight and calculated hydrodynamic radius (Table 1).

In the presence of F-actin, there was a 4.5-fold reduction in D_{aq}, as well as a 36% bound fraction, demonstrating binding of Rh–aldolase to F-actin. In the presence of its substrate, fructose-1,6-bisphosphate (FBP), both the mobile fraction and D_{aq} of aldolase returned to control levels, indicating competition between substrate- and actin-binding activities. A similar series of experiments with a fluorescein-labeled analog of enolase (Fl–enolase) demonstrated a lack of actin-binding activity for enolase (Pagliaro et al., 1989; Table 1). Even in the presence of an F-actin-aldolase gel, enolase remained 100% mobile, and showed no significant decrease in D_{aq}.

B. Experiments *In Vivo*

We microinjected Rh–aldolase and Fl–enolase into living 3T3 cells and performed ratio-imaging, to map enzyme distribution, and FRAP, to measure D_{cyto}, and spatial differences in D_{cyto}. The image of aldolase distribution shown in Figure 8 is relatively uniform, and there is no apparent localization of aldolase on stress fibers. However, ratio-imaging of cells coinjected with Rh–aldolase and a size-fractionated fluorescein labeled dextran indicated that the dextran was relatively excluded from domains around stress fibers, while aldolase entered this region. The ratio image shown in Figure 9 reveals a series of thin, light lines along stress fibers; aldolase is increased relative to dextran in these zones. These data suggest that a fraction of aldolase incorporates into a microdomain closely associated with stress fibers in 3T3 cells, and exists partially in the solid phase of the cytoplasm. The partitioning of aldolase and other glycolytic enzymes between the

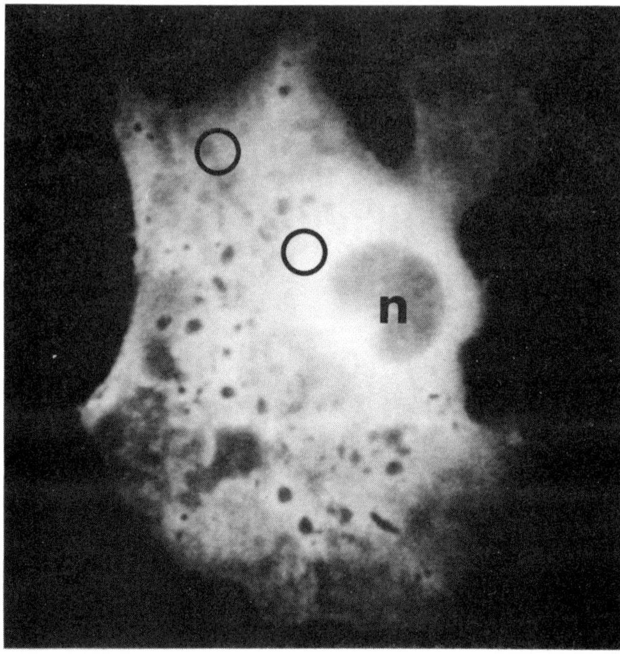

Figure 8. Image of aldolase distribution *in vivo.* Rhodamine-labeled aldolase was microinjected into a living 3T3 cell in tracer amounts. Enzyme distribution appears relatively uniform, except for exclusion from the nucleus (n) and vesicles. The size and location of typical peripheral (upper) and perinuclear (lower) FRAP measurement spots are indicated by the circles. (Reproduced from Pagliaro and Taylor (1988), with permission.)

fluid and solid phases of cytoplasm could play fundamental roles in the regulation of glycolysis, the organization of cytoplasm, cell motility, and neoplastic transformation. Similar experiments with enolase revealed no such organization of enolase around stress fibers (Pagliaro et al., 1989).

FRAP measurements of Rh–aldolase and Fl–enolase *in vivo* demonstrated the existence of a bound fraction of aldolase, but not enolase (Table 2). In addition, there were significant spatial differences in the D_{cyto} and mobile fraction of aldolase, but not enolase. For aldolase, D_{cyto} in the perinuclear region of cells was 1.4×10^{-7} cm^2/sec with a 21% bound fraction, while D_{cyto} in the cell periphery was 8.1×10^{-8} cm^2/sec with 100% mobility. Enolase was 100% mobile in both perinuclear and peripheral regions, and it had smaller spatial differences in D_{cyto}. These data are consistent with models in which actin-binding activity is responsible for the bound fraction of aldolase *in vivo*.

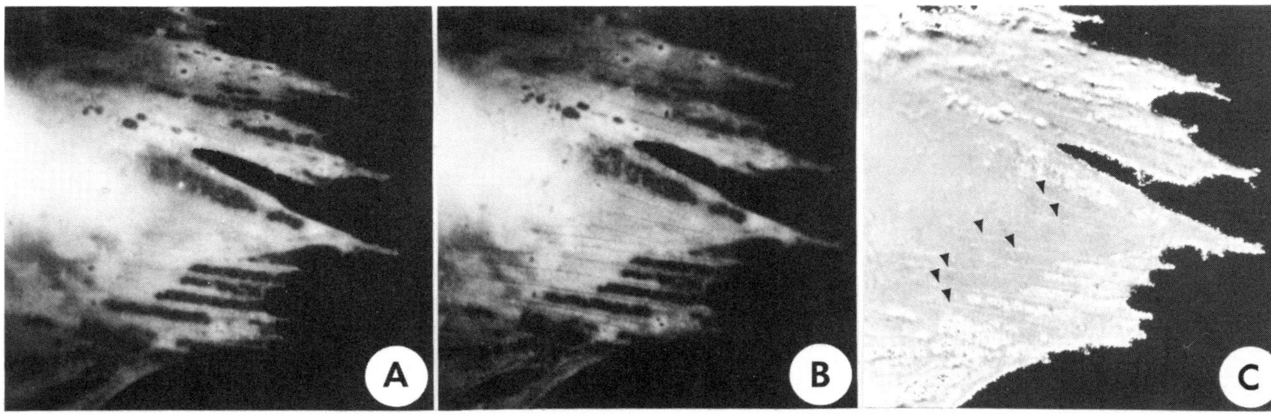

Figure 9. Ratio image map of aldolase distribution *in vivo*. In this experiment, both rhodamine-labeled aldolase and size-fractionated, fluorescein-labeled dextran were microinjected into a living 3T3 cell in tracer amounts. While the difference between the aldolase image (**A**) and the dextran image (**B**) is not immediately apparent, the ratio image (**C**) reveal a series of thin bright lines in the stress fiber-rich lamellum (arrowheads), indicating a relative concentration of aldolase around stress fibers. These data are consistent with models of enzyme organization such as those shown in Figure 2. (Reproduced from Pagliaro and Taylor (1988), with permission.)

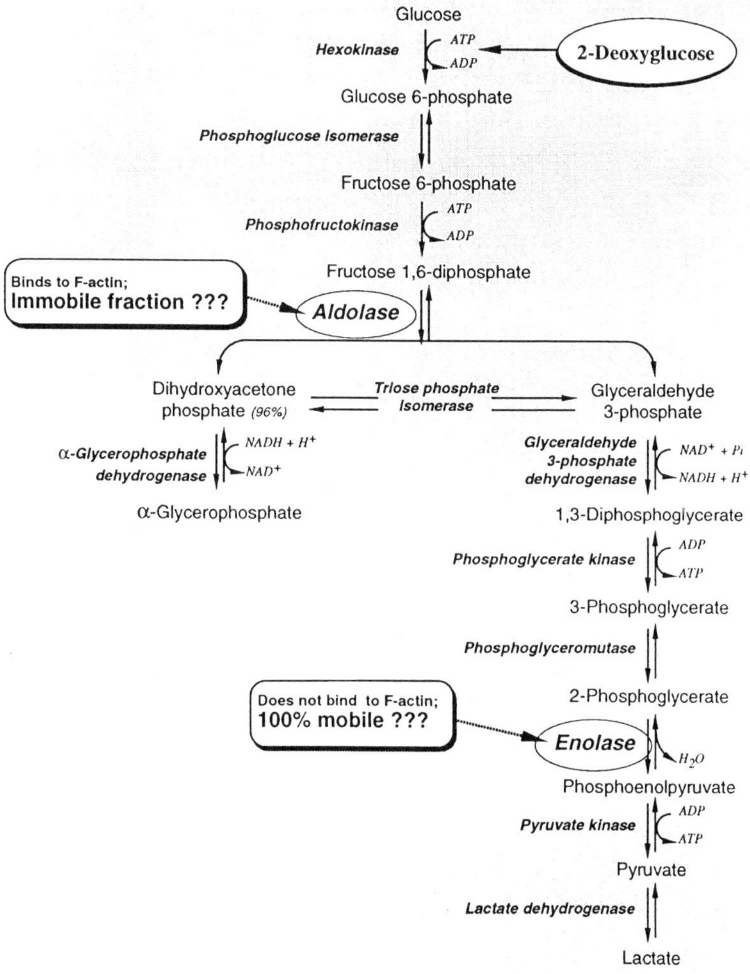

Figure 10. A metabolic experiment: if glycolysis is blocked with the competitive hexokinase inhibitor 2-deoxyglucose, will the mobility of glycolytic enzymes be affected? The results (shown in Table 2) demonstrate that enzyme organization is metabolically sensitive, and can be followed over time in individual living cells.

C. Metabolic Experiments: 2-Deoxyglucose and Cytochalasin D

Based on the above work, we hypothesized that if some glycolytic enzymes are preferentially localized around the actin cytoskeleton, and if such cytoskeletal localization is necessary for preferential use of glycolytic ATP for cell motility, then agents which inhibit glycolysis and/or cell motility might also affect enzyme localization. We used two approaches (Pagliaro and Taylor, 1992). First, depletion of 'downstream' glycolytic substrates with the competitive hexokinase inhibitor 2-deoxyglucose (2-DG) could decrease catalytic activity and significantly affect enzyme–enzyme and (if they exist) enzyme-structural protein interactions *in vivo*. The simplest prediction based on our hypothesis is that inhibition of glycolysis with 2-DG would result in the loss of the immobile fraction of aldolase *in vivo*, but that enolase would be relatively unaffected by 2-DG. We tested that prediction directly by performing FRAP experiments, both before and after perfusion with medium containing 2-DG, in cells coinjected with fluorescent analogs of aldolase and enolase (Table 2). Perfusion of medium containing 1 mg/ml 2-DG into the cell chamber rapidly released the ~ 21% bound fraction of Rh–aldolase in the perinuclear region of 3T3 cells, while the mobility of aldolase in the cell periphery, and the mobility of enolase, were not significantly affected.

Second, we performed FRAP experiments on cells before and after perfusion with the actin-capping agent cytochalasin D (CD; Cooper, 1987), in order to directly evaluate the role the actin cytoskeleton played in maintaining the immobile fraction of Rh–aldolase in cells. If the actin cytoskeleton acts as a scaffold for some glycolytic enzymes, then disruption of the scaffold should liberate the enzymes bound to it. Perfusion of medium containing 0.5 μM CD caused a rapid and complete release of the bound fraction of Rh–aldolase. Figure 11B shows the mobility of Rh–aldolase and Fl–enolase in the perinuclear region of two different

Table 2. *In Vivo* FRAP Data: Rhodamine–Aldolase and Fluorescein-Enolase in Living Swiss 3T3 Cells

	Control		*+ 2-DG*	
	$D_{37,cyto} \times 10^8$[a]	*Percent Mobile*	*$D_{37,cyto} \times 10^8$*	*Percent Mobile*
Rh-Aldolase				
Perinuclear (8)[b]	14.3 ± 6.7[c]	79.9 ± 7.6	9.11 ± 2.3	105.1 ± 6.1
Peripheral (7)	8.10 ± 3.2	101.8 ± 7.8	6.69 ± 3.9	101.8 ± 6.6
Fl-Enolase				
Perinuclear (3)	12.4 ± 9.0	98.3 ± 8.1	10.3 ± 4.1	102.4 ± 1.3
Peripheral (3)	9.63 ± 4.6	100.9 ± 3.3	7.61 ± 1.6	98.3 ± 5.1

Notes: [a]Aqueous diffusion coefficient, in cm^2/sec.
[b]Number of measurements used calculating the mean.

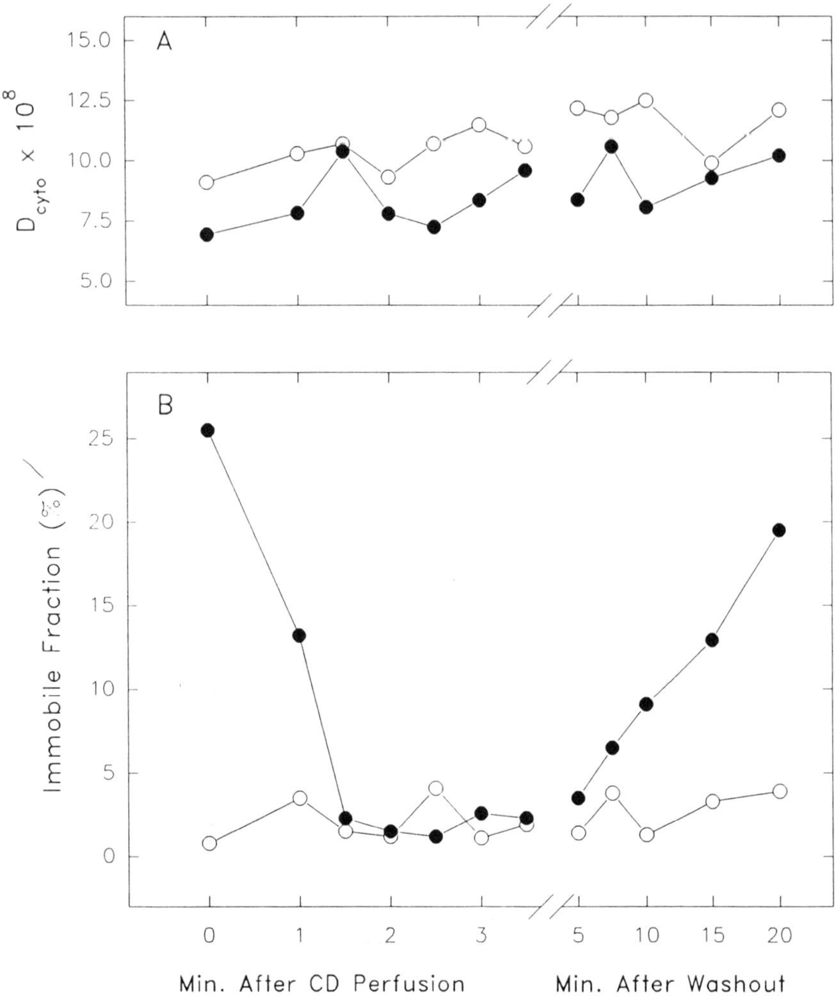

Figure 11. The effect of disrupting the actin cytoskeleton on the mobility of the bound fraction of aldolase, measured with FRAP of Rh–aldolase (solid circles) and Fl–enolase (open circles) in the perinuclear region of two different representative cells during perfusion with CD and subsequent washout. The D_{cyto} values of Rh–aldolase and Fl–enolase were not significantly affected by CD perfusion (**A**). The immobile fraction of Rh–aldolase decreased from 25.5% before CD perfusion to 2.3% by 90 seconds after perfusion; when CD was washed out of the chamber, 80%–90% of the control perinuclear immobile fraction of Rh–aldolase returned after 20 minutes (**B**). The immobile fraction of Fl–enolase was not significantly affected by either CD perfusion or washout. (Reproduced from Pagliaro and Taylor (1992), with permission.)

representative cells after perfusion with CD and subsequent washout. In the example shown, the immobile fraction of Rh–aldolase decreased from 25.5% before CD perfusion to 2.3% by 90 s after perfusion. When CD was subsequently washed out of the chamber, 80–90% of the control perinuclear immobile fraction of Rh–aldolase returned after 20 minutes. Fl–enolase was 100% mobile, and its mobility was not significantly affected by either CD perfusion or subsequent washout. The D_{cyto} values of Rh–aldolase and Fl–enolase were not significantly affected by CD perfusion (Figure 11A).

Together, these results are consistent with models in which aldolase binds to the actin cytoskeleton *in vivo*, and they demonstrate that the bound fraction of aldolase is metabolically sensitive. It is important to note that the changes we have seen are not "all or nothing" changes; like other delicate equilibria, the modulation of a 20% bound fraction of enzyme is a significant and reproducible result of a complex, large-scale series of weak interactions (Pagliaro, 1993).

VI. PROSPECTUS

Work from a number of laboratories over almost three decades has generated a substantial foundation of evidence which supports non-diffusion-limited schemes for glycolytic enzyme organization and metabolism. It is clear that glycolytic enzyme behavior in conditions which mimic the intracellular environment is significantly different from that in dilute aqueous environments (Walsh & Knull, 1988; Lakatos and Minton, 1991), and that considerations such as cellular water organization and volume occupancy cannot be neglected (Clegg et al., 1990). The evidence for organization of 'soluble' enzymes into functional units such as metabolons (Robinson and Srere, 1985), for functional compartmentation via metabolic channeling (Ovadi et al., 1989; Clegg and Jackson, 1990; Ovadi, 1991; Sumegi et al., 1991), and predictive metabolic modeling (Brooks and Storey, 1991), have built a solid foundation for real-time experiments *in vivo*. When fully exploited, fluorescence microscopy approaches will play an important role in this adventure.

ACKNOWLEDGMENTS

This work was supported by the following grants: Office of Naval Research N00014-92-J-1107, National Institutes of Health P30 DK17047, Department of Defense N00014-93-11052, National Science Foundation MCB 9403510, and the University of Washington Graduate School and Royalty Research Funds.

REFERENCES

Adams, S. R., Kao, J. P. Y., Grynkiewicz, G., Minta, A., & Tsien, R. Y. (1988). Biologically useful chelators that release Ca^{2+} upon illumination. J. Am. Chem. Soc. 110, 3212–3220.

Arnold, H. & Pette, D. (1968). Binding of glycolytic enzymes to structure proteins of the muscle. Eur. J. Biochem. 6, 163–171.

Arnold, H. & Pette, D. (1970). Binding of aldolase and triosephosphate dehydrogenase to F-actin and modification of catalytic properties of aldolase. Eur. J. Biochem. 15, 360–366.

Arnold, H., Nolte, J., & Pette, D. (1969). Quantitative and histochemical studies on the desorption and readsorption of aldolase in cross-striated muscle. J. Histochem. Cytochem. 17, 314–320.

Arnold, H., Henning R., & Pette, D. (1971). Quantitative comparison of the binding of various glycolytic enzymes to F-actin and the interaction of aldolase with G-actin. Eur. J. Biochem. 22, 121–126.

Axelrod, D., Koppel, D. E., Schlessinger, J., Elson, E., & Webb, W. W. (1976). Mobility measurement by analysis of fluorescence photobleaching recovery kinetics. Biophys. J. 16, 1055–1069.

Bergmeyer, H. U., Gawehn, K., & Grassl, M. (1974). Methods of Enzymatic Analysis, pp. 425–515, Academic Press, New York.

Bricknell, O. L., Daries, P. S., & Opie, L. H. (1981). A relationship between adenosine triphosphate, glycolysis and ischaemic contracture in isolated rat heart. J. Mol. Cell. Cardiol. 13, 941–945.

Bright, G. R., Rogowska, J., Fisher, G. W., & Taylor, D. L. (1987). Fluorescence ratio imaging microscopy: temporal and spatial measurements in single living cells. BioTechniques. 5, 1–6.

Bright, G. R., Fisher, G. W., Rogowska, J., & Taylor, D. L. (1989). Fluorescence ratio imaging microscopy. In: Methods in Cell Biology. Fluorescence Microscopy of Living Cells in Culture. Part B. Quantitative Fluorescence Microscopy—Imaging and Spectroscopy (Taylor, D. L. & Wang, Y. L., eds.), 30, 157–192, Academic Press, San Diego.

Brindle, K. M. (1991). Enzymology *in vivo*. Biochem. Soc. Trans. 19, 985–986.

Bronstein, W. W. & Knull, H. R. (1981). Interaction of muscle glycolytic enzymes with thin filament proteins. Can. J. Biochem. 59, 494–499.

Brooks, S. P. J. & Storey, K. B. (1991). A quantitative evaluation of the effect of enzyme complexes on the glycolytic rate *in vivo*: mathematical modeling of the glycolytic complex. J. Theor. Biol. 149, 361–375.

Clarke, F. M. & Masters, C. J. (1975). On the association of glycolytic enzymes with structural proteins of skeletal muscle. Biochim. Biophys. Acta. 381, 37–46.

Clarke, F. M. & Masters, C. J. (1976). Interactions between muscle proteins and glycolytic enzymes. Int. J. Biochem. 7, 359–365.

Clarke, F. M. & Morton, D. J. (1976). Aldolase binding to actin-containing filaments: formation of paracrystals. Biochem. J. 159, 797–798.

Clarke, F. M., Masters, C. J., & Winzor, D. J. (1974). Interaction of aldolase with the troponin-tropomyosin complex of bovine muscle. Biochem. J. 139, 785–788.

Clarke, F. M., Morton, D. J., Stephan, P., & Weidemann, J. (1985a). The functional duality of glycolytic enzymes: potential integrators of cytoplasmic structure and function. In: Cell Motility: Mechanism and Regulation (10th Yamada Conference; Ishikawa, H., Hatano, S., & Sato, H., eds.), pp. 235–250, University of Tokyo Press, Tokyo.

Clarke, F. M., Stephan, P., Morton, D. J., & Weidemann, J. (1985b). Glycolytic enzyme organization via the cytoskeleton and its role in metabolic regulation. In: Regulation of Carbohydrate Metabolism (Beitner, R., ed.), pp. 1–31, CRC Press, Boca Raton, Florida.

Clegg, J. S. (1988). The aqueous intracellular compartments. In: Microcompartmentation (Jones, D. P., ed.), pp. 1–16, CRC Press, Boca Raton, Florida.

Clegg, J. S. & Jackson, S. A. (1990). Glucose metabolism and the channeling of glycolytic intermediates in permeabilized L929 cells. Arch. Biochem. Biophys. 278, 452–460.

Clegg, J. S. Jackson, S. A., & Fendl, K. (1990). Effects of reduced cell volume and water content on glycolysis in L-929 cells. J. Cell Physiol. 142, 386–391.

Cooper, J. A. (1987). Effects of cytochalasin and phalloidin on actin. J. Cell Biol. 105:1473–1478.

de Duve, C. (1972). Is there a glycolytic particle? In: Structure and Function of Oxidation-Reduction Enzymes (Akeson, A. & Ehrenberg, A., eds.), pp. 715–728, Pergamon Press, Oxford.

Dolbeare, F. A. (1981). Fluorometric quantification of specific chemical species in single cells. In: Modern Fluorescence Spectroscopy (Wehry, E. L., ed.), Vol. 3, 251–293, Plenum, New York.

Dolken, G., Leisner, E., & Pette, D. (1975). Immunofluorescent localization of glycogenolytic and glycolytic enzyme proteins and of malate dehydrogenase isozymes in cross-striated skeletal muscle and heart of the rabbit. Histochem. J. 43, 113–121.

Edidin, M., Zagyansky, Y., & Lardner, T. J. (1976). Measurement of membrane protein lateral diffusion in single cells. Science 191, 466–468.

Gibbins, J. R. (1982). Epithelial cell motility: the effect of 2-deoxyglucose on cell migration, ATP production, and the structure of the cytoplasmic ground substance in lamellipodia of epithelial cells in culture. Cell Motil. Cytoskel. 2, 25–46.

Gurney, A. M. & Lester, H. A. (1987). Light-flash physiology with synthetic photosensitive compounds. Physiol. Rev. 67, 583–617.

Halling, P. J. (1989). Do the laws of chemistry apply to living cells? Trends Biochem. Sci. 14, 317–318.

Hardin, C. D., Raeymaekers, L., & Paul, R. J. (1992). Comparison of endogenous and exogenous sources of ATP in fueling Ca^{2+} uptake in smooth muscle plasma membrane vesicles. J. Gen. Physiol. 99, 21–41.

Harris, S. J. & Winzor, D. J. (1989). Effect of calcium ion on the interaction of aldolase with rabbit muscle myofibrils. Biochim. Biophys. Acta. 999, 95–99.

Huitorel, P. & Pantaloni, D. (1985). Bundling of microtubules by glyceraldehyde-3-phosphate dehydrogenase and its modulation by ATP. Eur. J. Biochem. 150, 265–269.

Jacobson, K., Elson, E., Koppel, D., & Webb, W. (1982). Fluorescence photobleaching in cell biology. Nature 295, 283–284.

Jongkind, J. F., Verkerk, A., & Sernetz, M. (1986). Detection of acid-β-galactosidase activity in viable human fibroblasts by flow cytometry. Cytometry 7, 463–466.

Kamiya, N., Nakajima, H., & Abe, S. (1957). Physiology of the motive force of protoplasmic streaming. Protoplasma 48, 94–112.

Kaplan, J. H., Forbush, B., & Hoffman, J. F. (1978). Rapid photolytic release of adenosine 5′-triphosphate from a protected analogue: utilization by the Na:K pump of human red blood cell ghosts. Biochemistry 17, 1929–1935.

Kohen, E., Kohen, C., Hirschberg, J. G., Wouters, A. W., Thorell, B., Westerhoff, H. V., & Charyulu, K. K. N. (1983). Metabolic control and compartmentation in single living cells. Cell Biochem. and Function. 1, 3–16.

Kropp, E. S. & Wilson, J. E. (1970). Hexokinase binding sites on mitochondrial membranes. Biochem. Biophys. Res. Commun. 38, 74–79.

Lakatos, S. & Minton, A. P. (1991). Interactions between globular proteins and F-actin in isotonic saline solution. J. Biol. Chem. 266, 18707–18713.

Lardy, H. A. (1965). On the direction of pyridine nucleotide oxidation-reduction reactions in gluconeogenesis and lipogenesis. In: Control of Energy Metabolism (Chance, B., Estabrook, R. W., & Williamson, J. R. eds.), pp. 245–248, Academic Press, New York.

Leamon, C. P. & Low, P. S. (1991). Delivery of macromolecules into living cells: a method that exploits folate receptor endocytosis. Proc. Natl. Acad. Sci. USA 88, 5572–5576.

Leamon, C. P. & Low, P. S. (1992). Cytotoxicity of momordin-folate conjugates in cultured human cells. J. Biol. Chem. 267, 24966–24971.

Luby-Phelps, K. (1989). Preparation of fluorescently labeled dextrans and ficolls. In: Methods in Cell Biology. Fluorescence Microscopy of Living Cells in Culture. Part A. Fluorescent Analogs, Labeling Cells, and Basic Microscopy (Wang, Y. L. & Taylor, D. L., eds.), 29, 59–73, Academic Press, San Diego.

Lynch, R. M., & Paul, R. J. (1983). Compartmentation of glycolytic and glycogenolytic metabolism in vascular smooth muscle. Science 222, 1344–1346.

Masters, C. J. (1984). Interactions between glycolytic enzymes and components of the cytomatrix. J. Cell Biol. 99, 222s–225s.

McCray, J. A., Herbette, L., Kihara, T., & Trentham, D. R. (1980). A new approach to time-resolved studies of ATP-requiring biological systems: laser flash photolysis of caged ATP. Proc. Natl. Acad. Sci. USA 77, 7237–7241.

McNeil, P. L. (1989). Incorporation of macromolecules into living cells. In: Methods in Cell Biology. Fluorescence Microscopy of Living Cells in Culture. Part A. Fluorescent Analogs, Labeling Cells, and Basic Microscopy (Wang, Y. L., & Taylor, D. L., eds.), 29, 153–173, Academic Press, San Diego.

McNeil, P. L., Murphy, R. F., Lanni, F., & Taylor, D. L. (1984). A method for incorporating macromolecules into adherent cells. J. Cell Biol. 98, 1556–1564.

Mercer, R. W. & Dunham, P. B. (1981). Membrane-bound ATP fuels the Na/K pump. Studies on membrane-bound glycolytic enzymes on inside-out vesicles from human red cell membranes. J. Gen. Physiol. 78, 547–568.

Morton, D. J., Clarke, F. M., & Masters, C. J. (1977). An electron microscope study of the interaction between fructose diphosphate aldolase and actin-containing filaments. J. Cell Biol. 74, 1016–1023.

Orosz, F., Christova, T. Y., & Ovadi, J. (1988). Modulation of phosphofructokinase action by macromolecular interactions. Quantitative analysis of the phosphofructokinase-aldolase-calmodulin system. Biochim. Biophys. Acta 957, 293–300.

Ottaway, J. H. & Mowbray, J. (1977). The role of compartmentation in the control of glycolysis. In: Current Topics in Cellular Regulation (Horecker, B. L. & Stadtman, E. R., eds.), pp. 107–208, Academic Press, New York.

Ovadi, J. (1991). Physiological significance of metabolic channelling. J. Theor. Biol. 152, 1–22.

Ovadi, J., Tompa, P., Vertessy, B., Orosz, F., Keleti, T., & Welch, G. R. (1989). Transient-time analysis of substrate-channeling in interacting enzyme systems. 1989 Biochem. J. 257, 187–190.

Pagliaro, L. (1993). Glycolysis revisited—a funny thing happened on the way to the Krebs cycle. News in Physiol. Sci. 8, 219–223.

Pagliaro, L. & Taylor, D. L. (1988). Aldolase exists in both the fluid and solid phases of cytoplasm. J. Cell Biol. 107, 981–991.

Pagliaro, L. & Taylor, D. L. (1992). 2-Deoxyglucose and cytochalasin D modulate aldolase mobility in living 3T3 cells. J. Cell Biol. 118, 859–863.

Pagliaro, L., Kerr, K., & Taylor, D. L. (1989). Enolase exists in the fluid phase of 3T3 cells. J. Cell Sci. 94, 333–342.

Paul, R. J., Hardin, C. D., Raeymaekers, L., Wuytack, F., & Casteels, R. (1989). Preferential support of Ca^{2+} uptake in smooth muscle plasma membrane vesicles by an endogenous glycolytic cascade. FASEB J. 3, 2298–2301.

Robinson, J. B. & Srere, P. A. (1985). Organization of Krebs tricarboxylic acid cycle enzymes in mitochondria. J. Biol. Chem. 260, 10800–10805.

Rotman, B. & Papermaster, B. W. (1966). Membrane properties of living mammalian cells as studied by enzymatic hydrolysis of fluorogenic substrates. Proc. Natl. Acad. Sci. USA 55, 134–141.

Shigekawa, K. & Dower, W. J. (1988). Electroporation of eukaryotes and prokaryotes: a general approach to the introduction of macromolecules into cells. Bio/Techniques 6, 742–751.

Sigel, P. & Pette, D. (1969). Intracellular localization of glycogenolytic and glycolytic enzymes in white and red rabbit skeletal muscle: a gel film method for coupled enzyme reactions in histochemistry. J. Histochem. Cytochem. 17, 225–237.

Simon, J. R., Gough, A., Urbanik, E., Wang, F., Lanni, F., Ware, B. R., & Taylor, D. L. (1988). Analysis of rhodamine and fluorescein-labeled F-actin diffusion in vitro by fluorescence photobleaching recovery. Biophy. J. 54, 801–815.

Srere, P. A. (1987). Complexes of sequential metabolic enzymes. Annu. Rev. Biochem. 56, 89–124.

Stephan, P., Clarke, F. M., & Morton, D. (1986). The indirect binding of triose-phosphate isomerase to myofibrils to form a glycolytic enzyme mini-complex. Biochim. Biophys. Acta 873, 127–135.

Stewart, M., Morton, D. J., & Clarke, F. M. (1980). Interaction of aldolase with actin-containing filaments: structural studies. Biochem. J. 186, 99–104.

Stryer, L. (1988). Biochemistry, third edition p. 370, W. H. Freeman and Co., New York.

Sumegi, B., Sherry, A. D., Malloy, C. R., Evans, C., & Srere, P. A. (1991). Is there tight channelling in the tricarboxylic acid cycle metabolon? Biochem. Soc. Trans. 19, 1002–1005.

Taylor, D. L. & Wang, Y. L. (1980). Fluorescently labeled molecules as probes of the structure and function of living cells. Nature 284, 405–410.

Taylor, D. L., Amato, P. A., Luby-Phelps, K., & McNeil, P. L. (1984). Fluorescent analog cytochemistry. Trends Biochem. Sci. 9, 88–91.

van Zoelen, E. J. J., Tertoolen, L. G. J., & de Laat, S. W. (1983). Simple computer method for evaluation of lateral diffusion coefficients from fluorescence photobleaching recovery kinetics. Biophys. J. 42, 103–108.

Walsh, J. & Knull, H. R. (1988). Heteromerous interactions among glycolytic enzymes and of glycolytic enzymes with F-actin: effects of polyethylene glycol. Biochim. Biophys. Acta. 952, 83–91.

Walsh, T. P., Winzor, D. J., Clarke, F. M., Masters, C. J., & Morton, D. J. (1980). Binding of aldolase to actin-containing filaments: evidence of interaction with the regulatory proteins of skeletal muscle. Biochem. J. 186, 89–98.

Walsh, J. L., Keith, T. J., & Knull, H. R. (1989). Glycolytic enzyme interactions with tubulin and microtubules. Biochim. Biophys. Acta 999, 64–70.

Wang, J., Reitz, F., Donaldson, T., Pagliaro, L. (1994). A computer-interfaced falling ball viscometer. J. Biophys. Biochem. Meth. 28, 251–261.

Wang, Y. L. (1989). Fluorescent analog cytochemistry: tracing functional protein components in living cells. In: Methods in Cell Biology. Fluorescence Microscopy of Living Cells in Culture. Part A. Fluorescent Analogs, Labeling Cells, and Basic Microscopy (Wang, Y. L., & Taylor, D. L., eds.), 29, 1–12, Academic Press, San Diego.

Weiss, J. N. & Lamp, S. T. (1987). Glycolysis preferentially inhibits ATP-sensitive K^+ channels in isolated guinea pig cardiac myocytes. Science 238, 67–69.

Welch, G. R. (1977). On the role of organized multienzyme systems in cellular metabolism: a general synthesis. Prog. Biophys. Molec. Biol. 32, 103–191.

Wilson, E. B. (1925). The Cell in Development and Heredity, third edition. The Macmillan Company, New York, 6, 637–638.

Wilson, J. E. (1978). Ambiquitous enzymes: variation in intracellular distribution as a regulatory mechanism. Trends Biochem. Sci. 3, 124–125.

Wilson, J. E. (1980). Brain hexokinase, the prototype ambiquitous enzyme. In Current Topics in Cellular Regulation (Horecker, B. L. and Stadtman, E. R., eds.) pp. 1–54, Academic Press, New York.

Wilson, J. E. (1988). Function of ambiquitous proteins in a heterogeneous medium. In: Microcompartmentation (Jones, D. P., ed.), pp. 171–190, CRC Press, Boca Raton, FL.

Wojcieszyn, J. W., Schlegel, R. A., Wu, E. S., & Jacobson, K. A. (1981). Diffusion of injected macromolecules within the cytoplasm of living cells. Proc. Nat. Acad. Sci. USA 78, 4407–4410.

Wolf, D. E. (1989). Designing, building, and using a fluorescence recovery after photobleaching instrument. In: Methods in Cell Biology. Fluorescence Microscopy of Living Cells in Culture. Part B. Quantitative Fluorescence Microscopy—Imaging and Spectroscopy (Taylor, D. L. & Wang, Y. L., eds.), 30, 271–306, Academic Press, San Diego.

Yguerabide, J., Schmidt, J. A., & Yguerabide, E. E. (1982). Lateral mobility in membranes as detected by fluorescence recovery after photobleaching. Biophys. J. 39, 69–75.

THE COOPERATIVE BEHAVIOR OF KREBS TRICARBOXYLIC ACID CYCLE ENZYMES

Paul A. Srere, Craig R. Malloy,

A. Dean Sherry, and Balazs Sumegi

Advances in Molecular and Cell Biology
Volume 11, pages 125–145.

ISBN: 1-55938-844-7

ABSTRACT

Studies of enzymes *in vitro* are usually carried out under conditions that are quite different from the *milieu* in which they usually operate. These differences usually include the concentration of enzymes, substrates, effectors, and other physical differences, including volume exclusion of the protein *milieu* and possible interactions of sequential enzymes.

Two recent technological advances, facile production of specific metabolic lesions by molecular biological techniques and metabolic studies with ^{13}C NMR spectroscopy, offer a new window for the observation of enzyme action *in vivo*. In this article we present data concerning the activity of the Krebs TCA cycle enzymes *in vivo* employing these two newer technologies.

I. INTRODUCTION

Modern biochemists assume that most metabolic pathways, their structure, regulation, and interaction are well understood. Yet at least two important areas of metabolism which have been studied for about twenty years, distributive regulation (Kacser and Burns, 1973) and enzyme interactions, are barely mentioned in any modern text of biochemistry. This may be due to the fact that workers in the field of metabolism are divided on the complete acceptance of these newer concepts of metabolism. It seems that the opposition to these newer paradigms do not question the data which has been presented as much as they oppose the idea that these new concepts play an important role in the metabolic processes of cells.

Many of our accepted current ideas concerning metabolic regulation are based on *in vitro* studies of the properties of enzymes and on the total metabolite concentration changes in an excised tissue after incubation of tissue pieces or homogenates. It is clear that such studies are inadequate in many ways to enable us to deduce an accurate assessment of metabolism *in vivo*.

It is important therefore to attempt the comparison of what we know about *in vitro* enzymology and enzymology as it must exist *in vivo*.

II. DIFFERENCES *IN VITRO* AND *IN VIVO*

A. Enzyme Concentration

When enzymes are assayed by most spectrophotometric techniques one typically uses an enzyme concentration in the range of 1–10 nM. However, the *in vivo* concentrations of a large number of enzymes has been estimated at 0.1–100 μM (Srere, 1967; 1969). Those enzymes that are involved in energy metabolism or other major metabolic pathways tend toward the high end of that range (10–100 μM in certain tissues) while others occur at the lower end of that concentration range.

B. Total Protein Concentration

In vitro enzyme studies are usually performed using purified or partially purified enzymes, the assay mixture of which may have as little as 1 μg/ml of protein. If homogenates are studied then it is possible that 1 mg/ml of total protein may be present. Even when homogenates are used without dilution for metabolic studies, at least 3- to 10-fold dilution of cellular protein will have occurred.

On the other hand, cellular protein concentrations are very high. Cytosolic protein concentrations have been estimated as 20–30% or about 200–300 mg protein per ml (Srere, 1981). Even higher concentrations have been reported for mitochondria. In the latter case, 40–60% protein (400–600 mg/ml) has been estimated for the concentration of protein in the mitochondrial matrix. Even the inner membrane of the mitochondrion, admittedly one of the most proteinaceous membranes known, has a protein content of 50%. To put another perspective on these protein concentrations one should remember that *protein crystals* have a protein content in the range of 40–60%.

C. Substrate Concentrations

In vitro, enzymes are studied over a range of substrate concentrations in order to determine the kinetic constants of each substrate. When some regulatory aspects of enzymes are studied, care is usually taken to use concentrations of substrates that are believed to exist *in vivo* (but see below).

In vivo concentrations of substrates have been difficult to determine. Recent techniques have used the procedure of "freeze-clamping" (Wollenberger et al., 1960) tissues. This procedure prevents substrate concentration changes which may occur when animals are sacrificed and tissues quickly become anaerobic. This procedure, however, only yields total metabolite content and cannot give results which reflect possible differences in organellar distribution of metabolites. Veech and his associates (personal communication) have obtained a few results for metabolite distributions in mitochondria and other regions of the cell. This required a herculean effort and the use of special equipment not generally available.

NMR techniques offer the possibility of noninvasively determining cytosolic metabolite concentrations. There are some problems in standardizing this method, but the technique promises to be a useful one (see below). Alternatively, Veech (1978) has presented calculations which can estimate the concentration of certain metabolites in the cytosol and mitochondria. These calculations depend on a near-equilibrium assumption for the enzyme reactions involved and a unique distribution of certain enzymes of the cytosol and mitochondria.

At best, therefore, one can assume a global total concentration for substrate in the cytosol and in the mitochondrial matrix. As a first approximation, this is a reasonable assumption when one considers diffusion coefficients versus enzyme turnover time.

D. Effector Concentrations

In vitro enzyme studies have identified effectors for many enzymes. Thus, the effect of pH, ionic strength, metal ions, and other metabolites on an enzyme's activity give perhaps the first indication of what may regulate an enzyme's activity *in vivo*. However, all of the difficulties that were noted above for knowing the concentrations of substrates in cells applies to knowing the concentration of effectors in cells.

E. Organizational Aspects

Almost all *in vitro* studies of enzyme behavior occurs with the enzymes and substrates randomly arrayed in dilute solutions. It is clear, however, that enzymes have distinct cellular locations, i.e. nucleus, mitochondria, cytosol, peroxisomes, Golgi apparatus, etc. and that in some organelles, such as mitochondria, they are even more specifically segregated, i.e. matrix, inner membrane, intermembrane compartment. Further, there exists substantial evidence that complexes of enzymes exists within each subcompartment (Srere and Mosbach, 1974; Srere, 1987).

In certain mitochondria, regular arrangements of protein complexes have been observed by electron microscopic techniques. Positive identification of all these structures (ATPase, etc.) have not been made as yet (Allen et al., 1989). Further, in certain cell types, mitochondria are seen to have specific locations within the cell usually adjacent to a site of known ATP utilization.

III. EFFECTS OF *IN VIVO* AND *IN VITRO* DIFFERENCES ON ENZYME BEHAVIOR

It is clear from the foregoing section that large and conceivably important differences exist between cellular conditions and the conditions used for the usual *in vitro* studies. Each of these differences could mislead one concerning metabolism and its control.

A. Enzyme Concentration

A number of enzymes have been studied over a wide range of concentrations which extend from usual *in vitro* conditions (~nM) to putative *in vivo* concentrations (~μM). For some enzymes, phosphofructokinase for example, no effects on kinetics are known (Hulme and Tipton, 1971), but for other enzymes, large effects are noted (Aragon and Sols, 1991). In the latter instances an important factor is the propensity for the enzyme to polymerize and whether or not the polymer has different kinetic characteristics from the minimum sized enzyme. It has been shown recently that when enzyme concentrations are high enough that a significant amount of total substrate is bound, sustained oscillations may occur (Ryde-Petterson, 1992).

B. Protein Concentration

Minton (1983) has elegantly shown that the total protein concentration of an assay can have a large effect on the activity of some enzymes. This is mainly due to the solvent exclusion effect of the macromolecules. Since cellular protein concentrations are quite high, it seems certain that this effect will exert some influence *in vivo*. Minton has shown that solvent exclusion leads to reduced activity coefficients which, in turn, will lead to protein aggregation. He has also shown that cellular structural proteins have an effect on protein behavior (Minton, 1990).

Higher protein concentration also leads to large changes in cellular viscosity so that reactions depending on the diffusion of larger molecules will be strongly affected. However, diffusion of small molecules does not seem to be greatly affected (diffusion coefficient change of one order of magnitude or less). This, however, is for cytosolic protein concentrations (20–30%). No estimate of diffusion coefficients for small molecules is available for a cellular compartment like the mitochondrial matrix where the protein concentration may be as high as 50%. At those protein concentrations, close packing of proteins exists with two consequences: (1) about 50% of the matrix water is protein bound, and (2) with the close protein packing, it is probable that molecules as small as 20 Å in diameter will have difficulty in diffusing through the resulting pores (Srere, 1981).

C. Substrate and Effector Concentrations

One would expect that within a cellular compartment, small molecules, such as most metabolic intermediates whose diffusion coefficients in aqueous media are large, would quickly equilibrate in a volume as small as an animal cell or a cellular organelle. The time for complete equilibration is expected to be orders of magnitude more rapid (~10^{-8} s) than is the turnover time of most enzymes (10^{-3} s). The diffusion coefficients for small molecules within cells (and within protein crystals) have been measured, and there is general agreement that they are not more than one order of magnitude smaller than those measured in water. A number of factors may influence this in the cellular *milieu* to a greater degree. Among these are tortuosity effects, unstirred layer effects, and ionic binding to the charged macromolecules. It is not known if these effects at the cellular level are great enough so that metabolite diffusion times are on the same order of magnitude as enzyme turnover. Not withstanding these uncertainties, unexpected gradients of ATP and O_2 have been proposed to exist in cells based on enzyme activities whose position in cells are established (Aw and Jones, 1985). Several laboratories have reported the existence of multiple pools of acetyl–CoA and oxalacetate within mitochondria.

One is led to the conclusion that the study of enzyme activities *in vitro* at "physiological" substrate and effector concentrations is at best an educated guess.

D. Organizational Aspects

It has been amply demonstrated that enzymes which are sequential in a metabolic pathway have a different kinetic behavior when they exist bound in complexes or even when they are immobilized together compared to the same enzymes free in solution (Robinson et al., 1987; Sumegi et al., 1991). Masters (1989) and his coworkers have shown that glycolytic enzymes bind to structural elements of muscle, and in doing so, the kinetics of the enzymes change. Differences have been observed in several other systems, including mitochondrial enzymes.

The remainder of this chapter will deal specifically with the Krebs TCA cycle, which has been studied *in vitro* with the newer techniques of molecular biology and ^{13}C NMR. Our results indicate quite clearly the difficulty of using data on isolated enzymes to explain the behavior of reactions associated with this cycle *in situ*.

IV. USE OF MOLECULAR BIOLOGY AND ^{13}C NMR TO STUDY THE KREBS TCA CYCLE *IN VIVO*

In recent years, yeast cells have been almost universally adopted as a model eucaryotic cell. The advantages of yeast are: (1) it is relatively fast growing, (2) a considerable amount of information is available on its genetics, and (3) it can be readily manipulated by molecular biological techniques. For the study of the Krebs

Table 1. Growth Pattern of TCA Cycle Mutants

		Growth in:	
Mutants	*Galactose*	*Glycerol*	*Acetate*
Wild type	+	+	+
mMDH mutant (A21–3)	+	+	–
ICDH1 mutant (A10–4)	+	+	–
KGDC (KGD1)	+	–	–
Fumarase (FUM1)	+	–	–

TCA cycle, a number of disruption mutants of enzymes of the cycle have been prepared, and their growth on various nutrients and the changes in metabolic profiles have been studied (Table 1). In this section we report the metabolic consequences in several TCA cycle mutant yeast cells which lack malate dehydrogenase (Tompson and McAlister-Henn, 1989), citrate synthase (Kim et al., 1986), (NAD^+)-isocitrate dehydrogenase (ICDH) (Keys and McAlister-Henn, 1990), ($NADP^+$)ICDH (Haselbeck and McAlister-Henn, 1991), and α-ketoglutarate dehydrogenase complex (αKGDC) (Repetto and Tzagoloff, 1989).

A. Mitochondrial Malate Dehydrogenase (MDH1) Mutant Yeast

Mitochondrial malate dehydrogenase mutant yeast cells were isolated from a yeast mutant collection which could not grow on acetate (Thompson and McAlister-Henn, 1989). The mutant cells lacked MDH1 activity and did not produce immunologically recognizable MDH protein. Transformation of this mutant with a plasmid containing the MDH1 gene restored its ability to grow on acetate indicating that the mutation was exclusively located on the MDH1 gene (M.T. McCammon, personal communication). It is an intriguing question as to why the MDH1 mutant yeast lost its ability to grow on acetate. This cell still retains an active glyoxylate cycle with active peroxisomal and cytosolic malate dehydrogenases. Therefore, the mutant yeast cell should be able to produce malate and oxalacetate from acetate. Since the yeast cells have both mitochondrial and cytoplasmic aspartate aminotransferase, oxalacetate can be converted to aspartate in the cytoplasm which can be transported back to the mitochondria where the mitochondrial aspartate aminotransferase can regenerate oxalacetate for the TCA cycle. The cycle can produce malate, which can be exported to the cytoplasm where cytoplasmic MDH and aspartate aminotransferase produce aspartate from malate. Aspartate can then reenter the mitochondria and shunt the missing MDH reaction. In this way, all intermediates can be synthesized, and the yeast cells should be able to produce energy for growth (Figure 1). Therefore, there is no apparent metabolic reason to explain the lack of growth on acetate of MDH1 mutant yeast cells.

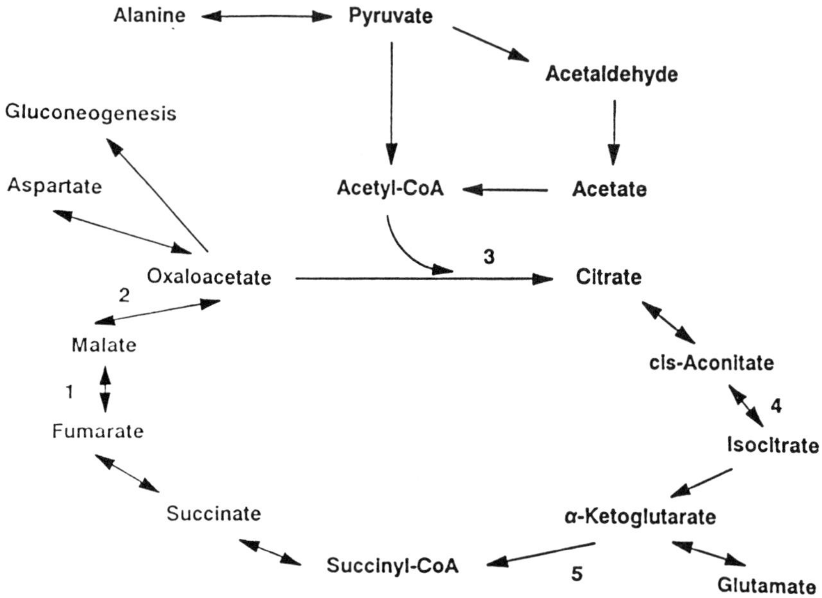

Figure 1. Pyruvate oxidation in yeast mutants. The following reactions are inhibited in the mutant: (**1**) fumarase (not discussed); (**2**) mitochondrial malate dehydrogenase; (**3**) mitochondrial citrate synthase; (**4**) mitochondrial isocitrate dehydrogenase; and (**5**) α-ketoglutarate dehydrogenase mutants.

We have studied this question in more detail by following the metabolism of [2-^{13}C] acetate and [3-^{13}C] pyruvate of the yeast MDH mutant with ^{13}C NMR (Table 2; also Sumegi et al., 1992). Surprisingly, the mutant cells oxidized acetate very poorly even in the presence of added aspartate or malate. In these mutant cells,

Table 2. Oxidation of [2-^{13}C] Acetate and [3-^{13}C] Pyruvate in Yeast Mutants

	^{13}C Incorporation into glutamate C4 from:	
Mutants	*[2-^{13}C] Acetate*[a]	*[3-^{13}C] Pyruvate*[b]
Wild type	1200	74
mMDH mutant	2	6.3
ICDH mutant	5	10
αKGDC mutant	3	20

Notes: [a]The ^{13}C content of glutamate C4 is given as percentage of ^{13}C content in the remaining acetate C2 signal. Very little acetate remains in the wild type cell so that this high value (1200) reflects that fact.

[b]The ^{13}C content of glutamate C4 is given as percentage of the major ^{13}C peaks of either acetate, pyruvate, or alanine. It was necessary to use different references because the pyruvate C3, acetate C2, and the alanine C3 were the major peaks in ICDH mutant, KGDC mutant, and malate dehydrogenase mutants, respectively.

[3-^{13}C] pyruvate was converted to alanine and acetate, and a portion of pyruvate was utilized by gluconeogenesis. Only a small fraction of it was oxidized in the TCA cycle. Since pyruvate can be converted to oxalacetate by pyruvate carboxylase and can be oxidized to acetyl–CoA by the pyruvate dehydrogenase complex or indirectly through pyruvate decarboxylase, aldehyde dehydrogenase, and acetate thiokinase, it is difficult to explain why these mutant yeast cells do not have TCA cycle activity. McAlister-Henn and her colleagues (Steffan et al., 1992) have studied some growth and metabolic characteristics of yeast MDH1 mutants which contain either the rat or *E. coli* MDH. They found that rat MDH restored acetate growth of the yeast and O_2 consumption of mitochondria with pyruvate and malate as substrates. However, *E. coli* MDH did not restore acetate growth, but did give mitochondrial O_2 consumption of pyruvate and malate at 60% of the rate of wild type yeast mitochondria (Steffan et al., 1992).

From the data presented here and from that of McAlister-Henn, no simple explanation of the growth pattern of the yeast mutant lacking mitochondrial malate dehydrogenase can be made.

B. Citrate Synthase-Defective Yeast Cells

The mitochondrial citrate synthase mutant yeast ($CS1^-$) cells, generated by gene disruption (Kim et al., 1986), retain their ability to grow on nonfermentable compounds like glycerol, lactate, and pyruvate, but they are unable to grow on acetate. Interestingly, the deletion of the citrate synthase gene caused significant changes in the enzyme composition of the yeast cell. The activities of the pyruvate dehydrogenase complex, αKGDC, isocitrate lyase, phosphoenolpyruvate carboxykinase, and acetyl–CoA synthase were significantly lower in the $CS1^-$ cell line than in the wild type or in the $CS2^-$ (peroxisomal) mutant (Kispal et al., 1988; 1989). Furthermore, the $CS1^-$ mitochondria were less active in oxidizing various Krebs TCA cycle intermediates than wild type or $CS2^-$ cells. ^{13}C NMR studies of acetate metabolism also indicated the presence of a less active TCA cycle in the mutant cells than in the wild type (Kispal et al., 1989). The citrate concentration was found to be higher in the $CS1^-$ cells than in the wild type so that the decreased TCA cycle activity must be due to some other phenomenon. Citrate synthesis in the mutant cells can occur in the cytoplasm using peroxisomal citrate synthase. It is, therefore, possible that the decreased TCA enzyme activities (PDC and αKGDC) contribute to the less active TCA cycle. Alternately, the absence of a citrate synthase protein may cause a defect in the organization of TCA cycle enzymes which is then responsible for decreased TCA cycle activity. The disruption of organization is supported by the data showing that the expression of an inactive citrate synthase gene in a plasmid restores partial growth of the $CS1^-$ cells on acetate. That is, not only the activity of an enzyme is necessary, but the presence of its protein moiety may also be important in the restoration of an active TCA cycle.

C. (NAD^+)ICDH-Deficient Yeast Mutants

There are two isocitrate dehydrogenase isozymes in yeast mitochondria, (NAD^+)ICDH and ($NADP^+$)ICDH, and the activity of the two isozymes are approximately equal in the mitochondria of wild type cells. However, there has been no general agreement about the physiological function of these isozymes. One view is that the main function of the ($NADP^+$)ICDH in the mitochondria is to produce NADPH for mitochondrial synthetic processes (Haselbeck and McAlister-Henn, 1991) while the (NAD^+)ICDH enzyme is involved in the Krebs TCA cycle. However, it is possible that the ($NADP^+$)ICDH, along with a transhydrogenase, can participate in the Krebs TCA cycle since the activity of the ($NADP^+$)ICDH isozyme present in yeast mitochondria is similar to that of the (NAD^+)ICDH one. A yeast mutant defective in (NAD^+)ICDH has been described by McAlister-Henn's group (Keys and McAlister-Henn, 1991), and it was found that, similar to all Krebs cycle mutants, these yeast cells are not able to grow on acetate. Since the (NAD^+)ICDH reaction can be shunted by the mitochondrial ($NADP^+$)ICDH, we are again faced with a paradox.

The TCA cycle activity has been studied in these cells by ^{13}C NMR techniques, and these data were similar to that of other TCA cycle mutants in that deletion of (NAD^+)ICDH significantly decreased flux through the TCA cycle as judged from the incorporation of ^{13}C acetate or ^{13}C pyruvate into glutamate (Table 2). This supports the view that (NAD^+)ICDH is part of the TCA cycle and the ($NADP^+$)ICDH isozyme is not. During the oxidation of [3-^{13}C] pyruvate, there was an accumulation of labeled ^{13}C citrate in the yeast mutants (Sumegi et al., 1992), suggesting that the TCA cycle was inhibited at the isocitrate dehydrogenase level.

The expression of ($NADP^+$)ICDH increased when yeast cells were grown under anaerobic conditions, in contrast to the decreased expression of (NAD^+)ICDH, αKGDC, and several other TCA cycle enzymes. More recently, ($NADP^+$)ICDH mutant yeast cells were generated, and the growth studies indicated the ($NADP^+$)ICDH is necessary for NADPH generation for mitochondrial synthetic processes (Repetto and Tzagoloff, 1989). However, the question of why the Krebs TCA cycle prefers the (NAD^+)ICDH over the $NADP^+$ isozyme still remains.

One explanation of how the TCA cycle enzymes differentiate between the two isozymes is based upon *in vitro* interaction studies with pure enzymes. The (NAD^+)ICDH interacts with αKGDC (Sumegi et al., 1991) and mitochondrial aconitase (Tyiska et al., 1986), and thus binds to the inner membrane through the interaction of the ketoglutarate dehydrogenase complex with complex I (Sumegi et al., 1991). In permeabilized mitochondria (Sumegi et al., 1990; 1991) showed increased rates of isocitrate oxidation compared to a soluble system which was interpreted as being due to channeling of substrates in the intact system. These data indicated that the interaction between the TCA cycle enzymes make possible substrate channeling which, in turn, increases flux through the TCA cycle 5- to 6-fold over the unorganized system. In all of these experiments, ($NADP^+$)ICDH was also studied and it showed no interaction with TCA cycle enzymes, suggesting that the low TCA cycle flux in the (NAD^+)ICDH defective yeast cell could be due to a lack of organization and substrate channeling in the yeast mitochondria.

D. α-Ketoglutarate Dehydrogenase-Defective Yeast Mutants

Mutation or deletion of the first component of α-ketoglutarate dehydrogenase (KGD1) gene yields a yeast cell which is unable to grow on acetate or glycerol in spite of the fact that it has an active respiratory chain (Repetto and Tzagoloff, 1989). The lack of growth on glycerol indicates that there is either inadequate energy production or the mutation results in decreased production of a needed intermediate(s). The oxidation of glycerol to acetate should provide the mutant cells with enough energy for growth. The mutant cells should be able to produce glutamate since they have pyruvate carboxylase and the necessary enzymes converting citrate to α-ketoglutarate and to glutamate. In addition, the rest of the TCA cycle intermediates can be formed from oxalacetate by the reversal of the malate dehydrogenase–succinyl-CoA synthetase span of the Krebs TCA cycle (Figure 1). However, a study of the oxidation of [3-^{13}C] pyruvate or [2-^{13}C] acetate (in the presence of unlabeled aspartate or malate) gave the surprising result that the pyruvate to glutamate and acetate to glutamate conversions were significantly inhibited in the mutant cells (Table 2).

We were surprised to note only a very low incorporation of [3-^{13}C] pyruvate into glutamate in αKGDC mutant cells. These cells could form oxalacetate from pyruvate since they showed increased gluconeogenesis and they could form acetyl–CoA from pyruvate. One would expect, therefore, the facile production of citrate, isocitrate, α-ketoglutarate, and glutamate. The fact that this did not occur in these mutant cells could be explained by the postulate that disruption of organization of the Krebs TCA cycle enzymes results in a great reduction in the activity of the remaining limbs of the cycle (Halper and Srere, 1977; Moore et al., 1984; Tyiska et al., 1986; Sumegi et al., 1990, 1991a,b, 1992). We and others have given physical evidence for the interaction between functionally related sequential mitochondrial enzymes, and evidence has been given for substrate channeling in *in vitro*, *in situ*, and *in vivo* systems. Therefore, the deletion of α-ketoglutarate dehydrogenase which interacts with other TCA cycle enzymes (Sumegi et al., 1990, 1991) can cause a major disruption in the organization of TCA cycle enzymes, indirectly affecting parts of the cycle where it is not involved. It is also possible that some unknown intermediate(s) accumulates which inhibits the cycle, but in this case, it would be necessary to assume that a different inhibitor accumulates in each mutation, making this explanation less likely.

All of these mitochondrial mutants showed an increased gluconeogenic activity which cannot be explained purely by shifts in intermediate concentrations. It is possible that defective mitochondrial metabolism somehow signals the nucleus to affect the expression of enzymes in other metabolic pathways. This phenomenon was seen in the case of different petite mutant yeast cells (Parikh et al., 1987; Srere, 1990). Therefore, we have to recognize that even a simple mitochondrial mutation may have complicated pleiotropic effects on apparently unrelated metabolic activities of the mutant yeast cells.

V. ^{13}C ISOTOPOMER ANALYSIS OF METABOLIC SPECTRA TO DETECT RELATIVE FLUXES THROUGH THE GLYOXALATE AND TCA CYCLES

The examples illustrated above suggest that ^{13}C NMR studies yield considerably less metabolic information about mutant cell lines than about wild type cells. The observation that a single mutation of a TCA cycle enzyme disrupts ^{13}C flux through virtually all reactions associated with the cycle, although interesting from a cellular organization perspective, is disappointing in that direct information about relative flux through competing pathways is no longer available. With this in mind, perhaps it is instructive to review the type of metabolic data one can obtain directly from a single ^{13}C NMR spectrum of a cell extract and use these principles to propose further experiments which may allow us to learn more about the intracellular *milieu*.

In 1981, den Hollander et al. presented one of the first ^{13}C NMR studies of cell metabolism. They reported time dependent spectra of *Saccharomyces cerevisiae* in the presence of 100 mM [2-^{13}C] acetate as a sole carbon source and observed high levels of ^{13}C enrichment in glutamate, aspartate and trehalose. By comparing the ^{13}C fractional enrichments in the trehalose C3 and C4 carbons versus those in the C1, C2, C5, and C6 carbons, they concluded that ^{13}C flux from acetate to trehalose occurred about equally via the glyoxalate cycle and the Krebs citric acid cycle. They argued that any phosphoenolpyruvate (PEP) derived from a citric acid cycle intermediate should yield trehalose with a 0.5:1:1 enrichment ratio in the C4:C5:C6 positions, respectively, while PEP derived from the glyoxalate cycle would yield trehalose with a 0:1:1 enrichment ratio in these same carbons. Their observed enrichment of trehalose C4 was about 33% that of C5 or C6 (0.33:1:1), so they concluded that glyoxalate flux must have been about equal to TCA cycle flux. The glutamate and aspartate spectra were not interpreted in any quantitative terms.

The glutamate spectrum should, in principle, provide information about relative flux through the TCA versus glyoxylate cycles, providing intermediates from these two cycles mix. In 1984, Walsh and Koshland presented a thorough study of metabolic flux in *E. coli* using both radiotracer and ^{13}C NMR data. Since there are no physical compartments in bacteria, they assumed that all metabolites common to the various metabolic pathways would mix at steady-state. After presenting 25 mM [2-^{13}C] acetate to *E. coli* and extracting those cells and purifying the labeled glutamate, a high resolution ^{13}C spectrum of that sample showed that the relative ^{13}C fractional enrichments in the glutamate C1:C2:C3:C4:C5 were 0.4:1.0:0.9:1.0:0 (after appropriate corrections for T_1 and nuclear Overhauser enhancement differences). A simple relationship between proportional flux through isocitrate dehydrogenase, f_k, the observed ^{13}C fractional enrichment in glutamate C1 ($\overline{X}$), and the ^{13}C fractional enrichments of the methyl carbon (a) and the carboxyl carbon (b) of the acetate precursor (as reflected by glutamate C4 and C5 in the NMR spectrum) was derived by Walsh and Koshland (1984) and later modified by London (1988).

$$f_k = \frac{2(\overline{X} - b)}{a - b} \tag{1}$$

In deriving these expressions, both authors assumed that anaplerotic flux through TCA cycle intermediate pools was insignificant compared to TCA cycle flux, and that there was complete equilibration of label between fumarate and malate. Using this relationship and the fractional enrichment values reported by the ^{13}C spectrum, Walsh and Koshland (1984) determined that f_k was 0.8 and hence proportional flux through the glyoxalate cycle was 0.2. We have shown previously (Malloy et al., 1987) that entry of an unlabeled substrate via an anaplerotic pathway (pyruvate carboxylase, for example) decreases the ^{13}C fractional enrichment of glutamate C3 without altering the ^{13}C fractional enrichment of glutamate C4. This is most easily detected as a decrease in the glutamate C3/C4 ratio from unity to something less than 1. The observation by Walsh and Koshland (1984) that the ^{13}C fractional enrichments of glutamate C2, C3 and C4 were equal within experimental error indicated that anaplerosis was, in fact, low compared to TCA cycle flux in these cells and hence equation 1 is valid. If, however, under other metabolic circumstances when anaplerosis may not be insignificant compared to TCA cycle flux then, as we show below, equation 1 cannot be used to estimate relative glyoxalate cycle flux.

An alternative to measurement of ^{13}C fractional enrichments in metabolite spectra is to measure relative singlet, doublet, triplet, and/or other multiplet areas within a *single* resonance and use that data to obtain information about relative metabolic fluxes. This technique, referred to as isotopomer analysis, has been developed and implemented in our laboratory largely for examining metabolism in mammalian tissue (Malloy et al., 1988; Malloy et al., 1990; Jeffrey et al., 1991). Since these multiplet ratios are sensitive not only to the ^{13}C enrichment at each carbon site within a molecule but also sensitive to the enrichment levels at all neighboring carbon sites (one bond interactions only), the isotopomer approach offers the possibility of deriving considerably more metabolic information from a single spectrum. Figure 2 shows the glutamate C2, C3, C4 and trehalose C1–C6 regions of a ^{13}C spectrum of an extract of yeast which had been presented with [2-^{13}C] acetate for a period of two hours before perchloric acid extraction. These spectra, similar to those presented by den Hollander et al. (1981), indicate that both metabolite pools are highly enriched by the [2-^{13}C] acetate and that all resonances in glutamate and trehalose appear as multiplets due to ^{13}C–^{13}C spin–spin coupling.

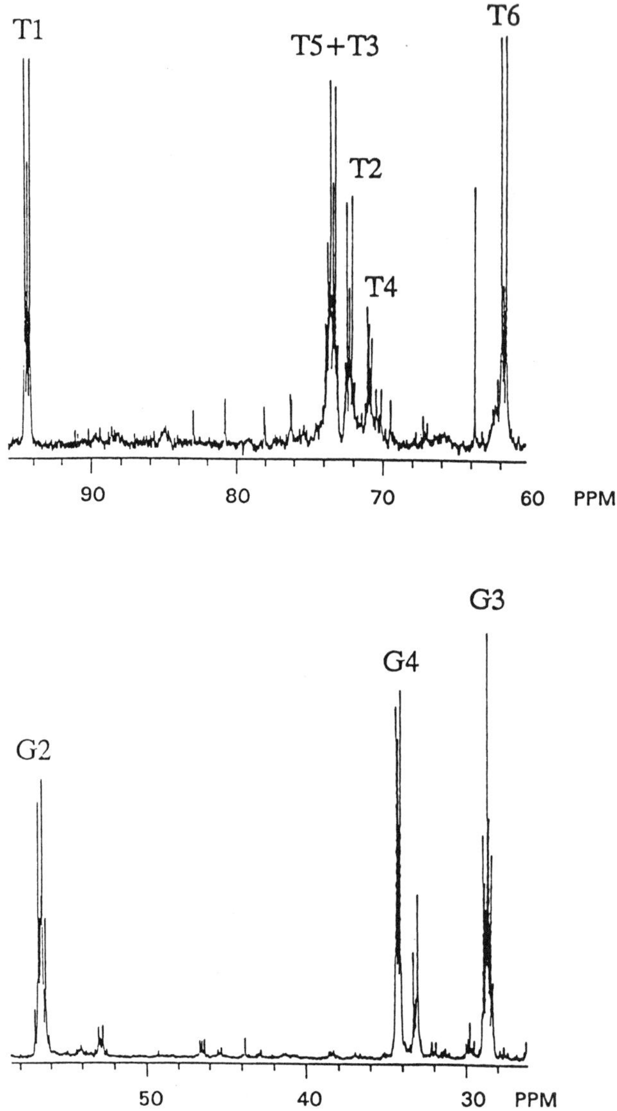

Figure 2. High resolution ^{13}C NMR spectrum of an extract of *Saccharomyces cerevisiae* after being presented with 10 mM [2-^{13}C] acetate for two hours before perchloric acid extraction. The top spectrum shows the enriched trehalose (T1–T6) carbon multiplets while the bottom spectrum shows the glutamate C2, C3, and C4 carbon multiplets.

The observation that the resonance areas of glutamate C2 and C3 are significantly less than that of C4 (experimental ratios are 0.63:0.71:1.0) indicates that anaplerosis was significant during utilization of [2-^{13}C] acetate by these cells (Malloy et al., 1987; 1988; 1990; Jeffrey et al., 1991). One cannot apply the steady-state equations presented previously for mammalian systems (Malloy et al., 1987) to solve for relative anaplerotic flux (y) in this system because any flux through the glyoxalate cycle will influence the observed glutamate C3/C4 when y is anything other than zero. If one models this metabolic system using the computer program, tcaSIM (Jeffrey et al., 1991), it is easy to demonstrate that the ^{13}C fractional enrichment in glutamate C1 (C1F) is dramatically affected by anaplerosis (y) *and* by flux through the glyoxalate cycle.

If one assumes, for example, that [2-^{13}C] acetate is the only carbon source available to the acetyl–CoA pool while flux through all combined anaplerotic reactions and flux through the glyoxalate cycle are turned on sequentially, the ^{13}C fractional enrichments shown in Table 3 would be observed. When glyoxalate flux is turned on, C1F is affected but the C3/C4 ratio remains unaffected. When anaplerosis is turned on without glyoxalate activity, both C1F and C3/C4 decrease but when both anaplerosis and glyoxalate flux are turned on, C1F decreases even more while C3/C4 increases back toward unity. Thus, given that C2F and C3F are less than C4F in the experimental spectrum, presented in Figure 2, equation 1 cannot be used to analyze this spectrum or any spectrum where y does not equal zero.

A complete isotopomer analysis of this same glutamate spectrum yields the following results: fraction of acetyl–CoA supplied by [2-^{13}C] acetate (F_{c2}) = 0.87 ± 0.02, y = 0.23 ± 0.08, and the glyoxalate flux = 0.34 ± 0.08. This result indicates that both pathways (anaplerosis and glyoxalate) were indeed *quite* active. If one examines the glutamate isotopomers present with and without flux through

Table 3. ^{13}C Fractional Enrichments of Glutamate[a]

Anaplerotic flux (y)[b]	*glyoxalate flux*[b]	*C1F*	*Glutamate ^{13}C fractional enrichment*[c] *C2F*	*C3F*	*C4F*
0	0	0.50	1.00	1.00	1.00
0	0.2	0.40	1.00	1.00	1.00
	(equation 1 applies)				
0.2	0	0.30	0.71	0.71	1.00
	(equation in Malloy et al., 1987 applies)				
0.2	0.2	0.25	0.76	0.76	1.00
	(neither equation 1 nor equation in Malloy et al., 1987 apply)				

Notes: [a]Calculated ^{13}C fractional enrichments of glutamate during entry of 100% [2-^{13}C] acetyl-CoA (methyl carbon enriched only) into the Krebs citric acid cycle and the glyoxalate cycle, with or without entry of unlabeled substrate into these pools via an anaplerotic pathway.

[b]Expressed as a percentage of TCA cycle flux

[c]At isotopic steady-state

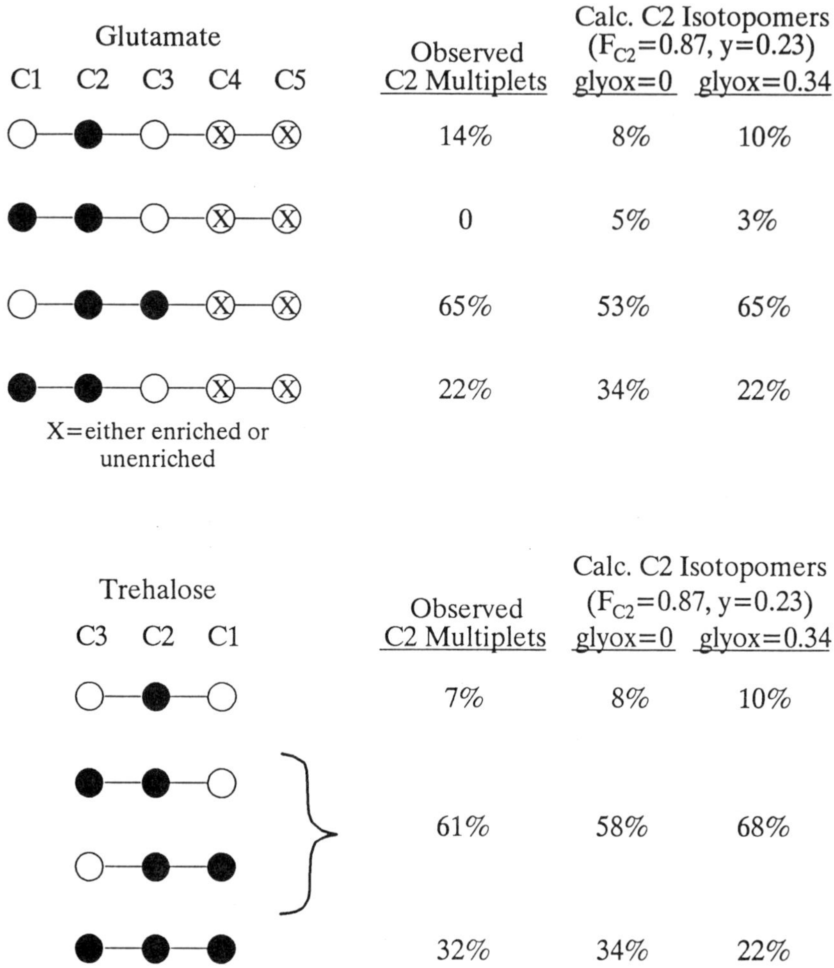

Figure 3. A comparison of experimental glutamate and trehalose C2 multiplet areas (expressed as a percent of the total resonance area) with the corresponding calculated multiplet areas for zero glyoxalate flux versus a glyoxalate flux of 34% relative to the TCA cycle flux. The multiplet areas in the first two right columns were calculated using tcaSIM (Jeffrey et al., 1991) with F_{C2} = 0.87 and y = 0.23 as input parameters (see text for definition of terms).

the glyoxalate cycle turned on, it is apparent that the C3 and C4 isotopomer populations (unlike the C3 and C4 fractional enrichments) are quite insensitive to glyoxalate flux while the C2 isotopomer population is quite sensitive. This point is illustrated in Figure 3 (top) which shows how the glutamate C2 isotopomer populations vary ± inclusion of glyoxalate cycle flux. The agreement between the observed glutamate C2 isotopomer populations and those calculated ± glyoxalate cycle activity shows quite clearly that the glyoxalate cycle is indeed active under these conditions. Furthermore, the isotopomer analysis reports that anaplerotic flux of unlabeled molecules into the citric acid cycle pools (for example, pyruvate → oxalacetate (mixing with TCA cycle pools) → malate (mixing with TCA cycle pools) → PEP) amounts to about 23% of the total flux through the citric acid cycle. As in the *E. coli* study (Walsh and Koshland, 1984), glutamate C5 did not become enriched so none of the labeled molecules which left the citric acid cycle as a gluconeogenic precursor reentered the cycle as acetyl–CoA (likely reflecting tight regulation of the pyruvate dehydrogenase complex during acetate utilization).

The observed ^{13}C fractional enrichment in trehalose C4, C5, and C6 in the spectrum shown in Figure 2 was approximately 0.39:0.67:1.00. One could use the same qualitative argument as presented by den Hollander et al. (1981) and Dickinson et al. (1983; 1991) to suggest that 0.39/0.50 or 78% of the trehalose C4, C5, and C6 carbons were derived from TCA cycle intermediates (i.e., about 22% were derived from the glyoxalate cycle). However, this analysis again assumes that anaplerotic flux through the citric acid cycle pools is negligible compared to TCA cycle flux and the glutamate spectrum shows that this is not true. As before, the ^{13}C fractional enrichments of the trehalose carbons will be sensitive to anaplerotic flux so the best approach once again might be an isotopomer analysis. However, there is less information available in the trehalose spectrum because of symmetry so only qualitative arguments can be presented at this time.

The expanded trehalose C2, C4, and C6 resonances are shown in Figure 4 and a summary of the C2 multiplet areas are given in Figure 3 (bottom). One may assume for the moment that trehalose C1–C3 have the identical isotopomer populations as C6–C4, respectively, and, since C1F was equal to C6F in this spectrum, flux through the pentose phosphate pathway is also negligible. As with glutamate C2, the trehalose C2 isotopomer population was found to be sensitive to *both* anaplerosis and glyoxalate flux. The calculated trehalose C2 isotopomers shown in Figure 3 are for the same metabolic conditions as reported by the glutamate spectrum ± glyoxalate cycle activity. Surprisingly, there appears to be better agreement between the observed trehalose multiplets and those calculated for zero glyoxalate flux than with a glyoxalate flux of 34%. This tends to suggest that *all* trehalose precursors were derived from metabolites which were in isotopomeric steady-state with the citric acid cycle and *none* were derived exclusively from the glyoxalate cycle.

This conclusion at first seems quite surprising, yet one must consider that the net product of the glyoxalate cycle is succinate and this metabolite must presumably reenter the mitochondria before being oxidized to fumarate. Thus, it may not be unreasonable to anticipate that the three carbon gluconeogenic precursors derived from these succinate molecules which ultimately lead to trehalose appear to have been derived from the Krebs citric acid cycle rather than the glyoxalate cycle. It is clear that further ^{13}C NMR experiments on mutants lacking isocitrate lyase activity

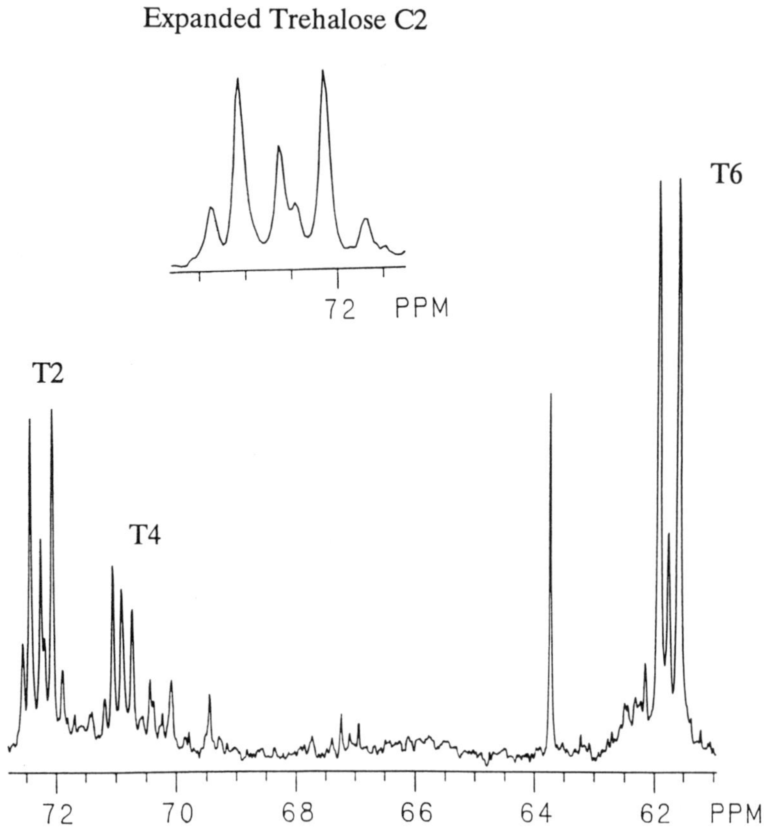

Figure 4. Expanded ^{13}C spectrum of trehalose showing the C2 multiplet intensities as in inset.

or mutants incapable of transporting succinate into mitochondria could provide new insights into the coordination of metabolic flux in eucaryotic cells.

VI. AFTERWORD

Great progress has been made toward the understanding of metabolism and its regulation from *in vitro* studies on individual enzymes and their behavior. It has always been clear, however, that the extrapolation of solution behavior to that in a crowded gel-like cell environment was perilous. Advances in understanding came only with advances in our ability to observe *in vivo* enzyme behavior more accurately, and this arrived only with technological advances. In this brief chapter, we have tried to illustrate some newer insights into the operation of the Krebs TCA cycle made possible by the application of molecular biological and ^{13}C NMR technologies. Both of these newer techniques carry with them newer problems which still need to be solved. However, an understanding of this metabolic path has emerged which is by no means complete but has relieved us of some simplistic ideas about control and organization which has persisted for years.

ACKNOWLEDGMENTS

The authors wish to thank Mrs. Penny Kerby for excellent secretarial assistance and Dr. Claudia Evans for scientific evaluation. These studies were funded by grants from the Department of Veterans Affairs (PAS), National Science Foundation (PAS), and NIDDK (PAS, CRM, ADS). Additional support provided by a Clinical Investigator Award of the Department of Veterans Affairs (CRM), The Robert A. Welch Foundation (ADS), and a Biomedical Research Initiative from the Meadows Foundation (ADS).

REFERENCES

Allen, R. D., Schroeder, C. L., & Fok, A. K. (1989). An investigation of mitochondrial inner membranes by rapid-freeze deep-etch techniques. J. Cell Biol. 108, 2233–2240.

Aragon, J. J. & Sols, A. (1991). Regulation of enzyme activity in the cell: effect of enzyme concentration. FASEB J. 5, 2945–2950.

Aw, T. Y. & Jones, D. P. (1985). ATP concentration gradients in cytosol of liver cells during hypoxia. Am. J. Physiol. 249, C385–C392.

den Hollander, J. A., Behar, K. L., & Shulman, R. G. (1981). ^{13}C NMR study of transamination during acetate utilization by *Saccharomyces cerevisiae*. Proc. Natl. Acad. Sci. USA, 78, 2693–2697.

Dickinson, J. R. & Hewlins, J. E. (1991). ^{13}C NMR analysis of a developmental pathway mutation in *Saccharomyces cerevisiae* reveals a cell depressed for succinate dehydrogenase. J. Gen. Microbiol. 137, 1033–1037.

Dickinson, J. R., Dawes, I. W., Boyd, A. S. F., & Baxter, R. L. (1983). ^{13}C NMR studies of acetate metabolism during sporulation of *Saccharomyces cerevisiae*. Proc. Natl. Acad. Sci. USA 80, 5847–5851.

Halper, L. A. & Srere, P. A. (1977). Interaction between citrate synthase and mitochondrial malate dehydrogenase in the presence of polyethylene glycol. Arch. Biochem. Biophys. 184, 529–534.

Haselbeck, R. J. & McAlister-Henn, L. (1991). Isolation, nucleotide sequence, and disruption of the *Saccharomyces cerevisiae* gene encoding mitochondrial NADP(H)-specific isocitrate dehydrogenase. J. Biol. Chem. 266, 2339–2345.

Hulme, E. C. & Tipton, K. F. (1971). The isotope-exchange reactions of ox heart phosphofructokinase. Biochem. J. 122:181–187.

Jeffrey, F. M. H., Rajagopal, A., Malloy, C. R., & Sherry, A. D. (1991). ^{13}C-NMR: a simple yet comprehensive method for analysis of intermediary metabolism. Trends Biochem. Sci. 16, 5–10.

Kacser, H. & Burns, J. A. (1973). The control of flux. In: Rate Control in Biological Processes, (D.D. Davis, ed.), pp. 65–104, Cambridge University Press, Cambridge, England.

Keys, D. A. & McAlister-Henn, L. (1990). Subunit structure, expression, and function of NAD(H)-specific isocitrate dehydrogenase in *Sacchromyces cerevisiae*. J. Bacteriol. 172, 4280–4287.

Kim, K. S., Rosenkrantz, M. S., & Guarente, L. P. (1986). *Saccharomyces cerevisiae* contains two functional citrate synthase genes. Mol. Cell. Biol. 6, 1936–1942.

Kispal, G., Rosenkrantz, M., Guarente, L., & Srere, P. A. (1988). Metabolic changes in *Saccharomyces cerevisiae* strains lacking citrate synthases. J. Biol. Chem. 263, 11145–11149.

Kispal, G., Evans, C. T., Malloy, C., & Srere, P. A. (1989). Metabolic studies on citrate synthase mutants of yeast: a change in phenotype following transformation with an inactive enzyme. J. Biol. Chem. 264, 11204–11210.

London, R. E. (1988). ^{13}C Labeling in studies of metabolic regulation. Prog. N.M.R. Spectros. 20, 337–383.

Malloy, C. R., Sherry, A. D., & Jeffrey, F. M. H. (1987). Carbon flux through citric acid cycle pathways in perfused heart by ^{13}C NMR spectroscopy. FEBS Lett. 212, 58–62.

Malloy, C. R., Sherry, A. D., & Jeffrey, F. M. H. (1988). Evaluation of carbon flux and substrate selection through alternate pathways involving the citric acid cycle of the heart by ^{13}C NMR spectroscopy. J. Biol. Chem. 263, 6964–6971.

Malloy, C. R., Sherry, A. D., & Jeffrey, F. M. H. (1990). Analysis of tricarboxylic acid cycle of the heart using ^{13}C isotope isomers. Am. J. Physiol. 259, H987–H995.

Masters, C. (1989). Interactions of glycolytic enzymes. Trends Biochem. Sci. 14, 361–364.

Minton, A. P. (1983). The effect of volume occupancy upon the thermodynamic activity of proteins: some biochemical consequences. Mol. Cell. Biochem. 55, 119–140.

Minton, A. P. (1990). Holobiochemistry: the effect of local environment upon the equilibria and rates of biochemical reactions. Int. J. Biochem. 22, 1063–1067.

Moore, G. E., Gadol, S. M., Robinson, J. B. Jr., & Srere, P. A. (1984). Binding of citrate synthase and malate dehydrogenase to mitochondrial inner membranes: tissue distribution and metabolite effects. Biochem. Biophys. Res. Commun. 121, 612–618.

Repetto, B. & Tzagoloff, A. (1989). Structure and regulation of KGD1, the structural gene for yeast alpha-ketoglutarate dehydrogenase. Mol. Cell. Biol. 9, 2695–2705.

Robinson, J. B. Jr., Inman, L., Sumegi, B., & Srere, P. A. (1987). Further characterization of the krebs tricarboxylic acid cycle metabolon. J. Biol. Chem. 262, 1786–1790.

Ryde-Petterson, V. (1992). Oscillations in coupled enzymic reactions at high concentration of enzyme. Biochim. Biophys. Acta 1119, 118–122.

Srere, P. A. (1967). Enzyme concentration in tissues. Science 158, 936–937.

Srere, P. A. (1969). Some complexities of metabolic regulation. Biochem. Med. 3, 61–72.

Srere, P. A. (1981). Protein crystals as a model for mitochondrial matrix proteins. Trends Biochem. Sci. 6, 4–6.

Srere, P. A. (1987). Complexes of sequential metabolic enzymes. Ann. Rev. Biochem. 56, 89–124.

Srere, P. A. (1990). Citric acid cycle redux. Trends Biochem. Sci. 15, 411–412.

Srere, P. A. & Mosbach, K. (1974). Metabolic compartmentation: symbiotic, organellar, multienzymic, and microenvironmental. Ann. Rev. Microbiol. 28, 61–83.

Steffan, J. S., Minard, K. I., and McAlister-Henn, L. (1992). Expression and function of heterologous forms of malate dehydrogenase in yeast. Arch. Biochem. Biophys. 293, 93–102.

Sumegi, B. & Porpaczy, Z. (1990). Structural and metabolic consequences of mitochondrial protein-protein interactions. In: Structural and Organizational Aspects of Metabolic Regulation, (P. A. Srere, M. E. Jones, and C. Mathews, eds.), UCLA Symp. on Molecular and Cellular Biology, New Series, vol. 133, pp. 229–244, Alan R. Liss, Inc., New York.

Sumegi, B., Porpaczy, Z., & Alkonyi, I. (1991a). Kinetic advantage of the interaction between the fatty acid beta-oxidation enzymes and the complexes of the respiratory chain. Biochim. Biophys. Acta 1081, 121–128.

Sumegi, B., Porpaczy, Z., Srere, P. A., and Sherry, A.D. (1991b). Role of α-ketoacid dehydrogenase complexes in the organization of mitochondrial matrix. In: Biochemistry and Physiology of Thiamine Diphosphate Enzymes, (H. Bisswanger and J. Ullrich, eds.), pp. 232–241, VCH Publisher, New York.

Sumegi, B., McCammon, M. T., Sherry, A. D., Keys, D. A., McAlister-Henn, L., & Srere, P.A. (1992). Metabolism of [3-^{13}C] pyruvate in TCA cycle mutants of yeast. Biochemistry 31, 8720–8725.

Thompson, L. M. & McAlister-Henn, L. (1989). Dispensable presequence for cellular localization and function of mitochondrial malate dehydrogenase from *Saccharomyces cerevisiae*. J. Biol. Chem. 264, 12091–12096.

Tyiska, R. L., Williams, J. S., Brent, L. G., Hudson, A. P., Clark, B. J., Robinson, J. B. Jr., & Srere, P. A. (1986). Interactions of matrix enzyme activities with mitochondrial inner membranes. In: The Organization of Cell Metabolism, (G. R. Welch and J. S. Clegg, eds.), NATO Series A: Life Sciences, vol. 127, pp. 177–189, Plenum Press, New York.

Veech, R. L. (1978). Regulation of coenzyme potential by near equilibrium reactions. In: Microenvironments and Metabolic Compartmentation, (P. A. Srere, and R. W. Estabrook, eds.), pp. 17–64, Academic Press, New York.

Walsh, K. & Koshland, D. E., Jr. (1984). Determination of flux through the branch point of two metabolic cycles. J. Biol. Chem. 259, 9646–9654.

Wollenberger, A., Ristau, O., & Schoffa, G. (1960). Eine einfache technik der extrem schnellen abkühlung grösserer gewebestücke. Pflügers Arch. Ges. Physiol. Menschen Tiere 270, 399.

NMR STUDIES OF ERYTHROCYTE METABOLISM

Hilary A. Berthon and Philip W. Kuchel

Advances in Molecular and Cell Biology
Volume 11, pages 147–205.

ISBN: 1-55938-844-7

ABSTRACT

The aim of this chapter is to survey phenomena relating to the regulation of metabolism in intact cells and how NMR spectroscopy can contribute to a greater understanding of these phenomena. The emphasis of the chapter is on isolated erythrocytes and hemolysates, so it is not a comprehensive review of *in vivo* NMR spectroscopy, and many of the examples which have been chosen come from our own particular work on human erythrocytes. The aspect of erythrocyte metabolism which has been subjected to most scrutiny by NMR spectroscopy is the pentose phosphate pathway, so work relating to this group of reactions is presented in greater detail than that from other areas; the work will serve as an example of the type of applications to which NMR can be put in studies of cells. The general physical characteristics of the intracellular *milieu* of the human erythrocyte, which can be measured using NMR spectroscopy, such as pH, viscosity, rotational and translational diffusion coefficients, and metabolite concentrations, are derived from generic experiments that can in principle be applied in studies of any cell type.

I. INTRODUCTION

Nuclear magnetic resonance (NMR) spectroscopy has been amply demonstrated to be the most versatile of all physical techniques for providing information on the microenvironment of enzymes, the nature of chemical transformations, and the rates at which these transformations occur in cells, tissues, and the whole body (e.g., Shulman, 1979; Gadian and Radda, 1981; Brindle and Campbell, 1987; Kuchel, 1989; Lundberg et al., 1990). It is a technique which, with ingenious application, can shed light on some of the postulated phenomena such as "substrate channeling" (Barun and Kirtley, 1977; Weber and Bernhard, 1982; Kurganov, 1986; Maretzki et al., 1989; Betts and Srivastava, 1991; Savageau, 1991; Shulman, 1991) which are alluded to in other chapters of this monograph. Indeed, the motivation for the NMR work carried out in many laboratories, including our own, has been to provide experimental confirmation of, and inspiration for, the large amount of quantitative theoretical work and much speculation, relating to the factors which influence metabolic and transport processes in intact cells.

In traditional studies of the kinetics of enzymic reactions, the enzyme is purified to homogeneity and the rate of the reaction determined at various pH values and with various buffer, substrate, inhibitor, and activator concentrations. By quantitative analysis of the data, using well-established theoretical approaches, it is possible to draw conclusions concerning the mechanism of the reaction catalyzed by the enzyme (e.g., Purich, 1983). A common *desideratum* is the order of addition of the substrates and release of the products from the enzyme during the reaction. Equally important though, is the provision of information which may be used to build computer-based models of metabolism of which the particular enzyme is but a small part. Thus, detailed kinetic studies of individual enzymes offer the potential for describing the operation of consecutive reactions involved in metabolic pathways, as exemplified by the early work of Chance and Higgins (Chance et al., 1960; Higgins, 1964) and the enormous contribution of Garfinkel and colleagues (e.g., Garfinkel et al., 1968).

With the use of computer simulations, based on the kinetic parameters evaluated *in vitro* for the individual enzymes, it has become possible to estimate the concentrations of metabolites in pathways. This type of procedure potentially offers a firmer basis than intuition in relating clinical and biochemical observations of metabolite concentrations to possible inborn errors which affect the pathway (Kuchel et al., 1977; Holzhütter et al., 1990). While computer models have reached a high level of complexity in terms of the number of reactants that can be realistically followed in simulated timecourses of metabolic experiments, there is also the need to be able to determine the essential relationships operating within the modeled system. It is in this context that metabolic control theory has assumed much importance (Kacser and Burns, 1973; Burns et al., 1985; Acerenza et al., 1989). Nevertheless, the speed with which calculations may be performed on modern computers is such that simplified 'phenomenological' models are to be avoided, especially when the experimental data indicate a greater complexity than can be readily expressed with simple equations. This approach, using detailed enzyme kinetic equations, was developed by Garfinkel (e.g., Garfinkel, 1980) and it allows greater scope for the extension of a model. Furthermore, it is possible to evaluate flux control coefficients (FCC or sensitivity coefficients; Burns et al., 1985) in these 'numerical' models (Kohn et al., 1979; Thorburn and Kuchel, 1985; Berthon et al., 1992), thus gaining insight that has been argued to be only available through the simple analytical models.

NMR spectroscopy is highly selective, enabling the identification of molecules in complex heterogeneous mixtures, such as those which constitute cells, and requires minimal specific sample preparation (Kuchel, 1989). This is in distinct contrast to such methods as high performance liquid chromatography (HPLC), for which it is often necessary to choose sample preparation conditions which allow the selective detection of a particular compound or class of compounds. This important characteristic of NMR spectroscopy was used by organic chemists for over a decade before the spectrometers reached a sufficient level of sensitivity in the late 1960's, for the spectroscopic study of cells. Eakin et al. (1972) first demonstrated the utility of ^{13}C NMR spectroscopy for monitoring the metabolism of ^{13}C-labeled glucose by intact yeast cells; Moon and Richards (1973) demonstrated the utility of ^{31}P NMR spectroscopy for measuring intracellular pH, and the rate of decline of 2,3-bisphosphoglycerate (2,3-BPG) in rabbit erythrocytes. Reports about different aspects of metabolism, such as the effects of treating human erythrocytes with inosine and pyruvate soon followed (Henderson et al., 1974).

The ubiquity of the highly receptive ^{1}H nucleus in biological samples meant that it only became amenable for use in metabolic studies after it was shown that the spin–echo technique could be used to selectively detect small mobile molecules while eliminating the signal from macromolecules and membranes (Brown et al., 1977). Thus, it became possible to monitor metabolic and transport events in intact cells with ^{1}H NMR. Due to their ready availability, relevance to many human clinical disorders, and simple metabolism and structure (lack of a nucleus, mitochondria, and endoplasmic reticulum), human erythrocytes have been extensively studied with NMR. They serve as a useful model system for developing NMR techniques, and raise intriguing questions, such as how the high intracellular protein concentration affects the rates of metabolic reactions.

Even though the human erythrocyte is a simple cell, it contains a large number of different enzymes and NMR spectroscopy provides a convenient means of studying the operation of these in the whole cell or in concentrated hemolysates. The latter enable the use of solutes which would otherwise not affect metabolism as a consequence of their membrane impermeability.

II. INTRACELLULAR ENVIRONMENT

A. Intracellular pH

The average pH in the human erythrocyte may be calculated from the transmembrane distribution ratio of radioactively labeled or fluorescent weak acids or bases (Kotyk and Slavik, 1989). The erythrocyte is too small for the convenient routine use of microelectrodes, but the variation of the chemical shift of the orthophosphate resonance, as well as the two phosphate ester moieties of 2,3-BPG, can be easily measured, providing a ^{31}P NMR probe of intracellular pH (Moon and Richards, 1973; Labotka, 1984). Greater precision in the estimates of pH can be obtained by use of metabolically innocuous probe compounds, such as methylphosphonate

(MeP), which exhibit a larger pH dependence of their chemical shift, and resonate at a frequency far removed from the peaks arising from endogenous compounds (Slonczewski et al., 1981; Labotka, 1984; Stewart et al., 1986). The ^1H NMR spectrum of human erythrocytes contains an envelope of the superimposed resonances with three or four peaks derived from the 12 histidine residues in the α and β chains of hemoglobin. The chemical shift of the maximum ordinate of the composite peaks varies with pH; thus after construction of a titration curve using a hemolysate, the pH can be 'read' from a spin–echo spectrum of whole cells by measuring the chemical shift of the relevant peak and with reference to the titration curve (Browne et al., 1976; Ferrige et al., 1979).

Less precise measurements of pH can be obtained using ^1H NMR and the chemical shift of the ^1H_2O resonance (Madden et al., 1991). However, since this chemical shift varies with temperature, precise thermal regulation of the sample is imperative.

$H^{13}CO_3^-$ added to human erythrocytes is rapidly dehydrated via carbonic anhydrase to yield an equilibrium mixture with $^{13}CO_2$; thus, the 'fully relaxed' ^{13}C NMR spectrum of the cell suspension yields the concentration ratio of the two solutes. This information, used in conjunction with the Henderson-Hasselbalch equation, gives an estimate of the mean pH of the sample (Kuchel et al., 1987); this uses the apparent pK_a under intracellular conditions of ~6.2. Rapid exchange of both bicarbonate and CO_2 across the erythrocyte membrane ensures that the measured pH value is the average of that in the intra- and extracellular regions, weighted according to their relative volumes.

The rate of glycolysis in human erythrocytes increases with an increase in pH: this effect has been attributed largely to the pH sensitivity of phosphofructokinase, which has a high FCC for this pathway. Whether the average pH reported by NMR spectroscopy is relevant to the local hydrogen ion concentration has not been determined, and whether the average value has direct bearing on the rate of glycolysis, via its effect on phosphofructokinase, is not yet established.

B. Intracellular Viscosity

The rate constant characterizing the longitudinal relaxation ($1/T_1$) of a given ^{13}C-nucleus that has been excited by a radiofrequency (r.f.) pulse, is determined primarily by the dipole–dipole interaction between the ^{13}C nucleus and the nuclei of directly bonded hydrogen atoms (Abragam, 1961). The dipole–dipole interactions are a function of the rotational motion of the molecule containing the ^{13}C nucleus, and thus the T_1 value of the nucleus is also a function of the rotational motion (Doddrell et al., 1972). The viscosity inside the unicellular blue-green alga, *Synechococcus* sp., and the unicellular green alga, *Dunaliella salina,* which are salt tolerant, was therefore able to be inferred from measurements of the T_1 of the ^{13}C nuclei at natural abundance in the endogenous osmolytes, 2-O-α-D-glucopyranosylglycerol and glycerol, respectively (Norton et al., 1982).

Similarly, the rotational correlation time of intracellular ^{13}C-glycine in human erythrocytes was measured after these cells had been incubated with the amino acid for ~1 h. With the appropriate incubation conditions, most notably the absence of glucose, the ^{13}C-glycine undergoes exchange with the unlabeled glycyl of glutathione (GSH), thus yielding ^{13}C-GSH (York et al., 1982). Again, the rotational correlation time of this tripeptide can be inferred from the T_1 values of its nuclei (Endre et al., 1983; Endre and Kuchel, 1986). In the most simple analysis, the rotational correlation time, τ_c, is given by the Debye equation:

Figure 1. Representation of the random packing of hemoglobin molecules in the human erythrocyte cytoplasm where the concentration is 330 g 1^{-1} (Dacie and Lewis, 1975). The macromolecules are drawn as cross-sections of spheres, and the space between two concentric circles represents two layers of water molecules in the protein-hydration shell. Note that the representation here has all the spheres in one plane; in reality a cross-section of a random array of spheres will have discs of various sizes present.

$$\tau_c = \frac{4\pi r^3 \eta}{3kT} \tag{1}$$

where r is the Stokes radius of the molecule, η is the viscosity of the medium, k is the Boltzmann constant, and T is the absolute temperature. This analysis, however, makes the tenuous assumption that the molecule is a sphere.

The correlation time of spin-labeled GSH in erythrocytes has been measured using electron spin resonance and the value has been calibrated in terms of viscosity by using the compound in sucrose solutions, whose bulk viscosities were measured with an Ostwald capillary viscometer (Daveloose et al., 1983). This interpretation is flawed because the rotational correlation time refers to a microviscosity and this is a function of the nature of the surrounding molecules (Hermann and Müller, 1986; Endre and Kuchel, 1986). For hemoglobin solutions with concentrations comparable to those inside the human erythrocyte, the bulk viscosity is substantially greater (~5 times) than the viscosity estimated from measurements of the rotational correlation time of ^{13}C-labeled hemoglobin in mouse erythrocytes (London et al., 1975a). In other words, the *micro*viscosity is much less than the bulk- or *macro*viscosity. This outcome can be appreciated by considering the qualitative aspects of flowing and jostling spheres and the manner in which their motion is impeded by the close proximity of neighboring spheres, relative to their ability to rotate even though their neighbors may be proximal. Thus, a small molecule like glycine is relatively unimpeded in its rotational motion within the interstices of the matrix of jostling hemoglobin molecules (Endre and Kuchel, 1986).

Figure 1 is a representation of the relative packing density of hemoglobin molecules in a human erythrocyte and gives an impression of the free volume available for the rotational and translational movement of small molecules. From NMR analysis we have deduced that only with the most rapid of enzyme catalyzed reactions, such as those catalyzed by glutathione peroxidase, catalase and carbonic anhydrase, would there be any diffusion control of the reactions (Endre and Kuchel, 1986).

C. Translational Diffusion of Small Molecules Inside Cells

While it is possible to infer the translational diffusion coefficient from a knowledge of the microviscosity and the translational correlation time of the molecule (e.g., London et al., 1975a), it is clearly much more satisfactory to make the measurement directly. Pulsed field gradient spin–echo NMR spectroscopy (PFGSE) is now recognized as the most versatile and probably the most convenient means of measuring diffusion coefficients of solute and solvent molecules (Stilbs, 1987; Callaghan, 1991), not only in homogeneous solutions, but also in heterogeneous systems including cells. It is probably the only method for measuring the microtranslational diffusion coefficients inside cells. These experiments have only recently been achieved due to the development of various means of avoiding the spectral peak distortion and dispersion that results from eddy currents; these are induced in conducting elements of the probe and surrounding magnet by the magnetic field gradient pulses, of short duration (a few ms), used in the experiment.

The translational diffusion coefficient of (2-^{13}C)glycine in human erythrocytes has been measured and it is evident that its diffusion coefficient closely corresponds to that inferred from the previously mentioned rotational correlation-time estimates (Price et al., 1989). Likewise, the translational diffusion coefficient of $H^{13}CO_3^-$ has been measured in human erythrocytes in which the transmembrane exchange of this ion, via band 3, had been blocked by an inhibitor, 4,4′-dinitrostilbene-2,2′-disulfonate (DNDS) (Price and Kuchel, 1990). By using ^{31}P PFGSE NMR spectroscopy, the translational diffusion coefficient of the exogenous compound, hypophosphite ($H_2PO_2^-$), was also measured. From all three measurements it was possible to infer a microviscosity in the intracellular milieu, and the value was in close agreement with that inferred from T_1 measurements on the respective compounds inside the cells (Price and Kuchel, 1990).

D. Translational Diffusion of Proteins Inside Cells

An early estimate of the diffusion coefficient of intra-erythrocytic hemoglobin was made by Everhart and Johnson (1982) using a PFGSE procedure with a spectrometer operating at a relatively low magnetic field strength (0.4 T). The analysis relied on the fact that the spin–echo signal attenuation of ^1H_2O, which is rapidly diffusing inside the cell, is much greater than that of the slowly diffusing hemoglobin. Thus, in the presence of large magnetic field gradients, the resultant ^{1}H NMR signal is dominated by the signal 'envelope' from the nonexchangeable protons of hemoglobin. A PFGSE analysis of the intensity of this signal envelope, as a function of the square of the intensity of the gradient pulses, yielded estimates of the diffusion coefficient of hemoglobin. More recent experiments have utilized the fact that the ^{1}H spin–echo spectrum of human erythrocytes, at 400 MHz, contains a relatively well defined envelope of resonances from the H2 and H4 nuclei of the twelve histidyls in each of the α and β chains of the hemoglobin (see pH section above). By detecting the diffusion via a small number of well defined amino acid residues in hemoglobin, the specificity of the measurement is obviously improved.

Analysis of the PFGSE data yielded an estimate of the diffusion coefficient of hemoglobin of $6.12 \pm 1.1 \times 10^{-12}$ m^2 s^{-1}. The Stokes radius of the hemoglobin A tetramer is 3.13 nm; therefore, the Einstein–Smoluchowky equation, incorporating the Stokes expression for the frictional coefficient of a sphere, yields an estimate of the translational diffusion coefficient under conditions of thermodynamic ideality of 6.34×10^{-11} s^{-1} (Kuchel and Chapman, 1991). This value is over ten times greater than that which was measured for the whole cells. It is concluded that the large difference is due to the high concentration of the protein, and the measured restriction to motion is in accordance with previous measurements that were made using frontal gel chromatography on columns of noncompressible porous glass beads, and also determined using sedimentation equilibrium analysis.

The translational diffusion coefficient of lysozyme at high concentration has also been measured (Haner and Schleich, 1989), and attempts to demonstrate protein–protein interaction using PFGSE NMR spectroscopy have recently been carried out in this laboratory (A. J. Lennon, N. R. Scott, B. E. Chapman and P. W. Kuchel, unpublished results). The simple idea underlying the experiments is that the measured *apparent* diffusion coefficient of a protein will decrease if it forms a complex with like or dissimilar proteins; the value of the apparent diffusion coefficient is a number average of the diffusion coefficients of all the possible complexes of the protein. The experiments are precursors of attempts to measure the protein–protein interactions which are known to occur (e.g., Maretzki et al., 1989) between, for example, glyceraldehyde-3-phosphate dehydrogenase and band 3.

E. Membrane Potential

Those solutes which are both charged and permeate the cell membrane will tend to establish a transmembrane concentration ratio which depends upon the bulk electrical potential across the membrane. Under normal conditions a human erythrocyte has a membrane potential of –10 mV inside the cell relative to outside (Rink and Hladky, 1982). The potential is brought about almost exclusively by the impermeable anions, hemoglobin, 2,3-BPG, and ATP, and is referred to as a Donnan potential. Because the state of ionization of these three species is pH dependent, then so too is the membrane potential (Raftos et al., 1990). Upon elevation of pH, there is a net efflux of cations, particularly K^+, with a consequent reduction in cell volume and a diminution in the membrane potential. This effect alone will bring about a change in the concentration of substrates, and will exert an effect on enzyme catalytic rates, both directly, via the groups important in the catalytic mechanism which are able to be pH-titrated, and indirectly, through alteration of the protein conformation.

The membrane potential of human erythrocytes can be measured with a conventional single-pulse ^{31}P NMR spectrum using the exogenous compound, hypophosphite (Kirk et al., 1988). The compound displays a 'split peak phenomenon' with separate resonances from the intra- and extracellular populations of the solute. Therefore the ratio of peak areas in a 'fully relaxed' spectrum corresponds to the ratio of the amounts of hypophosphite in each compartment; this ratio can be converted to a concentration ratio when the volumes accessible to the solute in each compartment are known. By use of the Nernst equation, the membrane potential can be determined.

An alternative method makes use of trifluoroacetate and trifluoroacetamide, which also exhibit the split peak phenomenon in ^{19}F NMR spectra of human erythrocytes (London and Gabel, 1989). The ratio of the intensities of the resonances of the uncharged solute, iodoacetamide, corresponds to the ratio of the compartment volumes, so the concentration ratio of the charged solute, iodoacetate, can be readily calculated. These NMR techniques are counterparts of those that use charged fluorescent dyes (Rink and Hladky, 1982); however, the latter techniques require the use of very dilute suspensions of cells and also procedures to 'calibrate' the membrane potential. NMR spectroscopy has the advantage of allowing the measurement to be carried out directly on cell suspensions of high hematocrit, during observations of other processes such as membrane transport and metabolism.

F. Cell Volume

As previously mentioned, changes in cell volume will directly affect the concentration of metabolites and effector molecules in the cell, and thus influence the rate of metabolic and transport processes. It was first demonstrated in 1985 that the highly water soluble phosphoryl compound, dimethyl methylphosphonate (DMMP), upon addition to a suspension of erythrocytes, gives rise to two resonances in the ^{31}P NMR spectrum. One resonance is assigned to the intracellular DMMP and the other to the extracellular DMMP (Kirk and Kuchel, 1985). The magnitude of the separation between the two resonances was shown to be an almost linear function of the cell volume, increasing as the cells were shrunken by the addition of high osmolality medium to the cell suspension. It was subsequently shown that the principal physical basis of this split-peak phenomenon is the reduction of the average extent of hydrogen bonding of the hydrogen atoms of water to the phosphoryl oxygen, resulting from the high concentration of protein inside the cells (Kirk and Kuchel, 1988a,b). The split-peak phenomenon is also apparent with a variety of fluorinated compounds including 3-fluoro-3-deoxy-D-glucose (Potts et al., 1990; Potts and Kuchel, 1992), difluorophosphate and monofluorophosphate (Xu et al., 1991; Xu and Kuchel, 1991) and F^- (Chapman and Kuchel, 1990). As a result of the transmembrane exchange of the solutes occuring on the subsecond time scale, they may be used to measure rapid changes in cell volume, and also their rates of exchange across the cell membrane can be measured using magnetization transfer procedures. The measurement of the rapid transmembrane exchange of solutes is an important developing area of biological NMR spectroscopy, which will not be discussed further here as it lies outside the theme of this chapter; for those interested, the area has recently been reviewed (Kirk, 1990; Kuchel, 1990; King and Boyd, 1991; Kuchel et al., 1994).

The general characteristics of the cytoplasm, discussed in this section, can obviously affect any metabolic reaction and could, therefore, have a bearing on the reactions of the pentose phosphate pathway, which are considered next.

III. PENTOSE PHOSPHATE PATHWAY

A. Background

The pentose phosphate pathway may be considered to consist of two major parts. In the oxidative part of the pathway, glucose 6-phosphate (Glc6P) is converted to the ketopentose phosphate, ribulose 5-phosphate (Ru5P), with the concomitant formation of one molecule of CO_2 and the reduction of two molecules of $NADP^+$ to NADPH (Figure 2). The formulation of the reactions comprising the non-oxidative pathway has been the subject of some controversy over the past two decades, but it is generally agreed to involve the initial transformation of three molecules of Ru5P to other pentose phosphates, followed by a series of reactions resulting in the net production of two hexose phosphate molecules and one triose phosphate molecule. The "classical" scheme for the non-oxidative pentose phosphate pathway is shown in Figure 2. All intermediates of the pentose phosphate pathway are phosphate esters.

It has been estimated that in human red blood cells under 'normal' conditions, ~5–10% of Glc6P metabolized enters the pentose phosphate pathway, with the remainder entering the glycolytic pathway (Sturman, 1967; Albrecht et al., 1971; Gaetani et al., 1974; Morelli et al., 1979; Magnani et al., 1983). However, under conditions of oxidative stress, which may be induced experimentally by oxidation of GSH, or addition of 0.1 – 0.5 mM methylene blue to a suspension of red cells, the proportion of Glc6P entering the pentose phosphate pathway can be as high as 80–100% (Thorburn and Kuchel, 1985). This high degree of metabolic flexibility enables the erythrocyte to function under the varying conditions of oxidative stress which accompany its circulation throughout the body.

NADPH performs its anti-oxidative function chiefly by maintaining the tripeptide, GSH, predominantly in the reduced state (the ratio of oxidized to reduced glutathione of 'normal' human red cells is ~1:500; Beutler, 1984). The Michaelis constant, K_m, of glutathione reductase for NADH is more than 20 times higher than that of NADPH (Scott et al., 1963), ensuring that NADPH serves as the major substrate. Hemoglobin, as well as many membrane proteins and enzymes, requires one or more intact sulfhydryl groups to retain its activity. Glutathione, the major non-protein thiol compound in red cells, affords the cell protection against free radicals thought to initiate chain reactions of membrane lipid peroxidation (Kosower and Kosower, 1974), and protects sulfhydryl groups in enzymes from oxidative attack. The combined action of glutathione reductase (EC 1.6.4.2) and glutathione peroxidase (EC 1.11.1.9) couples the pentose phosphate pathway to the removal of hydrogen peroxide (Figure 2). A wide variety of hydroperoxides, such as *t*-butyl hydroperoxide, lipid peroxides and steroid peroxides may also be reduced to their corresponding alcohols (Cohen and Hochstein, 1963; Little and O'Brien, 1968; Srivastava et al., 1974). Glyoxalase 1 (EC 4.4.1.5) and glyoxalase 2 (EC 3.1.2.6) use GSH as a cofactor to catalyze the conversion of ketoaldehydes such as

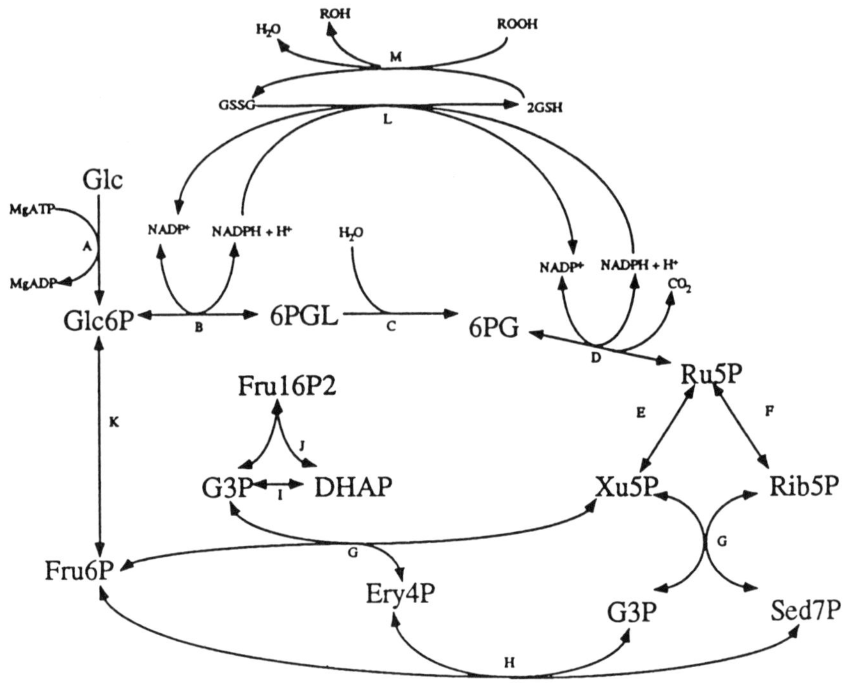

Figure 2. The reaction sequence of the classical pentose phosphate pathway, including the aldolase-mediated condensation of G3P and DHAP, and glutathione and hydroperoxide reduction reactions. The reactions are catalyzed by: (**A**) hexokinase (EC 2.7.1.1); (**B**) glucose-6-phosphate dehydrogenase (EC 1.1.1.49); (**C**) 6-phosphogluconolactonase (EC 3.1.1.31); (**D**) 6-phosphogluconate dehydrogenase (1.1.1.44); (**E**) ribulose-5-phosphate epimerase (EC 5.1.3.1); (**F**) ribose-5-phosphate isomerase (EC 5.3.1.6); (**G**) transketolase (EC 2.2.1.1); (**H**) transaldolase (EC 2.2.1.2.); (**I**) triosephosphate isomerase (EC 5.3.1.1); (**J**) aldolase (EC 4.1.2.13); (**K**) glucosephosphate isomerase (EC 5.3.1.9); (**L**) glutathione reductase (EC 1.6.4.2) and (**M**) glutathione peroxidase (EC 1.11.1.9). The aldolase-catalyzed reaction has been included although it does not constitute an integral part of this pathway. (See also Abbreviations.)

methylglyoxal into hydroxyacids such as lactate (see next section; Grimes, 1980; Rae et al., 1990).

The investigations of the pentose phosphate pathway in human erythrocytes described in this section are divided into several areas. First, a study designed to determine the factors which control the flux of metabolites through the pentose phosphate pathway under conditions of oxidative stress is described. Second, the alternative schemes of reactions for the non-oxidative pathway are examined in an approach combining ^{31}P NMR and computer simulations. Third, the nature of the control of the non-oxidative pentose phosphate pathway is explored. Recent studies employing ^{13}C NMR timecourse methods and isotopic isomer (isotopomer) analysis to examine the role of exchange reactions and the reversal of the reactions of the pentose phosphate pathway, are described. This is followed by examples of studies of metabolic flux through the pathway using multinuclear NMR spectroscopy.

An overriding aim of this work has been the summary of data pertaining to the operation of the pentose phosphate pathway in human erythrocytes (obtained by NMR methods) into a computer-based mathematical model. This model has evolved over the past 10 years and now includes data from the entire pathway and has been extended to discriminate between ^{13}C isotopomers (Berthon et al., 1993).

B. Control of the Oxidative Pentose Phosphate Pathway

In human erythrocytes, flux through the oxidative pentose phosphate pathway is limited mainly by the glucose-6-phosphate dehydrogenase reaction, which operates at less than 1% of its capacity, due to inhibition by the high cellular NADPH:$NADP^+$ ratio of at least 40:1 (Thorburn and Kuchel, 1985). There have been conflicting reports, however, about the enzyme exhibiting the highest degree of control under conditions of oxidative stress (Rose, 1961; Sturman, 1967; Albrecht et al., 1971; Gaetani et al., 1974; Morelli et al., 1979; Magnani et al., 1983). Therefore, studies in our laboratory have focused on elucidating the regulatory character of the oxidative pentose phosphate pathway under these conditions. A range of experimental systems including intact erythrocytes, hemolysates and a reconstituted metabolic system prepared from purified enzymes, were employed. The strategy adopted was to measure the kinetics of the regeneration of reduced glutathione (GSH) in samples initially containing high levels of oxidized glutathione (GSSG). As an additional aid to the analysis of this system, a computer-based mathematical model describing metabolism through this pathway, was constructed from data obtained from erythrocyte enzymes *in vitro*.

^1H NMR provided a convenient method for monitoring the reduction of GSSG to GSH in experimental systems (Brown et al., 1977). The most prominent difference in the ^1H spin–echo spectrum of GSH and GSSG is the diminution in amplitude of the GSH-glutamyl H4 resonance. The GSH-glutamyl H3 resonance was used as an internal reference in timecourses of GSH regeneration in human erythrocyte suspensions, since its amplitude does not change significantly during GSSG reduction. Experimentally, *t*-butyl hydroperoxide was used to convert cellular GSH to the oxidized form in the samples which were then incubated with 10 mM glucose (Thorburn and Kuchel, 1985).

Hemolysates were prepared by sonication of the cell suspensions. This procedure was preferable to freeze-thawing, as ATP levels were found to be maintained at concentrations comparable with the levels found in intact cells, for at least an hour after hemolysis by sonication; with freezing and thawing the ATP concentration falls to almost zero by ~20 min. The reason for this phenomenon is unknown but it may be associated with an alteration of the extent, or nature, of the association of the low and high molecular mass subunits of Na^+, K^+-ATPase. At low temperatures hydrophobic interactions between proteins are weakened and some multimeric proteins dissociate into monomers; thus if the small subunit of the ATPase is inhibitory the dissociation may enhance its catalytic rate. Nicotinamide and $NADP^+$ were added to the hemolysates in order to prevent the rapid decline in concentration of NAD and NADP occurring due to the action of NAD(P) glycohydrolase (see section below). This enzyme is bound to the extracellular surface of the erythrocyte membrane and, following hemolysis, it gains access to intracellular NAD and NADP, which it hydrolyzes (Kuchel and Chapman, 1985). Hemolysates supplemented with 25 mM nicotinamide and 1 mM $NADP^+$, exhibited initial GSH regeneration rates which were identical to those obtained in suspensions of whole cells (10.32 ± 0.29 mmol (l cell water)$^{-1}$ (min)$^{-1}$; Thorburn and Kuchel, 1985).

When glucose 6-phosphate (Glc6P) is replaced by (1-^{2}H)Glc6P, human erythocyte glucose-6-phosphate dehydrogenase exhibits a kinetic isotope effect (KIE) of between 1.4 and 2.4 (Rose, 1961). This observed deuterium KIE was used to probe the regulation of flux *in situ* by the enzyme, glucose-6-phosphate dehydrogenase. The absence of a detectable KIE, when GSH regeneration was measured with both ^{1}H NMR and a chemical assay in an erythrocyte suspension, hemolysate, or a reconstituted enzyme system, demonstrated that this enzyme had a low FCC for flux under the experimental conditions.

That hexokinase exhibited a high degree of control was indicated by the very rapid rate of GSH formation which occurred when Glc6P was added to a hemolysate sample prepared from cells which had been treated with *t*-butyl hydroperoxide and supplied with $NADP^+$, thus bypassing hexokinase (Thorburn, 1987). The rate of GSH regeneration, which was greater than 20 times faster than the rate observed in lysates supplied with glucose, resulted from the obviation of the rate-limiting hexokinase-catalyzed reaction. Furthermore, the addition of the hexokinase inhibitor, glucose 1,6-bisphosphate (Glc16P2), to a hemolysate sample, supplied with $NADP^+$, resulted in a significant reduction in both the observed rate of GSH regeneration and the measured hexokinase activity of the sample. Some previous investigations had supported the notion that hexokinase is ‘controlling’ under conditions of oxidative stress: by studying the effect of “loading” intact erythrocytes with hexokinase or glucose-6-phosphate dehydrogenase, Morelli et al. (1979) found that while no effect was observed with erythrocytes supplemented with the latter enzyme, additional hexokinase dramatically enhanced pentose phosphate pathway flux in the presence of 0.1 mM methylene blue.

In order to investigate the *in situ* metabolic character of the oxidative pentose phosphate pathway, a reconstituted metabolic system was constructed from semi-purified human erythrocyte enzymes (Thorburn and Kuchel, 1985), and results obtained from this system were compared with data from hemolysates and cell suspensions. The enzyme activities were chosen to correspond to one fifth of the ‘normal’ assayed levels, and a corrected value for the rate of GSH production was made by multiplying the measured rate by five (Thorburn and Kuchel, 1985). The corrected rate of GSH formation in the reconstituted system was 4.23 ± 0.14 times the value determined in intact cells and hemolysates (Thorburn and Kuchel, 1985). While using a physiological concentration of MgATP in the reconstituted system would serve to increase the activity of hexokinase by only 13.4%, the pathway flux measured in cells and hemolysates would be expected to be markedly reduced by the presence of the hexokinase inhibitor, Glc16P2, in the 0.1–0.3 mM range. Interestingly, when this effector was added to the reconstituted system, the rate of GSH regeneration was reduced to a greater degree than in hemolysates. This inhibition could be partially relieved by the presence of P_i, at physiological concentration (Thorburn and Kuchel, 1985). Halving the hexokinase concentration in the reconstituted system reduced the rate of GSH regeneration to 56 ± 5%; an identical reduction in the concentration of glucose-6-phosphate dehydrogenase, caused no significant change in the rate of GSH production.

Our approach to modeling the pentose phosphate pathway has entailed the detailed description of each enzymic reaction of the pathway using the *in vitro* kinetic data previously reported for each enzyme. The individual elementary rate equations were then derived from the various enzyme reaction mechanisms relevant to the constituent enzymes (e.g., Kuchel et al., 1977). In other words, a set of non-linear algebraic relationships between the steady-state kinetic parameters and the unitary rate constants was written and unitary rate constants, which were reasonably consistent with the published steady-state parameters, were calculated. This approach offers a means of determining how well the available kinetic data for each enzyme reflects the metabolic character of the system observed *in situ.* In our earlier work, computer simulations were carried out by the numerical integration of the resultant set of differential equations in the BIOSSIM computing package (Roman and Garfinkel, 1978; obtained from the SHARE program library agency, Triangle Universities Computation Centre, NC, USA).

Computer simulations of GSH production exhibited good agreement with experimental data. A quantitative assessment of the contribution of each reaction to the overall flux of reducing equivalents from Glc6P to GSSG under the chosen experimental conditions was carried out using simulations from the model. Flux control was shown to reside almost entirely in hexokinase, which had an FCC of 0.939, at normal enzyme concentrations (Thorburn, 1987).

Having obtained satisfactory agreement between the experimental data and simulations from the computer model, computer simulation methods were used to test the proposal that the severity of hemolysis observed in persons having a deficiency of glucose-6-phosphate dehydrogenase, may result from an additional deficiency of 6-phosphogluconolactonase (Beutler et al., 1985). Glucose-6-phosphate dehydrogenase deficiency is one of the most common disease-producing enzyme abnormalities, affecting over 100 million people worldwide. The problems associated with this disease are known to arise following the exposure of susceptible persons to various oxidative challenges. Simulations at 100, 1, 0.1, and 0.01% of the normal glucose-6-phosphate dehydrogenase activity and 100, 50, 1, and 0% of the normal 6-phosphogluconolactonase activity, were carried out (Thorburn and Kuchel, 1987a). The oxidative stress likely to be experienced by prehemolytic cells was mimicked by rendering all the glutathione to the oxidized form. When 6-phosphogluconolactonase activity was set to 1% of its normal activity, no significant effect on the rate of GSH regeneration at any of the glucose-6-phosphate dehydrogenase activities examined, was observed. Simulations using the kinetic data for the *Jalisco* variant of glucose-6-phosphate dehydrogenase, which possesses a higher affinity for the δ-lactone, yielded a similar result. Interestingly, as glucose-6-phosphate dehydrogenase was reduced from 100 to 0.01% of its normal activity, the effect of 6-phosphogluconolactonase deficiency on the GSH-regeneration time-course became proportionally less, due to the spontaneous hydrolysis of the δ-lactone becoming better able to keep up with the slower pathway flux.

In contrast with time courses in hemolysates, or those simulated by the computer model, time courses of GSH regeneration in intact cell suspensions were apparently biphasic, "tapering off" in rate when GSSG reduction was 60–70% complete (Thorburn and Kuchel, 1987b). It was hypothesized that this difference could be due to the cell-age related heterogeneity of erythocyte suspensions: as the younger cells (with higher hexokinase activity) completed GSH regeneration, the overall rate would be expected to progressively decline. In order to study this effect, freshly drawn erythrocytes (free from leukocytes) were separated into 10 fractions of increasing density, by centrifugation. This technique for separating cells according to their age, is based on the observation that erythrocytes increase in density as they age, due to an increasing concentration of hemoglobin caused by loss of cellular water and electrolytes (Murphy, 1973). The bottom fraction was found to have a hexokinase activity which was 38.1% lower than the top fraction, although the levels of this enzyme's inhibitors, 2,3-BPG and Glc16P2 were also substantially lower (Thorburn and Kuchel, 1987b). While computer simulations predicted that the rate of GSH regeneration in the bottom fraction of erythrocytes should be 27.5% lower than in the top fraction, experimentally, the difference was found to be only 11.5 ± 9.6%. However, this discrepancy may be explained if the measured hexokinase activity was not representative of the activity exhibited by the *bulk* of the cells in each fraction. Specifically, it was shown that the density-fractionation of erythrocytes yielded fractions which were still very heterogeneous with respect to the hexokinase activity of the individual cells in each fraction, with a substantial proportion of the hexokinase activity in the least dense fractions being located in the reticulocytes. Simulations based on the hexokinase activity of non-reticulocyte cells (reticulocytes were assumed to possess a hexokinase activity seven times that of mature cells; Rijksen et al., 1977), showed good agreement with the experimental data from cell suspensions. Thus, no decline in enzyme activities with cell age was detected: rather, loss of activity was correlated with a decreasing proportion of reticulocytes.

C. Classical versus L-type Scheme Controversy

Of the various alternative schemes which have been proposed for the non-oxidative pentose phosphate pathway (e.g., Sable, 1966; Wood, 1974; Severin and Stepanova, 1981; Susskind et al., 1982), the 'L-type' scheme (so named because it was first proposed to operate in liver) has evoked the most discussion. This formulation (Figure 3) ascribes important roles to four additional intermediates: arabinose 5-phosphate (Ara5P), *ido*-octulose 8-phosphate (*i*Oct8P), *ido*-octulose 1,8-bisphosphate (*i*Oct18P2) and sedoheptulose 1,7-bisphosphate (Sed17P2); and invokes the additional action of the following three enzymes: D-arabinose-5-phosphate isomerase, D-*glycero*-D-*ido*-octulose 1,8-bisphosphate:D-*altro*-heptulose-7-phosphotransferase, and aldolase. Transaldolase, essential in the classical scheme, has no role in the L-type scheme, which has been proposed to operate in a very wide range of tissues (Williams et al., 1987).

In order to determine the scheme for the non-oxidative pathway operating in human erythrocytes, ^{31}P NMR was used to follow the concentrations of pathway intermediates arising from the metabolism of various putative pentose phosphate pathway intermediates in hemolysates. These experimental data were visually

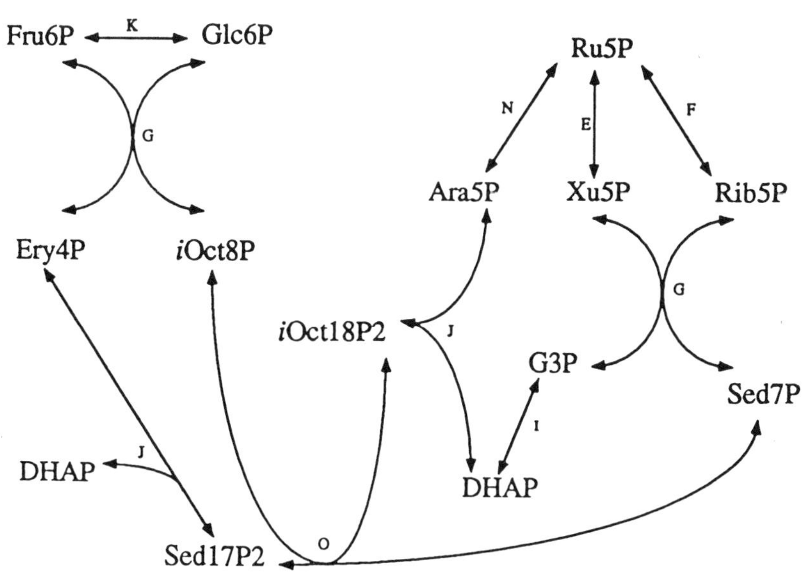

Figure 3. The reaction sequence of the L-type non-oxidative pentose phosphate pathway. The reactions are catalyzed by: (**N**) arabinose-5-phosphate isomerase (EC 5.3.1.13); (**O**) D-*glycero* D-*ido* octulose 1,8-bisphosphate:D-*altro*-heptulose 7-phosphotransferase (EC number not assigned); assignments for the remaining enzymes are as given in Figure 2. (See also Abbreviations.)

compared with simulations obtained from mathematical models comprising the reactions of each of the alternative schemes (McIntyre et al., 1989). The computer-based models were constructed from the *in vitro* kinetic data and mechanisms reported for each enzyme in a manner similar to that described for the construction of the model of the oxidative pathway (see previous section). Where the appropriate data were not available, or consistency between various steady-state kinetic data was not achievable, rational estimates of these parameters were made.

An important simplification of the experimental system in this study, and subsequent studies of the non-oxidative pentose phosphate pathway in our laboratory, was the metabolic 'isolation' of the non-oxidative pentose phosphate pathway, achieved by the preincubation of hemolysates at 37 °C for 40–45 min prior to use in metabolic experiments (Kuchel and Chapman, 1985). This results in the depletion of endogenous NAD(P)(H) (see section below), thus preventing 'recycling' through the oxidative pathway or 'removal' of sugar phosphates by entry into the glycolytic pathway.

Using a range of substrates to 'probe' the reactivity of the non-oxidative pentose phosphate pathway, a much more satisfactory agreement between experimental data and simulations using the classical reaction scheme than the L-type scheme, was obtained (McIntyre et al., 1989). The observed accumulation of Glc6P, following the addition of hemolysate to a buffered solution of ribose 5-phosphate (Rib5P), was predicted by the classical model but not by the L-type model. Additionally, the very high sedoheptulose 7-phosphate (Sed7P) and arabinose 5-phosphate (Ara5P) levels predicted by the L-type model were not observed experimentally. However, the identification of sedoheptulose 1,7-bisphosphate (Sed17P2) in the incubation mixture, arising from the addition of hemolysate to Rib5P, demonstrated the inadequacy of the classical scheme to provide a complete description of all the metabolic events occurring in the non-oxidative pentose phosphate pathway of the human erythrocyte.

No accumulation of *ido*-octulose 1,8-bisphosphate (*i*Oct18P2) was observed following the incubation of Ara5P (20 mM) and dihydroxyacetone phosphate (DHAP; 20 mM) with hemolysate, implying that the activity of aldolase for *i*Oct18P2 synthesis in a hemolysate is very low.

Transaldolase, proposed to play an essential role in the classical formulation of the pathway, may be inactivated by maintaining hemolysates at 55 °C for 5 min (Takeuchi et al., 1984). Thus, heat treatment of hemolysates offers a method of determining the importance of transaldolase to flux through the non-oxidative pathway (McIntyre et al., 1989). Following the addition of heat-treated hemolysates to Rib5P, the concentration of Sed7P rose to a very high level; the accumulation of Glc6P was observed to be very slight with ^{31}P NMR (McIntyre et al., 1989): in a subsequent experiment carried out under identical conditions (Berthon et al., 1992), no Glc6P accumulation was observed by ^{31}P NMR. A control experiment, designed to check the selectivity of the heat treatment for transaldolase inhibition entailed the addition of exogenous transaldolase to a heat treated hemolysate. The ^{31}P NMR time course of spectra acquired after the addition of this lysate to Rib5P, closely resembled that observed with a 'normal' hemolysate (McIntyre et al., 1989). Interestingly, inclusion of transaldolase-catalyzed reactions in the computer model of the L-type scheme, produced simulated metabolite concentrations which were in much closer agreement with those obtained experimentally, than simulations which excluded this reaction. However, the predicted high level of *ido*-octulose 8-phosphate (*i*Oct8P) was not observed experimentally (McIntyre et al., 1989).

In conclusion, ^{31}P NMR time course experiments with human hemolysates have provided cogent evidence for the involvement of transaldolase in the reaction scheme of the non-oxidative pentose phosphate pathway in human erythrocytes.

Conflicting results concerning the utilization of Ara5P via the non-oxidative pentose phosphate pathway have been reported. Radioactivity from Ara5P has been observed to be incorporated into other pathway intermediates in a variety of tissues (Blackmore et al., 1982; Bleakley et al., 1984; Butler et al., 1990). However, in a variety of cytosols from different tissues, the rate of Ara5P utilization never exceeded that of Rib5P (Williams et al., 1978b). Using chromatographic techniques to detect metabolites, Wood and Gascon (1980) found Ara5P to be virtually inert in rat liver extracts and in rat muscle. Our experiments showed that in human erythrocytes, the rate of Ara5P utilization, detected using ^{31}P NMR, was very slow and very much less than the normal carbon flux through the pathway (McIntyre et al., 1989). Although the intensity of the resonance from this compound diminished with time, there was a concomitant rise in the intensity of the P_i peak, suggesting that Ara5P was degraded by phosphohydrolase activity.

Williams (1980) has reported that Ara5P utilization is increased if Rib5P, erythrose 4-phosphate (Ery4P) or glyceraldehyde 3-phosphate (G3P) are also included in the incubation mixture. However, in a time course in which a dilute hemolysate was added to Rib5P and Ara5P (in a 4:1 molar ratio), the rate of formation of Glc6P was negligible (Figure 4A). These data are consistent with the previously reported highly inhibitory effect of Ara5P on transaldolase (Williams et al., 1978b), together with the essential role of transaldolase for the operation of the non-oxidative pentose phosphate pathway (McIntyre et al., 1989). The inhibition of transaldolase by Ara5P has been proposed to play a role in directing flux through the L-type scheme of the pathway in the initial stages of pentose phosphate metabolism in a rat liver enzyme preparation (Williams et al., 1978b). While the inhibition of yeast transketolase by Ara5P has been reported (Wood and Gascon, 1980), Williams (1980) finds no evidence to suggest inhibition of the liver enzyme, and ascribes the low observed pathway flux in the presence of Ara5P (Williams et al., 1978b) to an unspecified inhibition. The results obtained with human erythrocytes in our laboratory have shown no evidence to support the inclusion of Ara5P in the reaction scheme of the non-oxidative pentose phosphate pathway.

Williams and his co-workers have suggested that the importance of the L-type pathway in metabolism may be contingent upon a high aldolase activity in the tissue under examination (Williams et al., 1978b). The demonstration, by an immuno-

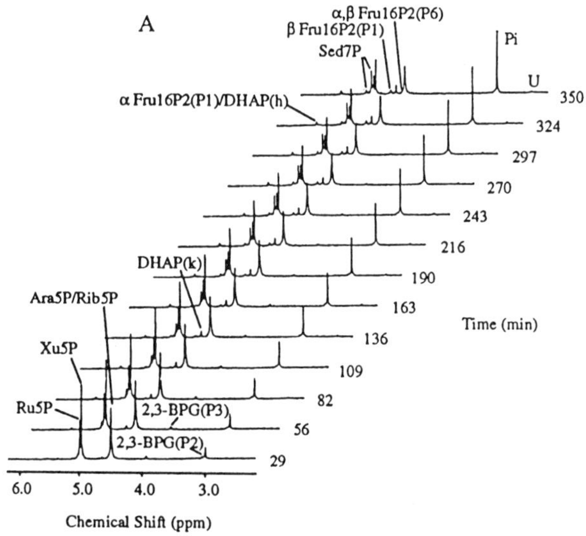

Figure 4. Time course of ^{31}P NMR spectra acquired following the addition of haemolysates to Rib5P and Ara5P (**A**), and Rib5P only (**B**). Sample composition: (**A**), Rib5P (17.4 mM), Ara5P (4.3 mM), TeP and MeP, hemolysate (200 μl, hematocrit, 0.92; final hematocrit equiv of sample, 0.10) in triethanolamine buffer (triethanolamine, 50 mM; KCl, 90 mM; NaCl, 10 mM; $MgCl_2$, 2 mM; (D,L)-dithiothreitol, 1 mM; EDTA, 0.5 mM; ^2H_2O, 10% v/v); final sample pH, 7.30; (**B**), Rib5P (20 mM), TeP and MeP, hemolysate (200 μl, hematocrit, 0.92; final hematocrit equiv of sample, 0.10) in triethanolamine buffer (same composition as given above); final sample pH, 7.30. (continued)

chemical procedure, of the failure of pathway flux to occur in a rat liver enzyme preparation in the absence of aldolase (Bleakley et al., 1984), has also supported this scheme. Interestingly, while muscle aldolase was ineffective in restoring the normal rate of pathway flux, the readdition of pure liver aldolase, restored pathway activity (Bleakley et al., 1984). Thus, the absence of NMR-detectable levels of octuloses may be a consequence of the presence of Class A aldolase in human erythrocytes (Penhoet et al., 1969; Yeltman and Harris, 1977). In contrast, liver aldolase demonstrates the kinetic, electrophoretic and immunological properties of the Class B type of the mammalian enzyme. A 13-fold difference in the K_m values of *i*Oct18P2 between each kind of aldolase has been reported (Paoletti et al., 1979).

Much of the evidence used in the debate surrounding the nature of the formulation of the non-oxidative pentose phosphate pathway has arisen from experiments conducted with radioisotopes and the observation of labeling patterns present in product molecules (e.g., Williams et al., 1978a,b; Scofield et al., 1985; Williams et al., 1987). Recently, we found Glc6P produced by the metabolism of (1-^{13}C)Rib5P in hemolysates, to be isotopically labeled at carbon atoms 1 and 3, with the degree

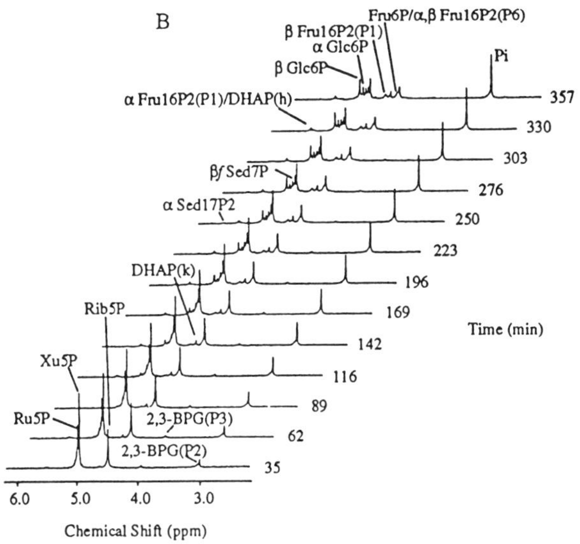

Figure 4. (continued) Acquisition parameters: (**A**) and (**B**), spectrometer frequency, 242.94 MHz; acquisition time, 1.0 s; pulse angle, $\pi/2$; recycle time, 20 s; number of transients per spectrum, 80. Proton decoupling (WALTZ-16) was applied continuously; the sample temperature was maintained at 37°C. The times given on the right hand side are the calculated midpoint times of each spectral accumulation. 'U', unassigned peak. (See also Abbreviations.)

of labeling at the former position being approximately double that in the latter position. The distribution of isotopic label was followed with ^{13}C NMR spectroscopy. These results are consistent with the classical mechanism, rather than the L-type mechanism of the non-oxidative pentose phosphate pathway. Operation of the pathway according to the classical scheme, assuming irreversibility of all reactions and no influence from side-reactions, would be predicted to result in the formation of product hexose phosphate with the ratio of labeling in carbon atoms 1 and 3 (C1:C3) being exactly 2:1 (Figure 2). Although deviation of the "identity badge" (Williams et al., 1986; Williams et al., 1987) C1:C3 ratio from this value, or the theoretical range if recycling is taken into account (Katz and Wood, 1960), has been used as evidence that the pentose phosphate cycle does not operate according to the classical scheme (e.g., Williams, 1980; Williams et al., 1987), this assertion has been repudiated by others (e.g., Landau and Wood, 1983; Scofield et al., 1985) who have asserted that the bidirectional nature of the reactions catalyzed by the pathway enzymes would, in particular, result in deviations from this value.

The presence of the L-type pathway intermediate, octulose monophosphate, as well as octulose and sedoheptulose bisphosphates has been detected in human erythrocytes, following incubation with inosine (Bartlett and Bucolo, 1960; Bucolo and Bartlett, 1960; Bartlett and Bucolo, 1968). The presence of Sed17P2 and

Oct18P2 in normal fresh erythrocytes (Vanderheiden, 1965) has also been reported, providing circumstantial evidence for the operation of this pathway of metabolism in human erythrocytes. However, our data, obtained with human hemolysates, has supported the classical scheme. Given the very specific physiological role of the erythrocyte, there is little justification for the extrapolation of results obtained with red cells to other tissues. The enzymes, transaldolase, transketolase and aldolase, exhibit a broad substrate specificity (Williams et al., 1987), potentially permitting a wide range of reactions to occur. Thus, any attempt to summarize the reactions of the non-oxidative pentose phosphate pathway in a computer model must, necessarily, represent an approximation to the intricate array of metabolic processes occurring *in vivo.* Nevertheless, the semiquantitative comparison between simulations and observations offers a very valuable "global" method for studying the interactive nature of the the reactions of the non-oxidative pathway which often give rise to complex, 'counterintuitive' behavior (Garfinkel, 1980). Although the ^{31}P NMR spectrum of the sugar phosphate metabolites of the pentose phosphate pathway is fairly crowded, many of the resonances are at least partially resolved and are able to be assigned (McIntyre et al., 1989). Offsetting the difficulties associated with the limited sensitivity of the ^{31}P NMR technique, is its capacity for monitoring a range of metabolites simultaneously, rendering it very suitable for the multivariant characterization of this metabolic pathway.

D. Control of the Non-Oxidative Pentose Phosphate Pathway

The degree of control exerted by the enzymes, transketolase and transaldolase, over flux through the non-oxidative pentose phosphate pathway in human erythrocytes, has been investigated using ^{31}P NMR data, in conjunction with computer simulations. ^{31}P NMR time course experiments were carried out with hemolysates having various activities of transketolase and transaldolase. The transketolase and transaldolase activities of hemolysates were altered using an antitransketolase serum and a heating-inactivation procedure, respectively (see above; Berthon et al., 1992). Although studies of the nature of the control of the non-oxidative pentose phosphate pathway in other tissues have led to conflicting postulates concerning the main rate-limiting enzyme of this pathway, transketolase has most often been ascribed a rate-limiting role (Wood, 1985). D-*glycero*-D-*ido*-octulose 1,8-bisphosphate:D-*altro*-heptulose-7-phosphotransferase, an enzyme of the L-type scheme of the non-oxidative pathway, has been proposed to fulfill a rate-limiting role in tissues where this pathway operates (Williams et al., 1978b; Williams, 1980; Arora et al., 1985; Butler et al., 1990).

Transketolase catalyses the reversible transfer of an aldol moiety to a suitable aldehydic acceptor and displays a broad substrate specificity (Williams et al., 1987). It has been proposed that it plays a role in both the classical and the L-type pathways with two primary transketolase reactions in the classical pathway, and two in the L-type pathway, and one proposed reaction in common (c.f., Figures 2 and 3).

Transaldolase catalyzes the transfer of a three-carbon ketol unit to an aldehydic acceptor.

Flux through the non-oxidative pentose phosphate pathway may be monitored by following, with ^{31}P NMR, the flux of pathway metabolites after the addition of a dilute hemolysate to Rib5P (Figure 4B). Rib5P rapidly equilibrates with the other pentose phosphates, Ru5P and Xu5P. Sed7P emerges with time, as well as DHAP which in aqueous solutions exists in both hydrate (*gem*-diol) and ketone forms (Kuchel et al., 1990). Glc6P gives rise to two resonances in the ^{31}P NMR spectrum, arising from the α- and β-anomeric forms (Figure 4B). The linear rate of the rise in the concentration of Glc6P, used as an index of non-oxidative pentose phosphate pathway flux, shows considerable dependence upon the transketolase activity of the hemolysate added to the sample (Berthon et al., 1992). In contrast, the transaldolase activity of the hemolysate is shown to have only a low degree of control over pathway flux (low FCC) at normal enzyme concentrations.

^{31}P NMR data have been used to refine a computer model describing the changes in concentration of pathway metabolites with time. This mathematical model consists of equations describing the formation and breakdown of all enzyme-substrate complexes formed by the reactions of the non-oxidative pentose phosphate pathway (Berthon et al., 1992) with corresponding unitary rate constants for each of these reactions. Computer simulations were carried out using the SCoP computer package (National Biomedical Simulation Resource, Duke University Medical Center, Durham, NC, USA). The concentration profiles of the metabolites: Sed7P, Glc6P, and Ru5P plus Xu5P, determined with NMR, were compared with the levels of these metabolites in computer simulations in which all enzymes were set to normal concentrations. These metabolites were chosen because they may readily be quantified from the ^{31}P NMR spectra and are suitable indicators for the kinetic characterization of the non-oxidative pentose phosphate pathway. The rate constants within the computer model were adjusted to match selected metabolite concentration profiles derived from NMR timecourses obtained from hemolysates with 'normal' enzyme concentrations. Computer simulations with transketolase activity modulated to the same values as those in the NMR samples with enzyme-depleted hemolysates, were carried out. The plots of simulated metabolite flux versus transketolase activity resembled the nearly linear relationship obtained experimentally (Berthon et al., 1992). Control coefficients for transketolase and transaldolase were calculated by fitting a phenomenological polynomial of degree 4 to the flux versus enzyme activity curves obtained from the computer simulations. Estimates of the FCCs derived from the model were 0.64 and 0.06 for transketolase and transaldolase, respectively.

The experimental and computer methods of examining the non-oxidative pentose phosphate pathway represent two extreme approaches. The experimental data, derived by the perturbation of the whole system (the modulation of the enzyme concentrations), yield information on the role of its parts (the enzymes, transketolase and transaldolase). The experiment *in numero* is performed by carrying out simulations with a mathematical model constructed from *in vitro* experiments on isolated enzymes, with modifications to achieve a closer visual fit (or match) to experimental time courses. The experimental system used in this work differed from the situation which would occur *in vivo* in that it was not in a steady-state. Thus, the summation theorem of metabolic control theory (Kacser and Burns, 1973) cannot be strictly applied: however, the general conclusions regarding the higher control exerted by transketolase remain valid because of the approximation to steady-state conditions observed while the rate of Glc6P remained almost constant (i.e., between 18 and 32 min; time-adjusted for a 100% hematocrit equiv.).

While the set of rate constants in the computer model represent the best match to experimental data which is achievable using the unitary rate constant modulation method, the simulations showed Glc6P reaching a higher level than was observed experimentally. Evidence of the hydrolysis of intermediates of the pathway was apparent in the growth of the inorganic phosphate peak throughout the ^{31}P NMR time courses (Figure 4), which could only partially be accounted for by the degradation of 2,3-BPG. A lack of information on the specificity of the phosphohydrolases present in the red cell, prevents the inclusion of this reaction in the computer model at this stage.

In the future, the incorporation of the hydrolysis of sugar phosphates into the model may result in its closer description of the rate of formation of Glc6P. The specificity of the antitransketolase serum was very high but not absolute, and yet it appeared not to have any effect on the other enzymes in the system (Berthon et al., 1992). In other words, the accumulation of intermediates in either transketolase or transaldolase-depleted experimental mixtures in NMR experiments gave no evidence for the decreased activity of any of the enzymes of the tested pathway apart from the enzyme that was deliberately removed.

E. Exchange Reactions and the Reversibility of the Reactions of the Non-Oxidative Pentose Phosphate Pathway

We have recently observed, using ^{13}C and ^{31}P NMR spectroscopy, metabolic flux in the 'reverse' direction of the non-oxidative pentose phosphate pathway (i.e., from hexose phosphates to pentose phosphates). These time course experiments provided data for the further kinetic characterization of the pathway, which were incorporated into the mathematical model. The ability of NMR to yield information on the degree of correlation between adjacent ^{13}C nuclei, through spin–spin coupling, enables discrimination between isotopomers of Glc6P to be made, thus providing a method of monitoring 'isotopic exchange' reactions. We have shown how this elegant method may be used to detect the reactions which result in the catabolism and reformation of Glc6P via the reactions of the non-oxidative pentose phosphate pathway.

In order to study metabolic flux through the non-oxidative pentose phosphate pathway in the 'reverse' direction, time courses of the utilization of (1-^{13}C)Glc6P,

Glc6P and Fru16P2 in hemolysates were carried out (Figure 5). The initial composition of the sample in time course experiments was chosen to provide optimal conditions for the reversal of the pentose phosphate pathway. The products of the pathway are hexose phosphates along with G3P (Figure 2). As Fru16P2 serves as a "reservoir" of G3P, experimental samples which contained both Glc6P and Fru16P2 were prepared. This approach has also been used by Williams and Blackmore (1983); theoretically, however, the formation of pentose phosphate may occur in the absence of additional metabolites such as Fru16P2 (Bonsignore et al., 1962).

The equilibria of the reactions comprising the pathway lie heavily in favor of the formation of hexose phosphates, resulting in relatively slight changes in the composition of the sample. (1-^{13}C)Glc6P was rapidly isomerized to (1-^{13}C)Fru6P; pentose phosphates (Ru5P, Xu5P and Rib5P) labeled at C1 were also rapidly formed and remained present in the sample at a low concentration. The randomization of isotopic label due to the dynamic nature of the equilibria of the pathway was evident in the slow accumulation of isotopic labelling of Glc6P at C3. (1,3-^{13}C)Glc6P may arise from the formation of (1,3-^{13}C)Sed7P by the transketolase-mediated reaction of (1-^{13}C)Xu5P with (1-^{13}C)Rib5P, followed by the transaldolase-catalyzed transfer of the "top" three-carbon unit of Sed7P onto G3P to give (1,3-^{13}C)Fru6P, viz.,

$$(1\text{-}^{13}\text{C})\text{Xu5P} + \text{TK} \rightleftharpoons \text{TK-}(1\text{-}^{13}\text{C})\text{Xu5P} \tag{2}$$

$$\text{TK-}(1\text{-}^{13}\text{C})\text{Xu5P} \rightleftharpoons \text{TK-}(1\text{-}^{13}\text{C})\text{Gla} + \text{G3P} \tag{3}$$

$$\text{TK-}(1\text{-}^{13}\text{C})\text{Gla} + (1\text{-}^{13}\text{C})\text{Rib5P} \rightleftharpoons \text{TK-}(1\text{-}^{13}\text{C})\text{Gla-}(1\text{-}^{13}\text{C})\text{Rib5P} \tag{4}$$

$$\text{TK-}(1\text{-}^{13}\text{C})\text{Gla-}(1\text{-}^{13}\text{C})\text{Rib5P} \rightleftharpoons \text{TK} + (1,3\text{-}^{13}\text{C})\text{Sed7P} \tag{5}$$

$$(1,3\text{-}^{13}\text{C})\text{Sed7P} + \text{TA} \rightleftharpoons \text{TA-}(1,3\text{-}^{13}\text{C})\text{Sed7P} \tag{6}$$

$$\text{TA-}(1,3\text{-}^{13}\text{C})\text{Sed7P} \rightleftharpoons \text{TA-}(1,3\text{-}^{13}\text{C})\text{Grn} + \text{Ery4P} \tag{7}$$

$$\text{TA-}(1,3\text{-}^{13}\text{C})\text{Grn} + \text{G3P} \rightleftharpoons \text{TA-}(1,3\text{-}^{13}\text{C})\text{Grn-G3P} \tag{8}$$

$$\text{TA-}(1,3\text{-}^{13}\text{C})\text{Grn-G3P} \rightleftharpoons \text{TA} + (1,3\text{-}^{13}\text{C})\text{Fru6P} \tag{9}$$

$$(1,3\text{-}^{13}\text{C})\text{Fru6P} \rightleftharpoons (1,3\text{-}^{13}\text{C})\text{Glc6P} \tag{10}$$

(catalyzed by glucosephosphate isomerase)

The various names are defined in the Abbreviations, but for convenience note that TK and TA refer to transketolase and transaldolase respectively, and Gla and Grn to the glycolaldehyde and dihydroxyacetone moieties transferred by these enzymes. Williams and Blackmore (1983) similarly found label redistribution to be slight in their investigations with a rat liver enzyme extract.

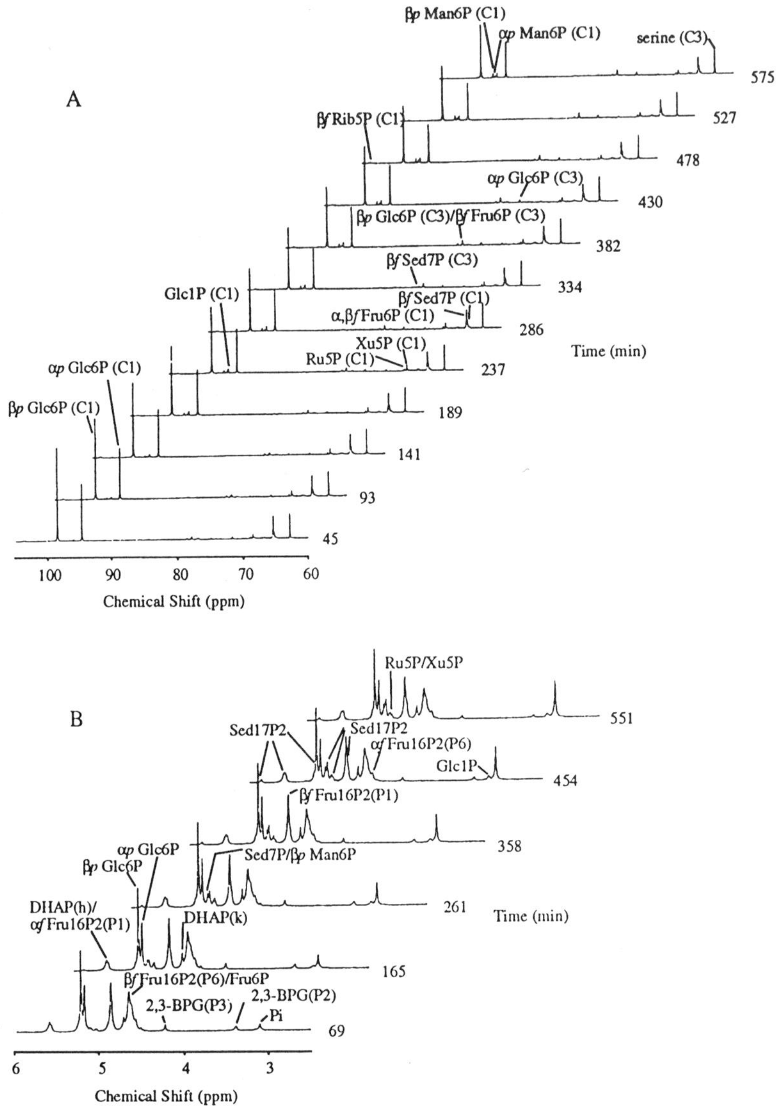

Figure 5. Time courses of ^{13}C (**A**) and ^{31}P (**B**) NMR spectra acquired after the addition of hemolysates to a sample containing (1-^{13}C)Glc6P, Glc6P and Fru16P2. Sample composition: (1-^{13}C)Glc6P (30 mM); Fru16P2 (30 mM); Glc6P (20 mM); (3-^{13}C)serine (2.7 mM); hemolysates (900 μl; final hematocrit equiv. 0.46); TeP and MeP; triethanolamine buffer (triethanolamine, 50 mM; KCl, 90 mM; NaCl, 10 mM; $MgCl_2$, 2 mM; D,L-dithiothreitol, 1 mM; EDTA, 0.5 mM; ^2H_2O, 10 percent v/v); pH, 7.28. Acquisition parameters: (**A**) spectrometer frequency, 150.919 MHz; number of transients per spectrum, 160; acquisition time, 0.6 s; pulse angle, π/2; recycle time, 10 s; spectral width, 31 240 Hz; spectra were processed into 64 k data points (continued)

The combined action of phosphoglucomutase and phosphomannose isomerase was evident in the formation of (1-^{13}C)Glc1P and (1-^{13}C)Man6P, respectively. The attainment of equilibrium between the glucose phosphates takes place rapidly and lies very much in favor of Glc6P (K_{eq} = 17.2; Bergmeyer, 1974); Man6P was formed slowly due to the low activity of the catalyzing enzyme in the hemolysate. Fru16P2 undergoes an aldolase-mediated scission to yield G3P and DHAP. None of the reactions in the scheme of Figure 2 can result in the labeling of any of these three molecules, which are not observed in the ^{13}C NMR spectra but may be followed in the ^{31}P NMR time course (Figure 5B). DHAP and Ery4P (also unlabeled by the reaction scheme of Figure 2) condense via aldolase to form Sed17P2, which was observed to accumulate slowly (Figure 5B).

Time courses of ^{13}C and ^{31}P NMR spectra acquired following the addition of a dilute hemolysate to (1-^{13}C)Glc6P and Sed17P2 showed similar features to the time courses obtained with (1-^{13}C)Glc6P and Fru16P2. It has been suggested that Sed17P2 may serve as a "sequestering agent" for Ery4P (Horecker and Mehler, 1955; Duke et al., 1981) as relatively low concentrations of the tetrose phosphate are known to inhibit a range of enzymes of the pentose phosphate pathway, and by sequestering it, this effect is obviated. The ^{31}P NMR time course of spectra revealed the reduction, with time, of the level of Sed17P2 and a concomitant increase in the concentration of Fru16P2. Interestingly, the equilibrium positions and rates of the reactions of the pentose phosphate pathway are "poised" so as to maintain the total bisphosphate concentration at a relatively constant value.

The computer model describing the non-oxidative pentose phosphate pathway was heuristically refined using time course data derived from the NMR experiments described above as well as additional data (H. A. Berthon and P. W. Kuchel, unpublished results). This model was elaborated to include all the isotopomers resulting from the metabolism of (1- ^{13}C)Glc6P by the reactions of the non-oxidative pentose phosphate pathway. The metabolite concentration profiles obtained *in numero* exhibited a satisfactory agreement with the experimental data. The computer-based model was further elaborated to include all isotopic isomers resulting from the metabolism of universally-labeled Glc6P (e.g., in principle, there are 2^7 = 128 possible ways ^{13}C label may be distributed in Sed7P, but only 32 ways in practice, according to the reaction scheme of Figure 2), and an experiment in which

---

Figure 5. (continued) and an exponential line-broadening factor of 3 Hz was used; (**B**) spectrometer frequency, 242.943 MHz; number of transients per spectrum, 64; acquisition time, 0.8 s; pulse angle, $\pi/2$; recycle time, 20 s; spectral width, 7 353 Hz; spectra were processed into 16 k data points; broadband proton decoupling (WALTZ-16) was applied continuously for both ^{13}C and ^{31}P NMR time courses; sample temperature, 37°C. The times given represent the calculated midpoint time of each spectral accumulation. (See also Abbreviations.)

(U-^{13}C)Glc6P was incubated with unlabeled Glc6P and Fru16P2 was carried out. The aim of this experiment was to detect the isotopic exchange reactions, reflected in carbon-carbon bond scissions of product hexose phosphate molecules. As these effects are fully described by the reversible equations which comprise the computer model, an additional aim of this work was to apply a rigorous estimation of the degree of their effect upon labeling patterns using computer simulation methods.

Isotopic exchange effects have been noted to occur in reactions catalyzed by transaldolase (Ljungdahl et al., 1961), transketolase (Clark et al., 1971), and aldolase (Bartlett et al., 1989). Along with reversal of flux through the non-oxidative pathway, followed by reformation of hexose 6-phosphate, the operation of exchange reactions has been used to explain labeling patterns which are not in accordance with those predicted, assuming irreversibility of the reactions of the non-oxidative pentose phosphate pathway by both the classical (e.g., Ljungdahl et al., 1961; Landau and Bartsch, 1966; Katz et al., 1975; Landau and Wood, 1983; Scofield et al., 1985) and the L-type mechanism (e.g., Williams et al., 1971; Williams et al., 1978b; Longenecker and Williams, 1980a,b; Williams, 1980; Williams and Blackmore, 1983; Williams et al., 1986; Butler et al., 1990).

The isotopic exchange reactions arise as a consequence of the reversibility of the group-translocation enzymic reactions. For example, transaldolase catalyses the transfer of a three-carbon unit (dihydroxyacetone; Grn) from Fru6P to G3P, resulting in the formation of new Fru6P and G3P molecules which may have different labeling patterns, *viz.,*

$$\text{Fru6P} + \text{TA} \rightleftharpoons \text{TA-Grn-G3P} \qquad (11)$$

$$\text{TA-Grn-G3P} \rightleftharpoons \text{TA-Grn} + \text{G3P} \qquad (12)$$

The overall reaction is:

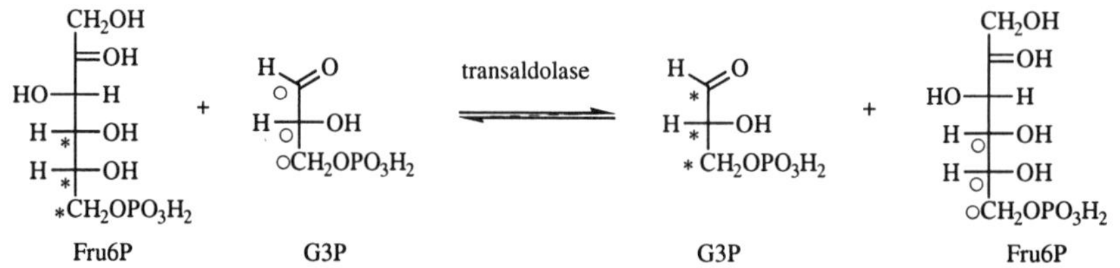

While in many respects metabolic NMR experiments with ^{13}C-labeled substrates resemble the studies which may be performed with ^{14}C-labeled compounds (Cohen et al., 1981), ^{13}C NMR has the additional capability of yielding information on the correlated nature of labeling via ^{13}C-^{13}C resonance splitting. Observation of the labeling of metabolites can yield information about the incorporation of complex structural units: the technique of diluting a multiply-labeled substrate with substrate at natural abundance introduces the means of determining which intact *groups* of carbon atoms are transferred into product molecules. One of the first quantitative studies using ^{13}C NMR for isotopomer analysis was published by London et al. (1975b) who investigated the glycolytic formation of lactate by *Lactobacillus delbrueckii,* and the photosynthetic incorporation of CO_2 into galactosyl glycerol by red algae. Other studies utilizing ^{13}C–^{13}C coupling for biochemical studies include those of Cohen (1983), Sherry et al. (1985), Kalderon et al. (1986), Malloy et al. (1987, 1990), London (1988; and references therein) and Jeffrey et al. (1991). The diagnosis and classification of some enzymic defects in humans by ^{13}C-isotopomer analysis has been described by Lapidot and co-workers (Kalderon et al., 1989; Gopher et al., 1990). The metabolic features of the diseases studied were indicated by the coupling patterns in ^{13}C NMR spectra of plasma samples, reflecting the presence of different metabolite isotopomers.

We employed the novel method of analyzing cross-peaks within a ^{13}C COSY spectrum to detect the distribution of isotopic forms of Glc6P present in a reaction mixture following the incubation of (U-^{13}C)Glc6P, Glc6P and Fru16P2 with hemolysates. This approach was pioneered by Jones and Sanders (1989) for biosynthetic investigations of the *Klebsiella* K3 serotype polysaccharide. In the experimental system, a mixture of universally ^{13}C-substituted and unsubstituted substrates was converted by the enzyme system into a range of isotopomers, which were identified from the coupling patterns exhibited in the ^{13}C COSY spectrum. Thus, we have a method for measuring the extent of carbon–carbon bond scissions occurring during the incubation time.

For a cross-peak derived from the coupling of two ^{13}C nuclei denoted by A and M, the nuclear coupling constant, J_{AM}, is termed the active coupling and results in the appearance of antiphase coupled resonances in the cross-peak of the COSY spectrum. If M is further coupled to a nucleus X, J_{MX}, the passive coupling is evident in the AM cross-peak (observed at the chemical shift of M in the frequency domain f_2) and is characterized by in-phase coupling.

The shorthand notation of Jones and Sanders (1989) has been adopted in the ensuing discussion of this experiment: e.g., the cross-peak observed between carbon atoms 1 and 2 of Glc6P, at the chemical shift of carbon 1 in f_2 is denoted by C^1_{12}, with the corresponding cross peak at the chemical shift of carbon 2 being denoted by C^2_{12}. The cross peaks are a combination of superimposed responses from the coupling patterns arising from the various isotopomers present in the sample.

^{13}C COSY spectroscopy was particularly well suited to the study of the 'reverse' operation of the non-oxidative pentose phosphate pathway because the equilibria of the enzymic reactions favor Glc6P formation, so the final mixture for analysis was not complicated by large amounts of other labeled metabolites. In one experiment, the reactions occurring on incubation of (U-^{13}C)Glc6P, Glc6P and Fru16P2 with hemolysates were stopped, after 10 h, by deproteinizing the sample by ultrafiltration. The βGlc6P C^1_{12} cross peak from the COSY spectrum (Figure 6) of the ultrafiltrate consisted of a simple coupling pattern resulting from active coupling between carbons 1 and 2 (C1 and C2). As might be expected, there was no evidence of additional coupling of C1 to any other carbon nuclei. The coupling

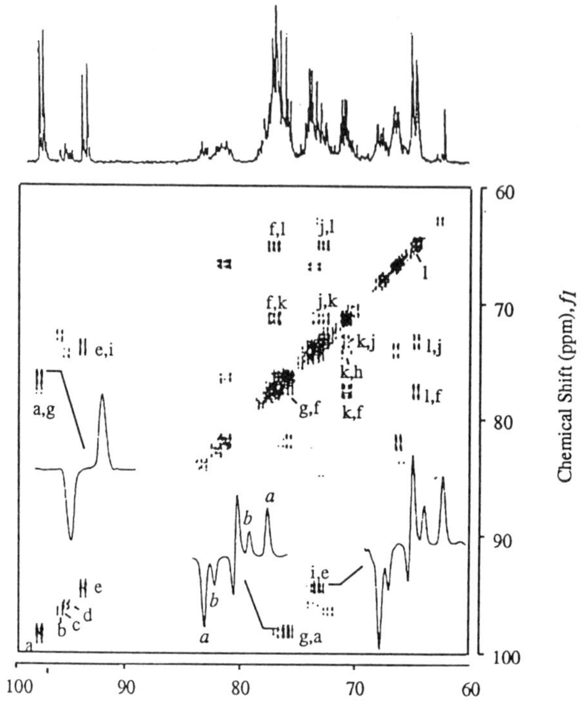

Figure 6. ^{13}C-^{13}C DQF-COSY of ultrafiltrate obtained following incubation of (U-^{13}C)Glc6P (28.5 mM), Glc6P (19 mM) and Fru16P2 (28.5 mM) with hemolysates (hematocrit equivalent, 0.46) for 10 h. Acquisition parameters: spectrometer frequency, 100.620 MHz; acquisition time, 0.205 s; pulse angle, ($\pi/2$) 15 μs; number of transients per increment, 160; 512 increments in f_1; 4 k data points in f_2; processing was with a phase-shifted sine function (shifted by $\pi/2$) applied in each dimension; f_1 was zero-filled to 1 k data points; broadband ^1H-decoupling (WALTZ-16) was applied continuously; set temperature, 298 K. Assignments: **a**, β Glc6P (C1); **b**, β Man6P (C1); **c**, α Man6P (C1); **d**, Glc1P (C1); **e**, α Glc6P (C1); **f**, β Glc6P (C3, C5); **g**, β Glc6P (C2); **h**, α Glc6P (C3); **i**, α Glc6P (C2); **j**, α Glc6P (C5); **k**, α,β Glc6P (C4); **l**, α,β Glc6P (C6); the cross-peaks are labeled with the f_2 assignment followed by the f_1 assignment; the following assignments of cross-peaks are explicitly given as there is some ambiguity due to co-resonance of peaks: **f**,**k**, β Glc6P (C3, C5), β Glc6P (C4); **f**,**l**, β Glc6P (C5), β Glc6P (C6); **g**,**f**, β Glc6P (C2), β Glc6P (C3); **j**,**k**, α Glc6P (C5), α Glc6P (C4); **j**,**l**, α Glc6P (C5), α Glc6P (C6); **k**,**f**, β Glc6P (C4), β Glc6P (C3, C5); **k**,**h**, α Glc6P (C4), α Glc6P (C3); **k**,**j**, α Glc6P (C4), α Glc6P (C5); **l**,**f**, β Glc6P (C6), β Glc6P (C5); **l**,**j**, α Glc6P (C6), α Glc6P (C5). Cross peaks were obtained by back-transformation of a selected row, followed by retransformation of the resultant fid with a digitization of 32 k points; the '*a*' and '*b*' on the $\beta Glc6PC^2_{12}$ peak refer to the AMX (doublet of doublets) and AM (doublet) coupled spins, respectively.

pattern from the cross-peak of βGlc6P C^2_{12} was more complex: a proportion of C2 nuclei coupled to C1 nuclei were also coupled to C3 nuclei; this was reflected in the presence of both active coupling (from coupling between C1 and C2) and passive coupling (from C2 and C3). Because the magnitudes of these coupling constants are similar, some cancellation of signal was observed in the centre of the multiplet. The peaks in the middle of the positive and negative portions of the multiplet (denoted *b*) resulted from the C2 nuclei which were coupled to C1 nuclei only; in this case no resultant passive coupling was observed, and this signal was superimposed upon the pattern derived from fully coupled C123 groups. The multiplet pattern observed with the β-anomer was also apparent in the α-anomeric form. The remaining cross-peaks arising from Glc6P may be similarly analyzed. The intensities of the peaks within the cross-peak multiplets were evaluated by integration using software (Bruker), following back-transformation of a selected row of the COSY spectrum to produce an fid which was then transformed with a greater extent of digitization (32 k points).

Computer simulations of the experiment predicted that the transaldolase "exchange" reaction dominates the 'mixing' of isotopic label, with (1,2,3-^{13}C)Glc6P rapidly rising in concentration, at an initial rate of 122 mmole l^{-1} min^{-1} (calculated by the fitting of a phenomenological polynomial of degree 3 to the initial 4 points in the 50-step computer simulation; Berthon et al., 1993). This "exchange" is due to the reactions:

$$(1,2,3,4,5,6\text{-}^{13}C)Fru6P + TA \rightleftharpoons TA\text{-}(1,2,3\text{-}^{13}C)Grn\text{-}(1,2,3\text{-}^{13}C)G3P \quad (14)$$

$$TA\text{-}(1,2,3\text{-}^{13}C)Grn\text{-}(1,2,3\text{-}^{13}C)G3P \rightleftharpoons$$
$$TA\text{-}(1,2,3\text{-}^{13}C)Grn + (1,2,3\text{-}^{13}C)G3P$$

$$G3P + TA\text{-}(1,2,3\text{-}^{13}C)Grn \rightleftharpoons TA\text{-}(1,2,3\text{-}^{13}C)Grn\text{-}G3P \quad (15)$$

$$TA\text{-}(1,2,3\text{-}^{13}C)Grn\text{-}G3P \rightleftharpoons TA + (1,2,3\text{-}^{13}C)Fru6P \quad (16)$$

with the overall reaction being:

$$(1,2,3,4,5,6\text{-}^{13}C)Fru6P + G3P \rightleftharpoons (1,2,3\text{-}^{13}C)Fru6P + (1,2,3\text{-}^{13}C)G3P \quad (17)$$

The simulated concentration of (1,2-^{13}C)Glc6P increased at an initial rate of 8.9 μmol l^{-1} min^{-1}. The scission of C2 – C3 bonds reflects the activity of transketolase, *viz.*,

$$Fru6P + TK \rightleftharpoons TK\text{-}Gla\text{-}Ery4P \quad (18)$$

$$TK\text{-}Gla\text{-}Ery4P \rightleftharpoons TK\text{-}Gla + Ery4P \quad (19)$$

$$(1,2,3,4,5,6\text{-}^{13}C)Fru6P + TK \rightleftharpoons TK\text{-}(1,2\text{-}^{13}C)Gla\text{-}(1,2,3,4\text{-}^{13}C)Ery4P \quad (20)$$

$$\text{TK-(1,2-}^{13}\text{C)Gla-(1,2,3,4-}^{13}\text{C)Ery4P} \rightleftharpoons$$

$$\text{TK-(1,2-}^{13}\text{C)Gla + (1,2,3,4-}^{13}\text{C)Ery4P} \quad (21)$$

$$\text{Ery4P+ TK-(1,2-}^{13}\text{C)Gla} \rightleftharpoons \text{TK-(1,2-}^{13}\text{C)Gla-Ery4P} \quad (22)$$

$$\text{TK-(1,2-}^{13}\text{C)Gla-Ery4P} \rightleftharpoons \text{TK + (1,2-}^{13}\text{C)Fru6P} \quad (23)$$

with the overall reaction being:

$$\text{Fru6P + (1,2,3,4,5,6-}^{13}\text{C)Fru6P + TK} \rightleftharpoons$$

$$\text{TK-Gla + (1,2,3,4-}^{13}\text{C)Ery4P + (1,2-}^{13}\text{C)Fru6P} \quad (24)$$

The simulation predicted that (4,5,6-^{13}C)Glc6P would be formed at an initial rate of 8.2 μmol l^{-1} min^{-1}; also the simulation predicted that (3,4,5,6-^{13}C)Fru6P, (3-^{13}C)Glc6P and (1,2,4,5,6-^{13}C)Glc6P would be produced at 8.7 μmol l^{-1} min^{-1}, 0.27 μmol l^{-1} min^{-1}, and 0.007 μmol l^{-1} min^{-1}, respectively. All other isotopomers formed were predicted to be of minor concentration. The semiquantitative data obtained from the experimentally obtained isotopomer ratios (Berthon et al., 1993) were compared with the ratios predicted using the mathematical model of the pentose phosphate pathway and found to be very similar.

^{13}C - ^{13}C COSY experiments provide a unique method for investigating isotopic exchange, a phenomenon which has, to a large extent, defied quantitative evaluation. An extension of the experiment described would be the acquisition of a series of COSY spectra on a single sample, which would enable the changes in the isotopomer composition of the sample to be monitored with time. The extent and significance of the randomization of isotopic label *in vivo* has been a contentious issue in the interpretation of data obtained from experiments employing radioactively labeled substrates, to detect flux through the pentose phosphate pathway (e.g., Katz, 1981; Blackmore et al., 1982; Landau and Wood, 1983; Williams et al., 1986). While the existence of isotopic exchange reactions, and flux reversal, have been invoked to explain experimental labeling patterns, mathematical models incorporating rate constants for the formation and breaking apart of each substrate–enzyme complex, remain the only rigorous way of analyzing the influence of these reactions. Thus, the combined ^{13}C NMR and computer simulation strategy provides a unique means for investigating the kinetic character of metabolism within the human erythrocyte, with potential for application to other cell types.

F. Multinuclear NMR Studies of the Pentose Phosphate Pathway

In our investigations of the pentose phosphate pathway in human erythrocytes, the three nuclei, ^{1}H, ^{13}C and ^{31}P, have served as complementary ‘probes’ (Styles et al., 1979) for monitoring the metabolism of ^{13}C-substituted metabolites. ^{31}P NMR may be used to yield information on net metabolite flux, while ^{13}C NMR spectra may be useful for this, as well as for elucidating the pathway of reactions by which metabolism proceeds by following isotope redistribution patterns. Due to the large number of ^{1}H nuclei with different magnetic environments in mixtures of sugar phosphates, the ^{1}H NMR spectra of these metabolite mixtures are crowded with resonances which are poorly resolved. The ^{1}H[^{13}C]spin–echo difference technique (Freeman et al., 1981), however, enables the spectrum to be simplified so that only protons directly bonded to ^{13}C nuclei are detected.

The simplest version of the ^{1}H[^{13}C]spin–echo difference experiment uses the following multipulse sequence (Freeman et al., 1981), in which the ^{1}H channel carries the spin-echo pulse sequence with signal acquisition, while the ^{13}C channel carries a single composite π pulse:

$$^1\text{H}: \pi/2_x\text{—}\tau\text{—}(\pi_x)\text{—}\tau\text{—Acquisition}$$

$$^{13}\text{C}: \qquad (\pi/2_{+x}\pi/2_{\pm x})$$

where the delay, τ, is chosen to be equal to $1/(2J_{CH})$ where J_{CH} is the ^{1}H - ^{13}C coupling constant in Hz. The two $\pi/2$ pulses applied to ^{13}C act together to give a π and 0 radian pulse in successive experiments. During the first time interval, τ, the proton magnetization vectors of the doublet build up a relative phase difference of π radians; the components can be viewed as being directed along the $+x$-axis and $-x$-axis in the rotating frame of reference. The π ^{13}C pulse causes the interchange of the spin-state labels on the ^{1}H vectors causing a reversal of their direction of precession so that they converge back onto the $+y$ axis during the second delay period, τ. Thus, at time 2τ, an echo is formed along the $+y$ axis, while protons uncoupled to ^{13}C nuclei are refocused along the $-y$ axis. Repetition of the entire pulse sequence without the ^{13}C π pulse results in all vectors being aligned along the $-y$ axis at time, 2τ. Subtraction of alternate transients yields a spectrum in which the signal from protons bonded to ^{12}C atoms is suppressed and that from protons coupled with ^{13}C atoms is reinforced. Similar ^{1}H[^{13}C] pulse sequences with ^{13}C broadband decoupling instead of a ^{13}C π pulse have also been employed (e.g., Rothman et al., 1985).

Also, a variant of the heteronuclear spin–echo method was used by Brindle et al. (1982a) in studies of the ^{13}C-label exchange between the methyl groups of alanine and pyruvate, catalyzed by alanine aminotransferase, both *in situ* in an erythrocyte suspension, and *in vitro*. Subsequently, the exchange was measured using a ^{1}H/^{15}N experiment in which a mixture of [2-^{15}N]- and [3-^{13}C] L-alanine (Brindle et al., 1984). ^{1}H detection of the fid is carried out with or without selective inversion of the ^{15}N spins after the spin–echo delay time τ is set equal to the $1/(2\,{}^3J_{^{15}N^1H})$, where the coupling constant is ~3 Hz. Thus as equilibrium exchange takes place, the ^{14}N-coupled peak components are inverted with respect to their ^{15}N-coupled counterparts and therefore the observed resonances decay to zero as the exchange proceeds. The ${}^2J_{^{13}C^1H}$ is 130 Hz, thus it is possible to add or subtract the two different spectra and then selectively display the signals from the methyl protons of [2-^{15}N,

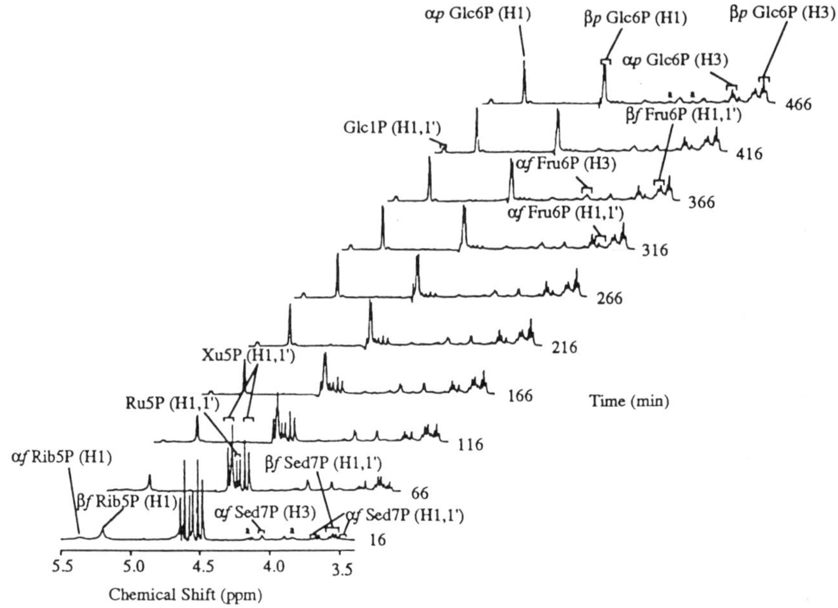

Figure 7. Time course of $^{1}H[^{13}C]$spin-echo difference spectra acquired following the addition of hemolysate to (1-^{13}C)Rib5P. NMR acquisition parameters: spectrometer frequency, 600.14 MHz; spectral width, 6 024 Hz; acquisition time, 0.6 s; H_2O presaturation time, 1.5 s; recycle time, 7.1 s; ^{13}C-decoupling (GARP) was applied during the acquisition; number of transients per spectrum, 192; temperature, 37 °C; 'a' denotes signal arising from protons not specifically coupled with ^{13}C nuclei but which were present at such high abundance that some residual signal was observed in the spin-echo difference spectrum. Sample composition: (1-^{13}C)Rib5P (15 mM), hemolysate (100 μl; final hematocrit equiv. 0.15) in imidazole buffer (100 mM). The time, in min, after the addition of hemolysate to the sample is given on the right-hand side of the spectra; every second spectrum is shown.

3-^{13}C] L-alanine or [2-^{14}N, 3-^{13}C] L-alanine. In these experiments the complication of having methyl peak-splitting due to coupling with the H2 protons can be obviated by using selective (1$\bar{3}$3$\bar{1}$) pulses on the H3 resonance. It is interesting to note that this was the first time that binomial pulses had been used to eliminate the effects of ^{1}H coupling (and excitation of the solvent water) (Brindle et al., 1984) and the experiment formed the basis of subsequent lactate-editing experiments. The groups of both Gadian and Shulman acknowledge this priority in their work on lactate metabolism *in vivo* (Hetherington et al., 1985; Williams et al., 1986).

Figure 7 shows a time course of $^{1}H[^{13}C]$spin–echo difference spectra acquired following the addition of a hemolysate to (1-^{13}C)Rib5P. The sample was contained within a 5 mm (outer diameter) NMR tube. Acquisition of the $^{1}H[^{13}C]$spin–echo difference spectra at 600.14 MHz was achieved with an 'inverse' probe, with inner coil double-tuned to ^1H and ^2H frequencies. ^{13}C-decoupling (GARP; Shaka et al., 1985) was applied continuously during the acquisition time. The proton spectra of most of the intermediates of the pathway were assigned, with the aid of two-dimensional techniques, enabling the assignment of resonances in the spin–echo difference time course. In the first spin–echo difference spectrum in this time course, the H1 resonances from both the anomeric forms of Rib5P were apparent. The H1,1′ doublet of doublets of the ketopentoses, labeled at C1 were also at their maximum intensity in the first spectrum of the time course. The formation of (1,3-^{13}C)Sed7P was evident from the appearance of H1,1′ and H3 peaks arising from this compound (α- and β-furanose forms were apparent). The reactions of the non-oxidative pentose phosphate pathway lead to labeling of Glc6P at C1 and C3, causing H1 and H3 resonances from this metabolite to emerge. The signals from H1,1′ and H3 of Fru6P were also detected, although these resonances are co-resonant with those arising from Sed7P.

The addition of hemolysate to samples containing (2-^{13}C)Glc6P and $NADP^+$, enabled flux through the oxidative, as well as the non-oxidative branch of the pentose phosphate pathway to be measured with $^1H[^{13}C]$, ^{13}C and ^{31}P NMR spectroscopy (H. A. Berthon, W. A. Bubb and P. W. Kuchel, unpublished data). The fall in the intensity of the resonances arising from Glc6P isotopically labeled at C2, was accompanied by a rise in the intensity of 6PG. Ru5P and Xu5P became labeled at C1, giving rise to overlapping doublet of doublets from the H1,1′ protons in the $^1H[^{13}C]$spin-echo difference spectrum. Also, in the ^{31}P NMR spectrum, the fall in the concentration of $NADP^+$ can be monitored by measuring the intensity of the doublet at lowest frequency assigned to one of the phosphorus atoms linking the two nucleoside moieties.

In contrast to the three-bond scalar coupling between protons and ^{31}P nuclei in sugar phosphate molecules (Cohen et al., 1984), the one-bond scalar coupling between protons and the ^{13}C nuclei is large; this means that when the delay τ is set to $1/J_{CH}$ there is minimal loss of signal through transverse relaxation during the echo, because τ is very small (a few ms). The potential enhancement in sensitivity (Alger and Shulman, 1984) of $^1H[^{13}C]$ spin-echo difference over proton-decoupled ^{13}C spectra is theoretically 1:62.5, which is the ratio of the relative receptivities of the two nuclei. However, the nOe consequent upon broadband proton decoupling of ^{13}C spectra reduces this advantage to about 1:20, and for practical purposes, rather smaller sensitivity enhancements have been reported (e.g., Rothman et al., 1985). In our experimental spectra, the high intensity of signals from protons not coupled to specifically labeled carbon atoms resulted in the appearance of residual signal in the spin–echo difference spectra, limiting the clarity, and thus the sensitivity with which intermediates of the pathway could be observed. Also, the co-resonance of signals in the proton spectrum, some of which have attached ^{13}C nuclei which are also unresolved in the ^{13}C spectrum (e.g., the H1,1′ and H3 resonances of Sed7P and Fru6P), represents a limitation of the technique. $^1H[^{13}C]$spin–echo difference spectroscopy is unsuitable for observing ^{13}C labeling at carbons which have no coupling to protons, such as carbonyl groups. The magnitude of the spin–echo delay time used for the time courses was optimized, in our experiments, for a typical coupling of C1 atoms to anomeric protons (~167 Hz). While the magnitude of the scalar coupling between ^{13}C and ^{1}H nuclei within carbohydrate molecules varies (e.g., in Glc6P, it is 170 Hz and 145 Hz for $^{1}J_{C1\text{-}H1}$ and $^{1}J_{C2\text{-}H2}$, respectively), satisfactory (although nonquantitative) spectra were obtained even when the spin–echo delay time was not optimal.

In summary, the multinuclear approach to our kinetic studies of the pentose phosphate pathway in human erythrocytes have proved to be an especially effective means of measuring metabolic fluxes, as well as isotopic redistribution reactions. Additionally, the spectral simplification afforded by the $^{1}H[^{13}C]$spin–echo difference technique has enabled metabolic flux through the pentose phosphate pathway to be followed by proton-detected NMR.

The metabolic link between the pentose phosphate pathway and glycolysis is well defined; Glc6P is diverted from the glycolytic pathway, passes via the pentose phosphate pathway, and returns to it as Fru6P and G3P. However, a less obvious link between the two pathways is that made up of electrons which flow from NADPH, a product of the oxidative part of the pentose phosphate pathway, to oxidized glutathione (GSSG) to yield GSH. GSH, in turn, donates electrons and then receives them back, in the glyoxalase reactions. These reactions are involved in the detoxification of a by-product (methylglyoxal) of two of the enzymic reactions (aldolase and triosephosphate isomerase) in the glycolytic pathway. Because this close metabolic link exists between the two major pathways, and it is mediated by the glyoxalase enzymes, it seemed relevant to discuss the glyoxalase system next.

IV. GLYOXALASE SYSTEM

Investigation of this system with NMR spectroscopy provides a good example of the ease with which the technique can be rapidly implemented to study enzymes whose reactants are difficult to assay by standard means; this advantage is in addition to that of being able to study the reactions in whole cells or lysates. The two glyoxalase enzymes have activities that rival the most active of the glycolytic enzymes, but the relevance and importance of this high activity is not fully understood (e.g., Thornalley, 1990).

Glyoxylases 1 and 2 catalyze the conversion of α-oxoaldehydes to the corresponding α-D-hydroxy acids. An essential cofactor for this reaction is reduced glutathione (GSH). Glyoxylase 1 appears to be present in all cells at all levels of evolution; in animal species, most tissues contain this enzyme (Carrington and Douglas, 1986). Although the enzymes have been known since 1913 (Dakin and Dudley, 1913), the kinetic characteristics of the whole pathway had not been thoroughly determined until the recent studies on human erythrocytes (Rae et al.,

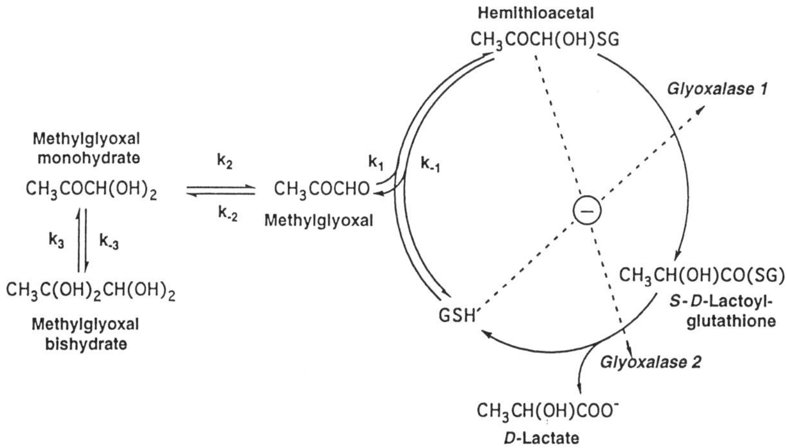

Figure 8. The glyoxylase reaction scheme. The k's denote unitary rate constants involved in the hydration of methylglyoxal and the formation of the hemithioacetal with GSH. The encircled sign indicates inhibition of the enzymes glyoxylase 1 and 2 by the two intermediates indicated at the base of each arrow.

1990). Despite the small number of reactions comprising the pathway, the glyoxylase system is mechanistically complicated, with both enzymes being subject to competitive inhibition by metabolites within the system (Figure 8). Glyoxylase 1 is inhibited by free GSH (Kermack and Matheson, 1957), and glyoxylase 2 is inhibited by the substrate of glyoxylase 1, the hemithioacetal (HPA), and to a lesser extent by free GSH (Uotila, 1973). Kinetic analysis of the system is also complicated by the necessity to consider the extent of hydration of the α-oxoaldehyde, methylglyoxal. This three-carbon α-oxoaldehyde, which may be converted to D-lactate via the glyoxylase system of reactions, is presumed to pass via the monocarboxylate transporter of the erythrocyte into plasma, and thence into the liver where it is oxidized in the peroxisomes of the hepatocytes. In aqueous solution, the dominant forms are the monohydrate (or aldehydrol) and the bishydrate (see Figure 9). The hemithioacetal is the true substrate of glyoxylase 1 (Racker, 1951) and is formed by the spontaneous reaction between GSH and the free (unhydrated) oxoaldehyde. Methylglyoxal itself is the product of a side reaction in the isomerization, catalyzed by triosephosphate isomerase, which interconverts DHAP and G3P (Browne et al., 1976; Webb et al., 1977; Richard, 1991) and the aldol scission/condensation catalyzed by aldolase (Grazi and Tombretta, 1978; Iyengar and Rose, 1981).

^1H NMR spectroscopy enables the direct measurement of the time-dependent changes in the concentration of the two hydrates of methylglyoxal which had not been previously directly measurable, as well as measurement of the substrates and products of the glyoxylase enzymes in human erythrocytes (Rae et al., 1990; Figure

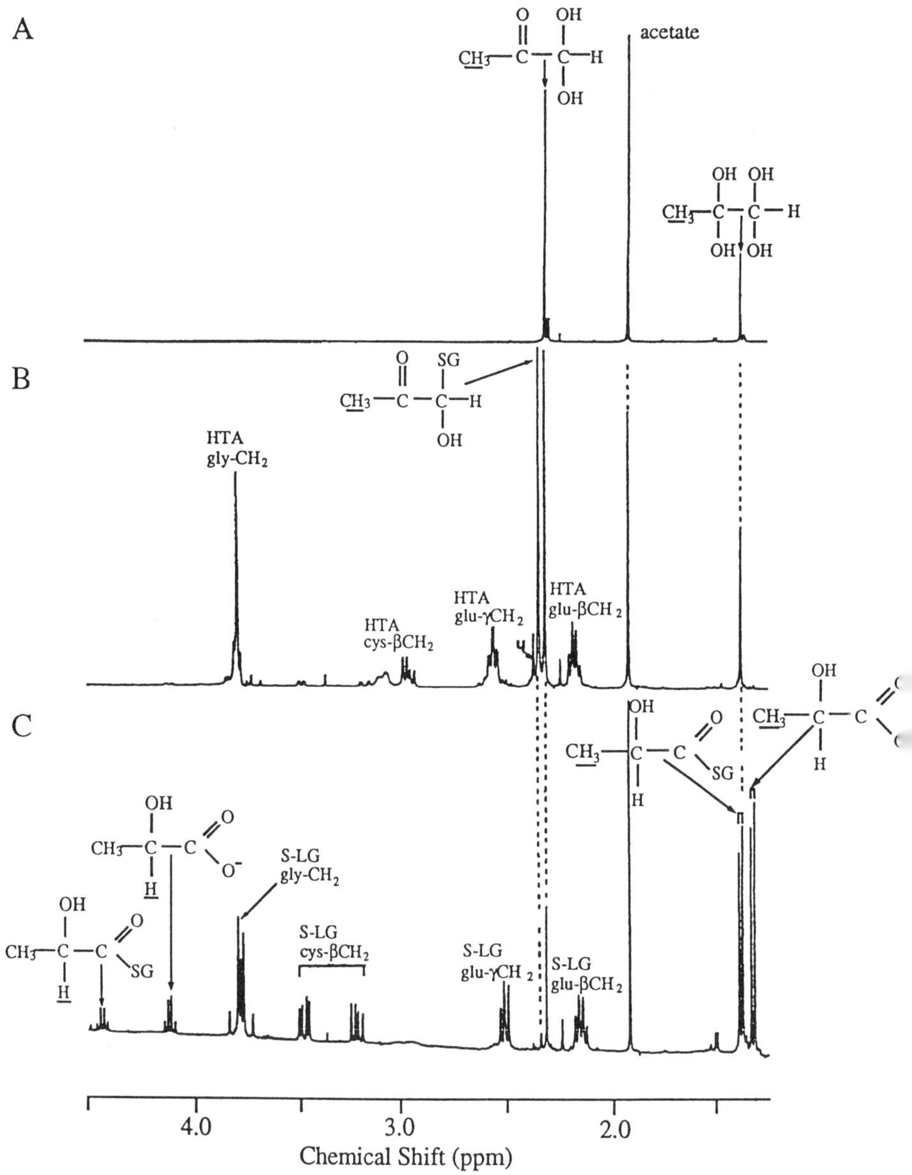

Figure 9. 400-MHz ^{1}H NMR spectrum showing the substrate, intermediate and products of the glyoxylase reactions using methylglyoxal as substrate. (**A**) ^{1}H-NMR spectrum of methylglyoxal showing $C\underline{H}_3$, resonances of the monohydrate (2.3 ppm) and bishydrate (1.4 ppm) of methylglyoxal, and acetate (1.9 ppm). (**B**) Methylglyoxal and GSH. The three major species present are methylglyoxal with assignments as in (**A**); (continued)

9). The rate constants describing the chemical equilibrium between the hydrated species of methylglyoxal, were determined using a magnetization transfer procedure and were found to be much slower (by several orders of magnitude) than the previously reported values derived using other experimental methods. The steady-state kinetic parameters describing the enzyme-catalyzed reactions were determined in dilute hemolysate samples, and these parameters were used in an overall kinetic model of the reaction scheme (Figure 8). The simulated time courses corresponded closely with those obtained experimentally using dilute hemolysates with various concentrations of reactants.

Erythrocytes obtained from a horse with a very low activity of glyoxylase 2, showed the anticipated accumulation of the intermediate compound, S-D-lactoyl glutathione (Rae et al., 1991). The long-term incubation of these cells with a high concentration of this intermediate did not lead to detectable levels of S-D-lactoyl glutathione in the extracellular fluid. This finding suggests the non-involvement of the glutathione-conjugate transporter which appears to carry out the transfer of glutathione conjugates generated via the enzyme, glutathione S-transferase, as part of the detoxification system of reactions acting upon xenobiotics in the cytoplasm (Board, 1981).

Fructose-bisphosphate aldolase (EC 4.1.2.13) catalyzes the aldol condensation of DHAP with G3P to give Fru16P2. This reaction operates in the reverse direction in glycolysis. The enzyme has long been known to be non-specific with respect to the aldehyde used in the condensation; both phosphorylated and unphosphorylated aldehydes can replace glyceraldehyde 3-phosphate (Meyerhof et al., 1936; Mehler and Cusic, 1967; Arora et al., 1988). Recently it was shown that methylglyoxal may also serve as a substrate and the product of the condensation reaction between DHAP and methylglyoxal is 6-deoxy-5-ketofructose 1-phosphate (Rae et al., 1992). The compound appears to exist in aqueous solution as a mixture of the free diketone, each of the possible monohydrates, and the bishydrate. The metabolic significance of the accumulation of such a compound is not known and its susceptibility to cytoplasmic phosphatases is also unexplored.

Although a definitive role has not been ascribed, the relatively high activity of the glyoxylase system enables its efficient scavenging of methylglyoxal and other oxoaldehydes. These compounds may potentially inactivate enzymes of the erythrocyte by the Schiff-base modification of them. Additionally, the likelihood of

---

Figure 9. (continued) free GSH with major assignments almost identical to those of the hemithioacetal; and the $C\underline{H}_3$ resonance of the hemithioacetal is a singlet to high frequency of the methylglyoxal monohydrate (1.35 ppm). **(C)** Spectrum recorded about 30 min after the addition of a hemolysate (2 percent final hematocrit equivalent value). (Other details are given in the paper by Rae et al. (1990) from which the Figure was derived.)

accumulation of compounds such as 6-deoxy-5-ketofructose 1-phosphate, which have no known metabolic roles, is minimized.

In concluding this section it can be reiterated that ^{1}H NMR spectroscopy has provided new insight into the quantitative aspects of the operation of the glyoxalase system and was instrumental in the discovery of a new metabolite related to it.

V. GLYCOLYSIS-SELECTED REACTIONS

A. Enzyme Organization

Glyceraldehyde-3-phosphate dehydrogenase in human erythrocytes appears as band 6 in electropherograms of proteins prepared by dissolving membranes in sodium dodecylsulfate, thus indicating that it binds to the membrane or cytoskeleton. This binding is not seen in erythrocytes from all animals and it only occurs in human erythrocytes when the membranes are prepared under conditions of low ionic strength (Maretzki et al., 1989). Earlier suggestions (Fossel and Solomon, 1977, 1978, 1979) that phosphoglycerate kinase and phosphoglycerate mutase form a complex *in situ*, either alone or together with glyceraldehyde 3-phosphate dehydrogenase, have largely been discredited (Momsen et al., 1979). On the other hand, interest in substrate channeling is linked to the very interesting observation that some parasites, including the insect trypanosomatid, *Crithidia luciliae*, have metabolic complexes in organelles known as a glycosomes (Grote et al., 1990). The metabolic product of the incomplete set of glycolytic reactions of a glycosome is 3-phosphoglycerate. Preliminary ^{31}P NMR experiments have demonstrated the accumulation of this compound in the cells but the signal-to-noise in the reported spectra of intact cells was not sufficient to unambiguously assign a resonance to 3-phosphoglycerate. Assignment was achieved by the use of cell extracts (Grote et al., 1990). The glycosome is a microorganelle with a single bilayer membrane which is impermeable to most intermediates, and it lacks the last three enzymes of glycolysis, enolase, pyruvate kinase, and lactate dehydrogenase, which are located in the cytoplasm. Interestingly, the membrane is permeable to phosphoenolpyruvate (PEP) which is produced not only in glycolysis but also via glycosomal PEP carboxykinase (Opperdoes, 1987).

Experiments with ^{32}P-phosphate ($^{32}P_i$) in human erythrocytes revealed that the specific radioactivity of cellular P_i stabilizes at a value below that of the extracellular P_i. This outcome can be explained on the basis of a single cellular P_i pool exchanging slowly with a large unlabeled pool of cellular organic phosphates. The very rapid initial phase of labeling (in a few seconds) of cellular contents probably occurs via glyceraldehyde-3-phosphate dehydrogenase which utilizes the P_i to produce 1,3-bisphosphoglycerate. The specific activity of the phosphorylated intermediates, though, is less than that of the mean P_i pool. This suggests that the enzyme does not derive its P_i from the extracellular pool, but that some form of compartmentation of the anion occurs in the cell (Kemp et al., 1989). A possible explanation for some of the inhomogeneity of the specific activity of P_i is the turnover of the phosphate esters of enzymes and cytoskeletal proteins about which little is known. It will be a challenge to devise NMR experiments to address the multitude of inferences possible from this interesting finding.

B. Lactate Dehydrogenase

The isotope exchange activity of lactate dehydrogenase has been measured using ^{1}H spin–echo NMR spectroscopy. In the spin–echo spectrum, the methyl resonance of natural lactate is inverted if τ in the pulse sequence is set to $1/(2\ J_{HH}) = 68$ ms; this is the consequence of the spin–spin coupling between the C2 (methenyl) and the C3 (methyl) protons. If the C2 proton is exchanged for a deuteron, then phase modulation due to the homonuclear coupling is removed and the methyl resonance adopts a positive (upright) phase. If the resonances in the spectra are broadened by exponential multiplication that is in excess of four times the value of J, this phase inversion appears as if it is symmetrical about the base line (Brindle et al., 1986); thus the addition of an inverted and an upright peak of equal absolute intensity will yield a 'null' peak. If lactate, uniformly labeled with deuterium ((U-^{2}H) L-lactate), is mixed with natural L-lactate, then a fully inverted methyl resonance is observed in the spin–echo ^{1}H NMR spectrum; but as equilibration of the reactants takes place via lactate dehydrogenase, in the presence of NAD^+, then the intensity of the inverted resonance reduces to almost zero over a time course of 30–60 min (Brindle et al., 1980). An analogous experiment can be performed using L-lactate labeled with deuterium in the C2 position ((2-^{2}H) L-lactate) mixed with an equimolar concentration of L-lactate labeled with deuterium in the C-3 position ((3-^{2}H) L-lactate). In this experiment the intensity of the methyl resonance declines over the time course. By analyzing such time courses in whole cells and hemolysates, and comparing them with those obtained with purified lactate dehydrogenase, Brindle et al. (1986) concluded that *in situ*, the enzyme is operating under conditions of competitive inhibition. The inhibitor was thought to be ADP-ribose (among several possible metabolites) which probably accumulated during the preparation of the cells, some samples of which had been metabolically manipulated (Tsuboi et al., 1966) to have variable levels of NAD^+. The enzyme also seems to display the effects of inhibition in perfused rat livers (Chung and Jue, 1982).

The above method of isotope exchange that is used to study the activity of lactate dehydrogenase *in situ*, supercedes a previous one which utilizes much higher concentrations of added lactate and pyruvate (Simpson et al., 1982a,b). A related method involves the measurement of deuteron/proton exchange on the C2 of lactate, which is evident from the change in phase of the ^{1}H spin–echo NMR spectrum, as a probe of the *in situ* activity of glyceraldehyde-3-phosphate dehydrogenase (Brindle et al., 1982b). The main conclusion in the paper was that there was no evidence that the enzyme is substantially bound to band 3 *in vivo*, since studies with purified enzyme and isolated erythrocyte membranes showed that the binding of the enzyme did inhibit it. Another conclusion was that the enzyme is unlikely to be rate limiting in glycolysis, as was also suggested much earlier in theoretical analyses, such as those of Heinrich et al. (1977).

C. Glyceraldehyde-3-Phosphate Dehydrogenase

The time course of lactate C2 proton/deuterium exchange was also studied as a measure of the activity of glyceraldehyde-3-phosphate dehydrogenase where that activity was modified by the known inhibitor iodoacetate (Foxall et al., 1984). The pattern of enzyme inhibition *in situ*, which is characterized by a binding step and a kinetic step, is similar to that *in vitro*. However, binding of the enzyme to the membrane in intact cells alters the inhibition pattern from that observed with the purified enzyme. This is an elegant demonstration of the use of an exchange reaction as a probe of the cellular environment of an enzyme. However, as with all indirect probes of the intracellular *milieu*, complications may exist with respect to the 'accessibility' of the alkylating agent (iodoacetate) to the enzyme, but in the present case other evidence had been provided to indicate that this was not a complication and that binding of the enzyme to band 3 was not evident *in situ* (see above; Brindle et al., 1982b).

VI. PROTEASES AND PEPTIDASES

Human erythrocytes contain proteolytic enzymes attached both to the outer surface of the membrane (Tökes and Chambers, 1975) and free in the cytoplasm (Pontremoli et al., 1980). The function of these enzymes has not been fully elucidated, although there is considerable information emerging on the role of the calcium-activated protease, calpain and of ATP-dependent proteases (Etlinger et al., 1983); the latter are involved in the degradation of proteins in reticulocytes as they differentiate to their end stage as erythrocytes (Rapoport et al., 1987).

^{1}H spin–echo NMR spectroscopy is a useful method for studying dipeptidase activity in intact erythrocytes and hemolysates. For example, the enzyme-catalyzed hydrolysis of the dipeptide glycylglycine yields the free amino acid glycine and the reaction is readily monitored over a ~20 min time course by recording a series of spectra (King et al., 1983). Similarly, L-cysteinylglycine is hydrolysed with a high maximal velocity (64.2 ± 3.2 mmol (l lysate)$^{-1}$ h^{-1}) and a K_m of 3.5 ± 0.6 mM. This process of dipeptide cleavage is involved in the turnover of GSH. The basis of the NMR analysis is that the H2 resonances of the glycyl of glycylglycine and L-cysteinylglycine have different resonance frequencies from those of the free amino acids, so that sequential spectra allow ready quantification of the concentration of the free amino acids and hence the determination of the reaction rate.

While the dipeptides discussed above were among the first studied with ^{1}H NMR, the dipeptide which became the most extensively studied by this technique was prolidase (King et al., 1986). This manganese-containing enzyme is highly active in the human erythrocyte; it operates on the *trans* isomer of dipeptides in which L-proline is the C-terminal residue (King et al., 1986). The conclusion relating to isomeric preference was reached by using magnetization transfer analysis to measure the rate of *cis-trans* isomerization of the dipeptide, and from time courses of NMR spectra with increasingly higher concentrations of the enzyme. The time courses demonstrated a more rapid decline of the resonances in the methyls of L-ala-L-pro that corresponded to the *trans*-isomer of the dipeptide, thus confirming the *trans*-isomer as the true substrate (King et al., 1986).

Both of the peptide hormones, enkephalin and bradykinin, are hydrolyzed by enzymes in human hemolysates (Vandenberg et al., 1985; Hyslop et al., 1987). By recording ^{1}H spin–echo NMR spectra of a hemolysate to which one or other of the hormones is added, it is possible to determine the degradation pathway of each. The pathways entail the operation of a series of peptidases with various specificities for terminal and internal amino acid residues. The sequential NMR spectra at 400 MHz reveal sufficiently high resolution to enable quantitative characterization of the various peptides that are released in the network of reactions. Thus, for enkephalin, a computer model was constructed to simulate the time course of the operation of the network of peptidolytic enzymes (Vandenberg et al., 1985).

VII. OTHER ENZYMES

A. Pyrimidine 5′-Nucleotidase

A deficiency of the activity of this enzyme is associated with a hereditary hemolytic anaemia. The defect was first described by Valentine et al. (1973) and by 1983, ~35 cases of the deficiency had been reported (Hansen et al., 1983). The activity of the enzyme in erythrocytes was first measured using a laborious assay involving the detection of inorganic phosphate released from CMP or UMP during a 2 h incubation. A radioactive method using ^{14}C-CMP as substrate is now commonly used (Torrence et al., 1977). It has recently been shown though, that ^{1}H NMR spectroscopy can, in principle, be used for assaying the enzyme under conditions suitable for use in clinical hematology (Kagimoto et al., 1986). However, because of the greater spectral complexity, ^{1}H NMR spectroscopy is not as amenable to the task as ^{31}P NMR spectroscopy. This spectral complexity is a consequence of the need to use spin–echo spectra to remove the broad ^{1}H-protein signals: the resulting J-modulation and the relatively crowded nature of these spectra make quantification of the reaction rate imprecise (Mendz et al., 1986a).

The basis of the ^{31}P NMR assay is the observation that the resonances corresponding to the 5′ phosphorus of CMP (~4.30 ppm; substrate) and inorganic phosphate

(~2.80 ppm; product) are clearly resolved in high resolution spectra (magnetic field strength 9.4 T; 400 MHz for ^{1}H). While both the decline in the substrate and increase in the product resonance can be readily detected, conservation of peak area (substrate + product = constant) is not observed because inorganic phosphate is consumed and released through various other reactions. Thus, an accurate estimate of this particular enzymic activity is only obtained from the decline of the CMP resonance. The assay was used in work which showed that the erythrocytes from multiple sclerosis patients had a normal activity of this enzyme; the activity in cells from normal controls and from the patients were 36 ± 2.9 and 32 ± 2.7 μmol (1 packed cells)$^{-1}$min^{-1}, respectively.

In the ^{1}H NMR spectra of neutralized perchloric acid extracts from normal and pyrimidine 5′-nucleotide deficient cells, the major difference lies in the high frequency region of the spectrum. In normal cells these resonances are assignable to the H2, H8 and H1′ protons of adenosine nucleotides, while in the defective cells there are additional resonances in this region, presumably from pyrimidine nucleosides and nucleotides.

The ^{31}P NMR spectra from the defective erythrocytes show that the intensities of signals from the α-, β- and γ-phosphate nuclei of nucleoside triphosphate are ~2 fold greater than in the normal cells. Additional resonances at −10.3, −11.0 and −11.9 ppm have been assigned to CDP-ethanolamine-phosphorus, CDP-choline-phosphorus and UDPG-phosphorus, respectively (Kagimoto et al., 1986). In earlier ^{31}P NMR work, Swanson et al. (1983) also noticed that the nucleoside triphosphate resonances were approximately twice the normal intensity in spectra from the defective cells. In addition, they noticed that the intracellular pH was ~0.2 units lower in the defective erythrocytes. The lower pH can be attributed to an elevation of intracellular impermeant negative charges which gives rise to a Donnan equilibrium, resulting in a correspondingly higher intracellular proton concentration (Raftos et al., 1990).

B. Enzymes of Phospholipid Turnover

Patients treated with lithium carbonate for manic depressive disorder demonstrate a sharp rise in erythrocyte choline concentration; this compound is a precursor of the important neurotransmitter, acetyl choline (Jope et al., 1978; Kuchel et al., 1980). The accumulation occurs in spite of the transport of choline across the erythrocyte membrane being *inhibited* by lithium ions (Lee et al., 1974).

In the course of NMR studies on GSH metabolism, an increase in choline concentration was seen to occur during the incubation of some hemolysates. This reaction did not occur in hemolysates prepared from cells washed in simple physiological saline (0.154 M NaCl), but did when the cells had been washed in Krebs bicarbonate medium. The latter medium contains Ca^{2+}. The most prominent choline resonance lies in the quaternary nitrogen methyl region of the spectrum, at a chemical shift that is slightly different from that of choline phosphate (2.95 ppm) and carnitine (Chapman et al., 1982). Because signals from choline, rather than choline phosphate, appeared in the ^1H spin–echo spectra during the time courses, it was concluded that phospholipase D was catalyzing the release. Until that time, this enzyme had been thought of as being found predominantly in the plant kingdom, although subsequent literature searches revealed some earlier work carried out by Hübscher (1962) which indicated the existence of the enzyme in rat liver. More recently, the enzyme has been partially purified from rat brain (Saito and Kanfer, 1975; Taki and Kanfer, 1979) and human eosinophils (Kater et al., 1976).

In recent studies of the Ca^{2+}-induced release of choline from phosphatidyl choline and L-α-glycerophosphocholine (GPC) in hemolysates, it became evident that there are different responses to activatory Ca^{2+}, and inhibitory Ba^{2+} and P_i ions. Thus, we proposed the existence of not only a phospholipase D but also a glycerophosphocholine phosphodiesterase (PDE; EC 3.1.4.2) in the cytoplasm of human erythrocytes (Selle et al., 1992). Also, by employing ^1H spin-echo and ^{31}P NMR spectroscopy, the release of glycerophosphocholine from lysophosphatidyl choline in hemolysates has been demonstrated. Thus, the existence of a lysophospholipase (EC 3.1.1.5) was confirmed using a much more direct method than the earlier radioactivity/chromatographic method (Mulder et al., 1965). In contrast with some earlier work, no evidence was found for a phospholipase A_2 activity in hemolysates, although it is known to exist in leukocytes. The failure to adequately remove leukocytes from the erythrocyte preparations is thought to be the basis of the artifact. Thus, it is concluded that lysophospholipase may constitute, with the earlier described glycerophosphocholine phosphodiesterase activity, a metabolic unit for the removal of hemolytic lysophosphatidylcholine which is formed in erythrocyte membranes as well as taken up from the plasma (Selle et al., 1993).

NMR spectroscopy, therefore, was instrumental in the discovery of phospholipase D, and the corresponding phosphodiesterase activity, in human erythrocytes. NMR provides a facile means of following the hydrolysis of membrane-associated lipids, such as lysophosphatidylcholine, that are difficult to detect, thus supplanting much more tedious procedures using radioactive substrates. Phospholipid metabolism in human erythrocytes appears to be more complex than has previously been thought, and defects in the constituent enzymes may be the basis of hemolytic anemias associated with, for example, high levels of plasma membrane phosphatidylcholine (Jaffé and Gottfried, 1968).

A computer model describing the rate of choline accumulation in human erythrocytes, in patients treated with lithium carbonate, has been devised. It is based upon the kinetics of choline release from phospholipid in human erythrocytes, the transmembrane flipping rate of lecithin, and the kinetics of the choline transport protein (Beilharz, et al., 1986). Thus, the model enables the prediction of choline concentrations in the erythrocytes of manic depressive patients, at various stages in their treatment with lithium carbonate.

C. NAD Glycohydrolase

Human erythrocyte NAD glycohydrolase (EC 3.2.2.5; NADase) is bound via a phosphatidyl inositide linkage (Kim et al., 1988) to the external surface of the cell membrane where it catalyzes the hydrolytic cleavage of NAD to yield ADP-ribose and nicotinamide. The enzyme is of relatively low activity in human erythrocytes but is of high activity in other mammalian species (Kuchel and Chapman, 1985). The physiological role of the enzyme is not known. NADase can be problematic in metabolic studies of hemolysates, since it catalyzes the degradation of NAD that is essential for glycolysis. However, it can also be exploited to bring about metabolic isolation of the non-oxidative pentose phosphate pathway from the reactions of the oxidative part, and from glycolysis (see earlier section). The enzyme is inhibited by its product, nicotinamide; this compound can therefore be used (in concentrations from 20 mM to 40 mM) in hemolysates to preserve glycolytic activity so that it is the same as in whole cells (Kuchel et al., 1982).

The ^{31}P NMR spectrum of NAD^+ at pH 7.4 contains a quartet at ~ −10 ppm; on incubating NAD^+ with human erythrocytes for 4 h, high frequency resonances arise with a concomitant decline of the original NAD^+. These new resonances were shown to correspond to ADP-ribose, by the addition of the authentic compound (Kuchel and Chapman, 1985). In addition, when ADP-ribose is incubated with whole cells no ^{1}H spin-echo NMR spectral changes occur in 2 h at 37°C. On the other hand, in a *hemolysate*, a dramatic diminution occurs in the adeninyl H8 resonance, with the concomitant elevation of two resonances in close proximity to it, one at high frequency (8.31 ppm) and the other at a lower frequency (8.24 ppm). These resonances are assigned to ADP and AMP respectively, and after several hours the dominant resonance is that of AMP-adeninyl H8 (Kuchel and Chapman, 1985). Thus, the ADP-ribose produced in the reaction with whole cells does not permeate the membrane of the erythrocytes at a rate that is readily detected, but in the hemolysates, it undergoes further degradation to yield adenine nucleoside phosphates.

D. Arginase

Human erythrocytes contain the urea cycle enzyme arginase in their cytoplasm. The activity is relatively high (Kuchel et al., 1984). The progress of the conversion of arginine to ornithine can be measured from the decline of the arginine Hδ resonance (2.99 ppm) and the emergence of the corresponding ornithine resonance (2.81 ppm) (Kuchel et al., 1984).

Because arginine is transported across the human erythrocyte membrane at a rate much slower than its hydrolysis to ornithine via arginase, its transport into the cells can be measured by acquiring sequential ^{1}H spin–echo NMR spectra. This convenient method of assaying the transport activity reveals a maximal velocity, V_{max} (0.07 ± 0.06 mmol (1 cell)$^{-1}$ min^{-1}) and Michaelis constant, K_m (6.3 ± 1.6 mmol 1^{-1}) for the transport protein that is consistent with previously reported values and values reported for other amino acids (Young and Ellory, 1977).

E. Miscellaneous

In principle, almost any enzyme can be assayed *in situ* in erythrocytes, hemolysates, or in buffer solutions, by NMR spectroscopy. Several enzymes that were among the first to be subjected to closer scrutiny by this technique, and therefore deserve special mention include: (1) alanine aminotransferase (Brindle et al., 1982a, 1984), and (2) malate dehydrogenase (Simpson et al., 1982c). (3) Quantitative analysis of defective triosephosphate isomerase in hemolysates from sufferers of a heritable hemolytic deficiency has been carried out with ^{1}H spin-echo NMR spectroscopy (Hyslop et al., 1991). (4) Also, the *in situ* equilibrium exchange kinetics of phosphoglyceromutase have been characterized by using magnetization transfer procedures (Robinson et al., 1984). (5) The analysis has been extended to include a 2D EXSY procedure which was also applied to adenylate kinase (Mendz et al., 1986b). (6) The reaction catalyzed by enolase has been studied using an inversion transfer procedure (Robinson et al., 1985) and the work has been extended to consider the reaction coupled to that catalyzed by (7) phosphoglyceromutase (Chapman et al., 1988).

The work on coupled enzyme-catalyzed reactions is part of a program to study the propagation of magnetization-labeling in sub-sequences of the glycolytic pathway; the work is at present incomplete.

VIII. CONCLUDING REMARKS

The utility of NMR spectroscopy for monitoring biochemical transformations in intact erythrocytes and hemolysates has been repeatedly demonstrated over the 20 years since the first report of the use of ^{31}P NMR to study these cells. Since the early work, applications of magnetization transfer procedures to biological systems have enabled the characterization of some fast equilibrium exchange reactions in the intact cell; however, perhaps, the most novel applications of magnetization transfer lie in the area of membrane transport, which is not the primary subject of this chapter. The work on solute transport in erythrocytes which grew out of the serendipitous finding of the split peak phenomenon has recently been reviewed (Kuchel et al., 1992; Kuchel et al., 1994).

The discovery of the exchange reactions associated with L-lactate in media constituted from D_2O provides a novel means of probing the environment of several of the glycolytic enzymes (Brown and Campbell, 1980). Isotopomer analysis has been applied successfully in studies of bacterial metabolism (Jones and Sanders, 1989) and it holds much promise for the further elucidation of reaction schemes such as the pentose phosphate pathway in the human erythrocyte (Berthon et al.,

1993). Pulsed field gradient spin echo (PFGSE) NMR spectroscopy, which has recently been used to estimate the translational diffusion coefficient of hemoglobin in human erythrocytes, at high resolution (Kuchel and Chapman, 1991), is poised to contribute to the characterization of heterogeneous protein-protein interactions. The lack of selectivity for detecting a particular protein by ^{1}H NMR raises the option of detecting other nuclei. The low sensitivity for detecting natural abundance levels of ^{13}C in proteins implies the need to use label enrichment with this nucleus. Thus, in certain situations cells can be 'fed' labeled amino acids, as was used by London et al. (1975a) with ^{13}C-histidine fed to rats to label erythrocyte hemoglobin, and as has recently been carried out using ^{19}F in 5-fluorotryptophan to label phosphoglycerate kinase in *Saccharomyces cerevisiae* (Brindle et al., 1989).

ABBREVIATIONS

a-, *altro*- configuration; α, when used with the structure of a sugar denotes the alpha anomeric form; Ara5P, arabinose 5-phosphate; β, when used with the structure of a sugar denotes the beta anomeric form; 2,3-BPG, 2,3-bisphosphoglycerate; cys, cysteine/cysteinyl; DHAP, dihydroxyacetone phosphate; DMMP, dimethylmethylphosphonate; Ery4P, erythrose 4-phosphate; *f*, furanose ring form; FCC, flux control coefficient; fid, free induction decay; Fru16P2, fructose 1,6-bisphosphate; Fru6P, fructose 6-phosphate; G3P, glyceraldehyde 3-phosphate; Glc1P, glucose 1-phosphate; Glc16P2, glucose 1,6-bisphosphate; Glc6P, glucose 6-phosphate; Gla, glycolaldehyde moiety transferred by transketolase; Glc, glucose; gly, glycine/glycyl; glu, glutamate/glutamyl; Grn, dihydroxyacetone moiety transferred by transaldolase; GSH, glutathione (reduced); GSSG, glutathione (oxidized); *i*-, *ido*- configuration; HTA, hemithioacetal; KIE, kinetic isotope effect; Man6P, mannose 6-phosphate; MeP, methylphosphonate; NADase, NAD glycohydrolase; $NADP^+$, nicotinamide adenine dinucleotide phosphate; NADPH, nicotinamide adenine dinucleotide phosphate (reduced); octulose 1,8-bisphosphate, Oct18P2; Oct8P, octulose 8-phosphate; 6PG, 6-phosphogluconate; 6PGL, 6-phosphogluconolactone; *p*, pyranose ring form; PDE, phosphodiesterase; PEP, phosphoenolpyruvate; PFGSE, pulsed field gradient spin-echo; P_i, inorganic phosphate; r.f., radiofrequency; Rib5P, ribose 5-phosphate; Ru5P, ribulose 5-phosphate; Sed17P2, sedoheptulose 1,7-bisphosphate; Sed7P, sedoheptulose 7-phosphate; S-LG, S-D-lactoylglutathione; TA, transaldolase; TeP, triethyl phosphate; TK, transketolase; Xu5P, xylulose 5-phosphate. All sugars and sugar phosphates were of the D-configuration unless otherwise stated.

ACKNOWLEDGMENTS

The Australian NH&MRC is thanked for project grant support to PWK. HAB is grateful for an Australian Postgraduate Research Award. Mr Bill Lowe is thanked for expert technical assistance in the preparation of this manuscript.

REFERENCES

Abragam, A. (1961). The Principles of Nuclear Magnetism. Clarendon Press, Oxford.

Acerenza, L., Sauro, H. M., & Kacser, H. (1989). Control analysis of time-dependent metabolic systems. J. Theor. Biol. 137, 423–444.

Albrecht, V., Roigas, H., Schultze, M., Jasobasch, G., & Rapoport, S. M. (1971). The influence of pH and methylene blue on the pathways of glucose utilization and lactate formation in erythrocytes of man. Eur. J. Biochem. 20, 44–50.

Alger, J. R. & Shulman, R. G. (1984). Metabolic applications of high-resolution ^{13}C nuclear magnetic resonance spectroscopy. Brit. Med. Bull. 40, 160–164.

Arora, K. K., Cortis, P., Bleakley, P. A., & Williams, J. F. (1985). Identification and measurement of D-*glycero* D-*ido* octulose 1,8-bisphosphate: D-*altro*-heptulose 7-phosphotransferase enzyme in tissues with L-type pentose phosphate pathway activity. Int. J. Biochem. 17, 1329–1337.

Arora, K. K., Collins, J. G., Macleod, J. K., & Williams, J. F. (1988). Rapid methods for the high yield synthesis of carbon-13 enriched intermediates of the pentose-phosphate pathway. Biol. Chem. Hoppe-Seyler 369, 549–557.

Bartlett, G. R. & Bucolo, G. (1960). Octulose phosphates from the human red blood cell. Biochem. Biophys. Res. Commun. 3, 474–478.

Bartlett, G. R. & Bucolo, G. (1968). The metabolism of ribonucleoside by the human erythrocyte. Biochim. Biophys. Acta 156, 240–253.

Bartlett, M. R. E., Collins, J. G., Flanigan, I., MacLeod, J. K., & Williams, J. F. (1989). CO_2 labeling of octulose bisphosphates during photosynthesis. An NMR study using intact spinach leaves. Biochem. Int. 18, 35–46.

Barun, K. De. & Kirtley, M. E. (1977). Interaction of phosphoglycerate kinase with human erythrocyte membranes. J. Biol. Chem. 252, 6715–6720.

Beilharz, R. G., Middlehurst, C. R., Kuchel, P. W., Hunt, G. E., & Johnson, J. F. S. (1986). An experimental study and computer simulation of the turnover of choline in erythrocytes of patients treated with lithium carbonate. Aust. J. Exp. Biol. Med. Sci. 64, 271–289.

Bergmeyer, H. U. (1974). In: Methods of Enzymatic Analysis. (Bergmeyer, H. U., ed.), 2nd ed., Verlag-Chemie GmbH, Weinheim.

Berthon, H. A., Kuchel, P. W., & Nixon, P. F. (1992). High control strength of transketolase in the non-oxidative pentose phosphate pathway of human erythrocytes: NMR, antibody and computer simulation studies. Biochemistry 31, 12792–12798.

Berthon, H. A., Bubb, W. A., & Kuchel, P. W. (1993). ^{13}CNMR isotoxomer and computer simulation studies of the non-oxidative pentose phosphate pathway of human erythrocytes. Biochem. J. 296, 379–387.

Betts, G. F., & Srivastava, D. K. (1991). The rationalization of high enzyme concentration in metabolic pathways such as glycolysis. J. Theor. Biol. 151, 155–167.

Beutler, E. (1984). Red cell metabolism. A manual of biochemical methods. 3rd ed., Grune and Stratton, Orlando.

Beutler, E., Kuhl, W., & Gelbart, T. (1985). 6-Phosphogluconolactonase deficiency, a hereditary erythrocyte enzyme deficiency-possible interactions with glucose-6-phosphate dehydrogenase deficiency. Proc. Natl. Acad. Sci. USA 82, 3876–3878.

Blackmore, P. F., Williams, J. F., Scofield, P. J., & Power, P. A. (1982). Mechanism and contribution of the pentose phosphate cycle to glucose metabolism in epididymal fat tissue. Int. J. Biochem. 14, 171–186.

Bleakley, P. A., Arora, K. K., & Williams, J. F. (1984). Evidence that aldolase and D-arabinose 5-phosphate are components of pentose pathway reactions in liver *in vitro*. Biochem. Int. 8, 491–500.

Board, P. G. (1981). Transport of glutathione S-conjugates from human erythrocytes. FEBS Lett. 124, 163–165.

Bonsignore, A., Pontremoli, S., Mangiarotti, A., De Flora, A., & Mangiarotti, M. (1962). A direct interconversion: D-fructose 6-phosphate $\rightleftharpoons$ sedoheptulose 7-phosphate and D-xylulose 5-phosphate catalyzed by the enzymes transketolase and transaldolase. J. Biol. Chem. 237, 3597–3602.

Brindle, K. M. & Campbell, I. D. (1987). NMR studies of kinetics in cells and tissues. Quart. Rev. Biophys. 19, 159–182.

Brindle, K. M., Brown, F. F., Campbell, I. D., Foxall, D. L., & Simpson, R. J. (1980). $^{1}H/^{2}H$ isotope exchange in intact erythrocytes. Biochem. Soc. Trans 8, 646–647.

Brindle, K. M., Boyd, J., Campbell, I. D., Porteous, R., & Soffe, N. (1982a). Observation of carbon labelling in cell metabolites using proton spin echo NMR. Biochem. Biophys. Res. Commun. 109, 864–871.

Brindle, K. M., Campbell, R. D., & Simpson, R. J. (1982b). A ^{1}H n.m.r. study of the kinetic properties expressed by glyceraldehyde phosphate dehydrogenase in the intact human erythrocyte. Biochem. J. 208, 583–592.

Brindle, K. M., Porteous, R., & Campbell, I. D. (1984). ^{1}H NMR measurements of enzyme-catalyzed ^{15}N-label exchange. J. Magn. Reson. 56, 543–547.

Brindle, K. M., Campbell, I. D., & Simpson, R. J. (1986). A ^{1}H-NMR study of the activity expressed by lactate dehydrogenase in the human erythrocyte. Eur. J. Biochem. 158, 299–305.

Brindle, K. M., Williams, S.-P. and Bolton, M. (1989). ^{19}F NMR detection of fluorine-labelled enzyme *in vivo*. FEBS Lett. 255, 121–124.

Brown, F. F. & Campbell, I. D. (1980). N.m.r. studies of red cells. Phil. Trans Roy. Soc. Lond. B. 289, 395–406.

Brown, F. F., Halsey, M. J., & Richards, R. E. (1976). Halothane interactions with haemoglobin. Proc. Roy. Soc. Lond. B. 193, 387–411.

Brown, F. F., Campbell, I. D., Kuchel, P. W., & Rabenstein, D. L. (1977). Human erythrocyte metabolism studies by ^{1}H spin echo NMR. FEBS Lett. 82, 12–16.

Browne, C. A., Campbell, I. D., Kierner, P. A., Phillips, D. C., Waley, S. G., and Wilson, I. A. (1976). Studies of the histidine residues of triosephosphate isomerase by proton magnetic resonance and X-ray crystallography. J. Mol. Biol. 100, 319–343.

Bucolo, G. & Bartlett, G. R. (1960). Sedoheptulose diphosphate formation by the human red blood cell. Biochem. Biophys. Res. Commun. 3, 620–624.

Burns, J. A., Cornish-Bowden, A., Groen, A. K., Heinrich, R., Kacser, H., Porteous, J. W., Rapoport, S. M., Rapoport, T. A., Stucki, J. W., Tager, J. M., Wanders, R. J. A., & Westerhoff, H. V. (1985). Control analysis of metabolic systems. Trends Biochem. Sci. 10, 16.

Butler, R. N., Arora, K. K., Flanigan, I., Lawson, M. J., Roberts-Thomson, I. C., & Williams, J. F. (1990). Pentose phosphate pathway in rat colonic epithelium. Biochem. Int. 22, 249–260.

Callaghan, P. T. (1991). Principles of NMR Microscopy. Clarendon Press, Oxford.

Carrington, S. J. & Douglas, K. T. (1986). The glyoxalase enigma—the biological consequences of a ubiquitous enzyme. IRCS Med. Sci. 14, 763–768.

Chance, B., Garfinkel, D., Higgins, J., Hess, B., & Chance, E. M. (1960). Metabolic control mechanisms. V. A solution for the equations representing interaction between glycolysis and respiration in ascites tumor cells. J. Biol. Chem. 235, 2426–2439.

Chapman, B. E. & Kuchel, P. W. (1990). Fluoride transmembrane exchange in human erythrocytes measured with ^{19}F NMR magnetization transfer. Eur. Biophys. J. 19, 41–45.

Chapman, B. E., Beilharz, G. R., York, M. J., & Kuchel, P. W. (1982). Endogenous phospholipase and choline release in human erythrocytes: A study using ^{1}H NMR spectroscopy. Biochem. Biophys. Res. Commun. 105: 1280–1287.

Chapman B. E., Stewart, I. M., Bulliman, B. T., Mendz, G. L., & Kuchel, P. W. (1988). ^{31}P magnetization transfer in the phosphoglyceromutase-enolase coupled enzyme system. Eur. Biophys. J. 16, 187–191.

Chung, Y. & Jue, T. (1982). ^{1}H NMR observation of redox potential in liver. Biochemistry 31, 11159–11165.

Clark, M. G., Williams, J. F., & Blackmore, P. F. (1971). The transketolase exchange reaction *in vitro*. Biochem. J. 125, 381–384.

Cohen, G. & Hochstein, P. (1963). Glutathione peroxidase: The primary agent for the elimination of hydrogen peroxide in erythrocytes. Biochemistry, 2, 1420–1428.

Cohen, J. S., Chen, C., & Bax, A. (1984). Selective observation of phosphate ester protons by ^{1}H{^{31}P} spin-echo difference spectroscopy. J. Magn. Reson. 59, 181–187.

Cohen, S. M. (1983). Simultaneous ^{13}C and ^{31}P NMR studies of perfused rat liver. J. Biol. Chem. 258, 14294–14308.

Cohen, S. M., Rognstad, R., Shulman, R. G., & Katz, J. (1981). A comparison of ^{13}C nuclear magnetic resonance and ^{14}C tracer studies of hepatic metabolism. J. Biol. Chem. 256, 3428–3432.

Dacie, J. V. & Lewis, S. M. (1975). Practical Haematology. 5th ed., Churchill Livingstone, Edinburgh.

Dakin, H. D. & Dudley, H. W. (1913). On glyoxalase. J. Biol. Chem. 14, 423–482.

Daveloose, D., Fabre, G., Berleur, F., Testylier, G., & Leterrier, F. (1983). A new spin label method for the measurement of erythrocyte internal microviscosity. Biochim. Biophys. Acta 763, 41–49.

Doddrell, D. M., Glushko, V., & Allerhand, A. (1972). Theory of nuclear Overhauser enhancement and ^{13}C-^{1}H dipolar relaxation in proton-decoupled carbon-13 NMR spectra of macromolecules. J. Chem. Phys. 56, 3683–3689.

Duke, C. C., MacLeod, J. K., & Williams, J. F. (1981). Nuclear magnetic resonance studies of D-erythrose 4-phosphate in aqueous solution. Structures of the major contributing monomeric and dimeric forms. Carbohydrate Res., 95, 1–26.

Eakin, R. T., Morgan, L. O., Gregg, C. T., & Matwiyoff, N. A. (1972). Carbon-13 nuclear magnetic resonance spectroscopy of living cells and their metabolism of a specifically labelled ^{13}C substrate. FEBS Lett. 28, 259–264.

Endre, Z. H. & Kuchel, P. W. (1986). Viscosity of concentrated solutions and of human erythrocyte cytoplasm determined from NMR measurement of molecular correlation times. Biophys. Chem. 24, 337–356.

Endre, Z. H., Chapman, B. E., & Kuchel, P. W. (1983). Intra-erythrocyte microviscosity and diffusion of specifically labelled [glycyl-α-^{13}C]glutathione by using ^{13}C n.m.r. Biochem. J. 216, 655–660.

Etlinger, J. D., Speiser, S., Marmorstein, S., & Wajnberg, E. (1983). Regulation of proteolysis in erythroid cells. Biomed. Biochim. Acta 42, S207–S211.

Everhart, C. H. & Johnson, C. S. (1982). The determination of tracer diffusion coefficients for proteins by means of pulsed field gradient NMR with applications to hemoglobin. J. Magn. Reson. 48, 466–474.

Ferrige, A. G., Lindon, J. C., & Paterson, R. A. (1979). High resolution proton nuclear magnetic resonance studies of interaction between deoxyhaemoglobin and small molecules, dithionite and diphosphoglycerate. J. Chem. Soc. Faraday Trans. 75, 2851–2864.

Fossel, E. T. & Solomon, A. K. (1977). Membrane mediated link between ion transport and metabolism in human red cells. Biochim. Biophys. Acta 464, 82–92.

Fossel, E. T. & Solomon, A. K. (1978). Ouabain-sensitive interaction between human red cell membrane and glycolytic enzyme complex in cytosol. Biochim. Biophys. Acta 510, 99–111.

Fossel, E. T. & Solomon, A. K. (1979). Effect of the sodium/potassium ratio on glyceraldehyde-3-phosphate dehydrogenase interaction with red cell vesicles. Biochim. Biophys. Acta 553, 142–153.

Foxall, D. L., Brindle, K. M., Campbell, I. D., & Simpson, R. J. (1984). The inhibition of erythrocyte glyceraldehyde 3-phosphate dehydrogenase *in situ* PMR studies. Biochim. Biophys. Acta 804, 209–215.

Freeman, R., Mareci, T. H., & Morris, G. A. (1981). Weak satellite signals in high-resolution NMR spectra: separating the wheat from the chaff. J. Magn. Reson. 42, 341–34.

Gadian, D. G. & Radda, G. K. (1981). NMR studies of tissue metabolism. Ann. Rev. Biochem. 50, 69–83.

Gaetani, G. D., Parker, J. C., & Kirkman, H. N. (1974). Intracellular restraint—a new basis for the limitation in response to oxidative stress in human erythrocytes containing low activity variants of glucose-6-phosphate dehydrogenase. Proc. Natl. Acad. Sci. USA 71, 3584–3587.

Garfinkel, D. (1980). Computer modeling, complex biological systems, and their simplifications. Am. J. Physiol. 239, R1–6.

Garfinkel, D., Frenkel, R. A., & Garfinkel, L. (1968). Simulation of the detailed regulation of glycolysis in a heart supernatant preparation. Comp. Biomed. Res. 2, 68–91.

Gopher, A., Vaisman, N., Mandel, H., & Lapidot, A. (1990). Determination of fructose metabolic pathways in normal and fructose-intolerant children: A ^{13}C NMR study using [U-^{13}C]fructose. Proc. Natl. Acad. Sci. USA 87, 5449–5453.

Grazi, E. & Tombetta, G. (1978). A new intermediate of the aldolase reaction, the pyruvaldehyde-orthophosphate complex. Biochem. J. 175, 361–365.

Grimes, A. J. (1980). In: Human Red Cell Metabolism, pp. 192–258, Blackwell Scientific Publications, Oxford.

Grote, R., Edwards, M. R., Norton, R. S., & O'Sullivan, W. J. (1990). A ^{31}P nuclear magnetic resonance study of *Crithidia luciliae*. Molec. Biochem. Parasitol. 42, 109–118.

Haner, R. L. & Schleich, T. (1989). Measurement of translational motion by pulse-gradient spin-echo nuclear magnetic resonance. In: Methods in Enzymology (Oppenheimer, N. J., and James, T. L., eds.), Vol. 176, pp. 418–446, Academic Press, San Diego.

Hansen, T, W. R., Seip, D., de Verdier, C. H., & Ericson, A. (1983). Erythrocyte pyrimidine 5′-nucleotidase deficiency. Scand. J. Haematol. 31,122–128.

Heinrich, R., Rapoport, S. M., and Rapoport, T. A. (1977). Metabolic regulation and mathematical models. Prog. Biophys. Molec. Biol. 32, 1–82.

Henderson, T. O., Costello, A. J. R., & Omachi, A. (1974). Phosphate metabolism in intact human erythrocytes: determination by phosphorus-31 nuclear magnetic resonance spectroscopy. Proc. Natl Acad. Sci. 71, 2487–2490.

Hermann, A. & Müller, P. (1986). Correlation of the internal microviscosity of human erythrocytes to the cell volume and the viscosity of hemoglobin solutions. Biochim. Biophys. Acta 885, 80–87.

Higgins, J. (1964). A chemical mechanism for oscillation of glycolytic intermediates in yeast cells. Proc. Natl Acad. Sci. USA 51, 989–994.

Hetherington, H. P., Avison, M. J., & Shulman, R. G. (1985). ^{1}H homonuclear editing of rat brain using semiselective pulses. Proc. Natl. Acad. Sci. USA 82, 3115–3118.

Holzhütter, H. G., Schuster, R., Buckwitz, D., & Jacobasch, G. (1990). Mathematical modelling of metabolic pathways affected by an enzyme deficiency. Biomed. Biochim. Acta 49, 791–800.

Horecker, B. L. & Mehler, A. H. (1955). Carbohydrate metabolism. Ann. Rev. Biochem. 24, 207–274.

Hübscher, G. (1962). Metabolism of phospholipids. VI. The effect of metal ions on the incorporation of L-serine into phosphatidylserine. Biochim. Biophys. Acta 57, 555–561.

Hyslop, S. J., King, G. F., & Kuchel, P. W. (1987). No evidence for bradykinin hydrolysis in human erythrocyte suspensions: ^{1}H NMR studies. Am. J. Hematol. 25, 183–189.

Hyslop, S. J., Beal, P., & Kuchel, P. W. (1991). ^{1}H NMR assay of triose-phosphate isomerase activity in erythrocytes. Anal. Biochem. 197, 178–181.

Iyengar, R. & Rose, I. A. (1981). Concentration of activated intermediates of the fructose 1,6 bisphosphate aldolase and triosephosphate isomerase reactions. Biochemistry 20, 1223–1229.

Jaffé, E. R. & Gottfried, E. L. (1968). Hereditary nonspherocytic hemolytic disease associated with an altered phospholipid composition of the erythrocytes. J. Clin. Invest. 47, 1375–1388.

Jeffrey, F. M. H., Rajagopal, A., Malloy, C. R., & Sherry, A. D. (1991). ^{13}C-NMR: a simple yet comprehensive method for analysis of intermediary metabolism. Trends Biochem. Sci., 16, 5–10.

Jones, D. N. M. & Sanders, J. K. M. (1989). Biosynthetic studies using ^{13}C-COSY: The *Klebsiella* K3 serotype polysaccharide. J. Am. Chem. Soc. 111, 5132–5137.

Jope, R. S., Jenden, B. J., Ehrlich, B. E., & Diamond, J. M. (1978). Choline accumulates in erythrocytes during lithium therapy. N. Engl. J. Med. 299, 833–834.

Kacser, H. & Burns, J. A. (1973). The control of flux. Symp. Soc. Exp. Biol. 27, 65–104.

Kagimoto, T., Shirono, K., Hagaki, T., Oda, T., Matsuzaki, H., Nagata, K. Nakagi, T., Morino, Y., & Takatsuki, K. (1986). Detection of pyrimidine 5′-nucleotidase deficiency using ^{1}H- or ^{31}P-nuclear magnetic resonance. Experientia 42, 69–72.

Kalderon, B., Gopher, A., & Lapidot, A. (1986). Metabolic pathways leading to liver glycogen repletion in vivo, studied by GC-MS and NMR. FEBS Lett. 204, 29–32.

Kalderon, B., Korman, S. H., Gutman, A., & Lapidot, A. (1989). Estimation of glucose carbon recycling in children with glycogen storage disease: A ^{13}C NMR study using [U-^{13}C]glucose. Proc. Natl. Acad. Sci. USA 86, 4690–4694.

Kater, L. A., Goetzl, E. J. and Austen, K. F. (1976). Isolation of human eosinophil phospholipase D. J. Clin. Invest. 57, 1173–1180.

Katz, J. (1981). The pentose cycle in liver. Trends Biochem. Sci. 6, XIV–XV.

Katz, J. & Wood, H. G. (1960). The use of glucose-C^{14} for the evaluation of the pathways of glucose metabolism. J. Biol. Chem. 235, 2165–2177.

Katz, J., Wals, P. A., Golden, S., & Rognstad, R. (1975). Recycling of glucose by rat hepatocytes. Eur. J. Biochem. 60, 91–101.

Kemp, G. J., Bevington, A., Khodja, D., Challa, A., & Russell, R. G. G. (1989). ^{31}P-labeling anomalies in human erythrocytes. Is there more than one pool of cellular P_i? Biochem. J. 264, 729–736.

Kermack, W. O. & Matheson, N. A. (1957). The effect of some analogues of glutathione on the glyoxalase system. Biochemistry 65, 48–58.

Kim, U-H., Rockwood, S. F., Kim, H-R., & Daynes, R. A. (1988). Membrane-associated NAD^{+} glycohydrolase from rabbit erythrocytes is solubilized by phosphatidylinositol-specific phospholipase C. Biochim. Biophys. Acta 965, 76–81.

King, G. F. & Boyd, C. A. R. (1991). Proton NMR studies of transmembrane solute transport. In: Cell Membrane Transport: Experimental Approaches and Methodologies (Yudilevich, D. L., Devés, R., Perán, S., & Cabantchik, Z. I., eds.), pp. 297–323, Plenum Press, New York.

King, G. F., York, M. J., Chapman, B. E., & Kuchel, P. W. (1983). Proton NMR spectroscopic studies of dipeptidase in human erythrocytes. Biochem. Biophys. Res. Commun. 110, 305–312.

King, G. F., Middlehurst, C. R., & Kuchel, P. W. (1986). Direct NMR evidence that prolidase is specific for the trans-isomer of imidodipeptide substrates. Biochemistry 25, 1054–1062.

Kirk, K. (1990). NMR methods for measuring membrane transport rates. NMR Biomed. 3, 1–16.

Kirk, K. & Kuchel, P. W. (1985). Red cell volume changes monitored using a new NMR procedure. J. Magn. Reson. 62, 568–572.

Kirk, K. & Kuchel, P. W. (1988a). The physical basis of the effect of hemoglobin on the ^{31}P NMR chemical shifts of some phosphoryl compounds. Biochemistry 27, 8803–8810.

Kirk, K. & Kuchel, P. W. (1988b). Characterisation of the transmembrane chemical shift differences in the ^{31}P NMR spectra of some phosphoryl compounds in erythrocyte suspensions. Biochemistry 27, 8795–8802.

Kirk, K., Kuchel, P. W., & Labotka, R. J. (1988). Hypophosphite ion as a ^{31}P NMR probe of membrane potential in erythrocyte suspensions. Biophys. J. 54, 241–248.

Kohn, M. C., Whitley, L. M., & Garfinkel, D. (1979). Instantaneous flux control analysis for biochemical systems. J. Theor. Biol. 76, 437–452.

Kosower, N. S. & Kosower, E. H. (1974). Protection of membranes by glutathione. In: Glutathione (Flohé, L., Beröhr, H.Chr., Sies, S., Waller, H.D., & Wendel, A., eds.), pp. 216–227, Georg Thieme, Stuttgart.

Kotyk, A. & Slavik, J. (1989). Intracellular pH and its measurements. CRC Press, Boca Raton.

Kuchel, P. W. (1989). Biological applications of NMR. In: Analytical NMR (Field, L. D. & Sternhell, S., eds.), pp. 157–219, John Wiley & Sons, Chichester.

Kuchel, P. W. (1990). Spin-exchange NMR spectroscopy in studies of the kinetics of enzymes and membrane transport. NMR Biomed. 3, 102–119.

Kuchel, P. W. & Chapman, B. E. (1985). Direct observation of the NAD glycohydrolase reaction in human erythrocytes using NMR spectroscopy. Experientia 41, 53–55.

Kuchel, P. W. & Chapman, B. E. (1991). Translational diffusion of haemoglobin in human erythrocytes and haemolysates. J. Magn. Reson. 94, 574–580.

Kuchel, P. W., Roberts, D. V., & Nichol, L. W. (1977). The computer simulation of the urea cycle: correlation of effects due to inborn errors in the catalytic properties of the enzymes with clinical-biochemical observations. Aust. J. Exptl. Biol. Med. Sci. 55, 309–326.

Kuchel, P. W., Singh, B. E., Hunt, G. E., Johnson, G. F. S., Begg, W. B., & Jones, A. J. (1980). Red cell choline in manic depressive patients taking lithium. New Engl. J. Med. 303: 705.

Kuchel, P. W., Jarvie, P. E., & Conyers, R. A. J. (1982). Erythrocyte glycolysis: A stimulation by nalidixic acid. Biochem. Med. 27, 95–100.

Kuchel, P. W., Chapman, B. E., Endre Z. H., King, G. F., Thorburn, D. R., & York, M. J. (1984). Monitoring metabolic reactions in erythrocytes using NMR spectroscopy. Biomed. Biochim. Acta 43, 719–726.

Kuchel, P. W., Bulliman, B. T., Chapman, B. E., & Kirk, K. (1987). The use of transmembrane differences in saturation transfer for measuring fast membrane transport: $H^{13}CO_3$-exchange across the human erythrocyte. J. Magn. Reson. 74, 1–11.

Kuchel, P. W., Berthon, H. A., Bubb, W. A., McIntyre, L. M., Nygh, N. K., & Thorburn, D. R. (1990). ^{13}C and ^{31}P NMR studies of the pentose phosphate pathway in human erythrocytes. Biomed. Biochim. Acta 49, S105–110.

Kuchel, P. W., Chapman, B. E., & Xu, A. S-L. (1992). Rates of anion transfer across erythrocyte membranes measured with NMR spectroscopy. In: The Band 3 Proteins: Anion Transporters, Binding Proteins and Senescent Antigens (Bamberg, E. & Passow, H., eds.), pp. 105–119, Elsevier, Amsterdam.

Kuchel, P. W., Kirk, K., & King, G. F. (1994). NMR methods for measuring membrane transport. In: Subcellular Biochemistry: Study of Biomembranes by Physico-Chemical Methods (Hilderson, H. J. & Ralston, G. B., eds.), Plenum Press, London, (in press).

Kurganov, B. I. (1986). The role of multienzyme complexes in integration of cellular metabolism. J. Theor. Biol. 119, 445–455.

Labotka, R. J. (1984). Measurement of intracellular pH and deoxyhemoglobin concentration in deoxygenated erythrocytes by phosphorus-31 nuclear magnetic resonance. Biochemistry 23, 5549–5555.

Landau, B. R. & Bartsch, G. E. (1966). Estimations of pathway contributions to glucose metabolism and the transaldolase reactions. J. Biol. Chem. 241, 741–749.

Landau, B. R. & Wood, H. G. (1983). The pentose cycle in animal tissues: evidence for the classical and against the 'L-type' pathway. Trends Biochem. Sci. 8, 292–296.

Lee, G., Lingach, C., Lyle, P. T., & Martiun, K. (1974). Lithium treatment strongly inhibits choline transport in human erythrocytes. Brit. J. Clin. Pharmacol. 1, 365–370.

Little, C. & O'Brien, P. J. (1968). An intracellular GSH-peroxidase with a lipid peroxide substrate. Biochem. Biophys. Res. Commun. 31, 145–150.

Ljungdahl, L., Wood, H. G., Racker, E., & Couri, D. (1961). Formation of unequally labeled fructose 6-phosphate by an exchange reaction catalyzed by transaldolase. J. Biol. Chem. 236, 1622–1625.

London, R. E. (1988). ^{13}C labeling in studies of metabolic regulation. Prog. NMR Spectr. 20, 337–383.

London, R. E. & Gabel, S. A. (1989). Determination of membrane potential and cell volume by ^{19}F NMR using trifluoroacetate and trifluoroacetamide probes. Biochemistry 28, 2378–2382.

London, R. E., Gregg, C. T., & Matwiyoff, N. A. (1975a). Nuclear magnetic resonance of rotational mobility of mouse hemoglobin labeled with [2-^{13}C]histidine. Science 188, 266–268.

London, R. E., Kollman, V. H., & Matwiyoff, N. A. (1975b). The quantitative analysis of carbon-carbon coupling in the ^{13}C nuclear magnetic resonance spectra of molecules biosynthesized from ^{13}C enriched precursors. J. Am. Chem. Soc. 97, 3565–3573.

Longenecker, J. P. & Williams, J. F. (1980a). Use of [2-^{14}C]glucose and [5-^{14}C]glucose for evaluating the mechanism and quantitative significance of the 'liver-cell' pentose cycle. Biochem. J. 188, 847–857.

Longenecker, J. P. & Williams, J. F. (1980b). Quantitative measurement of the L-type pentose phosphate cycle with [2-^{14}C]glucose and [5-^{14}C]glucose in isolated hepatocytes. Biochem. J. 188, 859–865.

Lundberg, P., Harmsen, E., Ho, C., & Vogel, H. (1990). Nuclear magnetic resonance studies of cellular metabolism. Anal. Biochem. 191, 193–222.

McIntyre, L. M., Thorburn, D. R., Bubb, W. A., & Kuchel, P. W. (1989). Comparison of computer simulations of the F-type and L-type non-oxidative hexose monophosphate shunts with ^{31}P-NMR experimental data from human erythrocytes. Eur. J. Biochem. 180, 399–420.

Madden, A., Leach, M. O., Collins, D. J., & Payne, G. S. (1991). The water resonance as an alternative pH reference: relevance to *in vivo* ^{31}P NMR localized spectroscopy studies. Magn. Reson. Med. 19, 416–421.

Magnani, M., Stocchi, V., Piatti, E., Dachà, M., Dallapiccola, B., & Fornaini, G. (1983). Red blood cell glucose metabolism in trisomy 10p: Possible role of hexokinase in the erythrocyte. Blood 61, 915–919.

Malloy, C. R., Sherry, A. D., & Jeffrey, F. M. H. (1987). Carbon flux through citric acid cycle pathways in perfused heart by ^{13}C NMR spectroscopy. FEBS Lett. 212, 58–62.

Malloy, C. R., Sherry, A. D., & Jeffrey, F. M. H. (1990). Analysis of tricarboxylic acid cycle of the heart using ^{13}C isotope isomers. Am. J. Physiol. 259, H987–H995.

Maretzki, D., Reimann, B., & Rapoport, S. M. (1989). A reappraisal of the binding of cytosolic enzymes to erythrocyte membranes. Trends Biochem. Sci. 14, 93–96.

Mehler, A. H. & Cusic, M. E. (1967). Aldolase reaction with sugar diphosphates. Science 155, 1101–1103.

Mendz, G. L., Middlehurst, C. R., Kuchel, P. W., & Crossie, P. A. (1986a). Determination of erythrocyte pyrimidine 5′-nucleotidase activity via ^{31}P nuclear magnetic resonance: Comparison of normal controls with multiple sclerosis patients. Experientia 42, 1016–1018.

Mendz, G. L., Robinson, G., & Kuchel, P. W. (1986b). Direct quantitative analysis of enzyme catalyzed reactions by 2-dimensional nuclear magnetic resonance spectroscopy: Adenylate kinase and phosphoglycerate mutase. J. Am. Chem. Soc. 108, 169–173.

Meyerhof, O., Lohman, K., & Schuster, P. (1936). Aldolase, a carbon-binding enzyme. I. Aldol condensation of dihydroxyacetone phosphate and acetaldehyde. Biochem. Z. 286, 301–318.

Momsen, G., Rose, Z. B., & Gupta, R. K. (1979). A reappraisal of ^{31}P NMR studies indicating enzyme complexation in red blood cells. Biochem. Biophys. Res. Commun. 91, 651–657.

Moon, R. B. & Richards, J. H. (1973). Determination of intracellular pH by ^{31}P magnetic resonance. J. Biol. Chem. 248, 7276–7278.

Morelli, A., Benatti, U., Salamino, F., Sparatore, B., Michetti, M., Melloni, E., Pontremoli, S., & De Flora, A. (1979). In vitro correction of erythrocyte glucose-6-phosphate dehydrogenase deficiency. Arch. Biochem Biophys. 197, 543–550.

Mulder, E., Van Den Berg, J. W. O., & Van Deenen, L. L. M. (1965). Metabolism of red-cell lipids II. Conversions of lysophosphoglycerides. Biochim. Biophys. Acta 106, 118–127.

Murphy, J. R. (1973). Influence of temperature and method of centrifugation on the separation of erythrocytes. J. Lab. Clin. Med. 82, 334–341.

Norton, R. S., MacKay, M. A., & Borowitzka, L. J. (1982). The physical state of osmoregulatory solutes in unicellular algae. A natural-abundance carbon-13 nuclear-magnetic-resonance relaxation study. Biochem. J. 202, 699–706.

Opperdoes, F. R. (1987). Compartmentation of carbohydrate metabolism in trypanosomes. Annu. Rev. Microbiol. 41, 127–151.

Paoletti, F., Williams, J. F., & Horecker, B. L. (1979). Synthesis and cleavage of octulose bisphosphates with liver and muscle aldolases. Arch. Biochem. Biophys. 198, 614–619.

Penhoet, E. E., Kochman, M., & Rutter, W. J. (1969). Molecular and catalytic properties of aldolase C. Biochemistry 8, 4396–4402.

Pontremoli, S., Melloni, E., Salamino, F., Speratore, B., Michetti, M., Benatti, A., Morelli, A., & de Flora, A. (1980). Identification of proteolytic activities in the cytosolic compartment of mature human erythrocytes. Eur. J. Biochem. 110, 421–430.

Potts, J. R. & Kuchel, P. W. (1992). Anomeric preference of fluoro-glucose exchange across human red cell membranes: ^{19}F-n.m.r. studies. Biochem. J. 281, 753–759.

Potts, J. R., Hounslow, A. M., & Kuchel, P. W. (1990). Exchange of fluorinated glucose across the red-cell membrane measured by ^{19}F NMR magnetization transfer. Biochem. J. 266, 925–928.

Price, W. S. & Kuchel, P. W. (1990). Restricted diffusion of bicarbonate and hypophosphite ions modulated by transport in suspensions of red blood cells. J. Magn. Reson. 90, 100–110.

Price, W. S., Chapman, B. E., Cornell, B. E., & Kuchel, P. W. (1989). Translational diffusion of glycine in erythrocytes measured at high resolution with pulsed field gradients. J. Magn. Reson. 83, 160–166.

Purich, D. L. (1983). Contemporary Enzyme Kinetics and Mechanism (Purich, D. L., ed.), Academic Press, Orlando.

Racker, E. (1951). The mechanism of action of glyoxalase. J. Biol. Chem. 190, 685–696.

Rae, C. D., Berners-Price, S. J., Bulliman, B. T., & Kuchel, P. W. (1990). Kinetic analysis of the human erythrocyte glyoxalase system using ^{1}H NMR and a computer model. Eur. J. Biochem. 193, 83–90.

Rae, C. D., Board, P. G., & Kuchel, P. W. (1991). Glyoxalase 2 deficiency in the erythrocytes of a horse: ^{1}H NMR studies of enzyme kinetics and transport of S-lactoylglutathione. Arch. Biochem. Biophys. 291, 291–299.

Rae, C. D., Bubb, W. A., & Kuchel, P. W. (1992). Aldolase-catalyzed diketonephosphate formation from oxoaldehydes. NMR studies and metabolic significance. J. Biol. Chem. 267, 9713–9717.

Raftos, J. E., Bulliman, B. T., & Kuchel, P. W. (1990). Evaluation of an electrochemical model of erythrocyte pH buffering using ^{31}P NMR data. J. Gen. Physiol. 95, 1183–1204.

Rapoport, S. M., Schewe, T., Thiele, B., Dubiel, W., & Schmidt, J. (1987). Maturational breakdown cascade of mitochondria in reticulocytes. Biomed. Biochim. Acta 46, S151–S155.

Richard, J. P. (1991). Kinetic parameters for the elimination reaction catalyzed by triosephosphate isomerase and an estimation of the reaction's physiological significance. Biochemistry 30, 4581–4585.

Rijksen, G., Schoop, I., & Staal, G. E. J. (1977). Properties of human erythrocyte hexokinase related to cell age. Clin. Chim. Acta 80, 193–202.

Rink, T. J. & Hladky, S. B. (1982). Measurement of red cell membrane potential with fluorescent dyes. In: Red Cell Membranes: A Methodological Approach (Ellory, J. C. & Young, J. D., eds.), pp. 321–334, Academic Press, London.

Robinson, G., Chapman, B. E., & Kuchel, P. W. (1984). ^{31}PNMR spin transfer in the phosphoglyceromutase reaction. Eur. J. Biochem. 143, 643–649.

Robinson, G., Kuchel, P. W., Chapman, B. E., Doddrell, D. M., & Irving, M. G. (1985). A simple procedure for selective inversion of NMR resonances for spin transfer enzyme kinetic measurements. J. Magn. Res. 63, 314–319.

Roman, G.-C. & Garfinkel, D. (1978). BIOSSIM—A structured machine-independent biological simulation language. Comput. Biomed. Res. 11, 3–15.

Rose, I. A. (1961). The use of kinetic isotope effects in the study of metabolic control. I. Degradation of glucose-1-d by the hexose monophosphate pathway. J. Biol. Chem. 236, 603–609.
Rothman, D. L., Behar, K. L., Hetherington, H. P., den Hollander, J. A., Bendall, M. R., Petroff, O. A. C., & Shulman, R. G. (1985). ^{1}H-observe/^{13}C-decouple spectroscopic measurements of lactate and glutamate in the rat brain *in vivo*. Proc. Natl. Acad. Sci. USA 82, 1633–1637.
Sable, H. Z. (1966). Biosynthesis of ribose and deoxyribose. Adv. Enzymol. 28, 391–460.
Saito, M. & Kanfer, J. N. (1975). Phosphatidohydrolase activity in a solubilized preparation from rat brain particulate fraction. Arch. Chem. Biophys. 169, 318–323.
Savageau, M. A. (1991). Metabolite channeling: implications for regulation of metabolism and for quantitative description of reactions *in vivo*. J. Theor. Biol. 152, 85–92.
Scofield, R. F., Kosugi, K., Chandramouli, V., Kumaran, K., Schumann, W. C., & Landau, B. R. (1985). The nature of the pentose pathway in liver. J. Biol. Chem. 260, 15439–15444.
Scott, E. M, Duncan, I. W., & Ekstrand, V. (1963). Purification and properties of glutathione reductase of human erythrocytes. J. Biol. Chem. 238, 3928–3933.
Selle, H., Chapman, B. E., & Kuchel, P. W. (1992). Release of choline by phospholipase D and a related phosphoric diester hydrolase in human erythrocytes. Biochem. J. 284, 61–65.
Selle, H., Chapman, B. E., & Kuchel, P. W. (1993). Glycerophosphocholine release in human erythrocytes. ^{1}H spin-echo and ^{31}P-NMR evidence for lysophospholipase. Eur. J. Biochem. 212, 411–416.
Severin, S. E. & Stepanova, N. G. (1981). Interrelationship between glycolysis and the anaerobic part of the pentose phosphate pathway of carbohydrate metabolism in the myocardium. Adv. Enzyme Regul. 19, 235–255.
Shaka, A. J., Barker, P. B., & Freeman, R. (1985). Computer-optimized decoupling scheme for wideband applications and low-level operation. J. Magn. Reson. 64, 547–552.
Sherry, A. D., Nunnally, R. L., & Peshock, R. M. (1985). Metabolic studies of pyruvate- and lactate-perfused guinea pig hearts by ^{13}C NMR. J. Biol. Chem. 260, 9272–9279.
Shulman, R. G. (1979). Biological Applications of Magnetic Resonance. Academic Press, New York.
Shulman, R. G. (1991). Is channeling a disprovable hypothesis? J. Theor. Biol. 152, 133–134.
Simpson, R. J., Brindle, K. M., Brown, F. F., Campbell, I. D., & Foxall, D. L. (1982a). Studies of lactate dehydrogenase in the purified state and in intact erythrocytes. Biochem. J. 202, 581–587.
Simpson, R. J., Brindle, K. M., Brown, F. F., Campbell, I. D., & Foxall, D. L. (1982b). A p.m.r. isotope-exchange method for studying the kinetic properties of dehydrogenase in intact cells. Biochem. J. 202, 573–579.
Simpson, R. J., Brindle, K. M., & Campbell, I. D. (1982c). Spin echo proton NMR studies of the metabolism of malate and fumarate in human erythrocytes. Dependence of free NAD levels. Biochim. Biophys. Acta 707, 191–200.
Slonczewski, J. L., Rosen, B. P., Alger, J. R., & Macnab, R. M. (1981). pH homeostasis in *Escherichia coli*: measurement by ^{31}P nuclear magnetic resonance of methylphosphonate and phosphate. Proc. Natl Acad. Sci. USA 78, 6271–6275.
Srivastava, S. K., Awasthi, Y. C., & Beutler, E. (1974). Useful agents for the study of glutathione metabolism in erythrocytes-organic hydroperoxides. Biochem. J. 139, 289–295.
Stewart, I. M., Chapman, B. E., Kirk, K., Kuchel, P. W., Lovric, V. A., & Raftos, J. E. (1986). Intracellular pH in stored erythrocytes. Refinement and further characterisation of the ^{31}P-NMR methylphosphonate procedure. Biochim. Biophys. Acta 885, 23–33.
Stilbs, P. (1987). Fourier transform pulsed-gradient spin-echo studies of molecular diffusion. Prog. NMR Spectr. 19, 1–45.
Sturman, J. A. (1967). Pentose phosphate pathway metabolism by normal and glucose-6-phosphate dehydrogenase-deficient human red cell haemolysates. Clin. Chim. Acta 18, 245–248.
Styles, P., Grathwohl, C., & Brown, F. F. (1979). Simultaneous multinuclear NMR by alternate scan recording of ^{31}P and ^{13}C spectra. J. Magn. Reson. 35, 329–336.
Susskind, B. M., Warren, L. G., & Reeves, R. E. (1982). A pathway for the interconversion of hexose and pentose in the parasitic amoeba *Entamoeba histolytica*. Biochem. J. 204, 191–196.

Swanson, M. S., Angle, C. R., Stohs, S. J., Wu., S. T., Salhany, J. M, Eliot, R. S., & Markin, R. S. (1983). ^{31}P NMR study of erythrocytes from a patient with hereditary pyrimidine-5′-nucleotide deficiency. Proc. Natl. Acad. Sci. 80: 169–172.

Takeuchi, T., Nishino, K., & Itokawa, Y. (1984). Improved determination of transketolase activity in erythrocytes. Clin. Chem. 30, 658–661.

Taki, T. & Kanfer, J. N. (1979). Partial purification and properties of a rat brain phospholipase D. J. Biol. Chem. 254, 9761–9765.

Thorburn, D. R. (1987). Biochemical and numerical studies of the kinetics of the hexose monophosphate shunt in human erythrocytes. Ph.D. Thesis, Department of Biochemistry, University of Sydney, Australia.

Thorburn, D. R. & Kuchel, P. W. (1985). Regulation of the human-erythrocyte hexose-monophosphate shunt under conditions of oxidative stress. A study using NMR spectroscopy, a kinetic isotope effect, a reconstituted system and computer simulation. Eur. J. Biochem. 150, 371–386.

Thorburn, D. R. & Kuchel, P. W. (1987a). Computer simulation of the metabolic consequences of the combined deficiency of 6-phosphogluconolactonase and glucose-6-phosphate dehydrogenase in human erythrocytes. J. Lab. Clin. Med. 110, 70–74.

Thorburn, D. R. & Kuchel, P. W. (1987b). Glutathione regeneration in density-fractionated erythrocytes. Biomed. Biochim. Acta 46, S248–252.

Thornalley, P. J. (1990). The glyoxalase system: new developments towards functional characterisation of a metabolic pathway fundamental to biological life. Biochem. J. 269, 1–11.

Tökes, Z. A. & Chambers, S. M. (1975). Proteolytic activity associated with human erythrocyte membranes. Biochim. Biophys. Acta 389, 325–338.

Torrance, J. D., West, C., & Beutler, E. J. (1977). A simple rapid radiometric assay for pyrimidine-5′-nucleotidase. J. Lab. Clin. Med. 90, 563–568.

Tsuboi, K. K., Allan, J. F., & Fukunaga, K. (1966). Cofactors and erythrocyte glycolytic capacity. J. Biol. Chem. 241, 1616–1620.

Uotila, L. (1973). Purification and characterisation of S-2-hydroxyacyl glutathione hydrolase (glyoxalase II) from human liver. Biochemistry 20, 3944–3951.

Valentine, W. N., Bennett, J. M., Krivit, W., Konrad, P. N., Lowman, J. T., Paglia, D. E., & Wakem, C. J. (1973). Nonspherocytic hemolytic anemia with increased red cell adenine nucleotides, glutathione and basophilic stippling and ribosephosphate pyrophosphokinase (RPK) deficiency: studies on two new kindreds. Brit. J. Haematol. 24, 157–167.

Vandenberg, J. I., King, G. F., & Kuchel, P. W. (1985). Enkephalin degradation by human erythrocytes and haemolysate studied using ^{1}H NMR spectroscopy. Arch. Biochem. Biophys. 242, 515–522.

Vanderheiden, B. S. (1965). Phosphate esters of human erythrocytes. IV. Sedoheptulose-1,7-diphosphate, octulose-1,8-diphosphate, inosine triphosphate and uridine diphosphate. Biochem. Biophys. Res. Commun. 21, 265–270.

Webb, M. R., Standring, D. N., & Knowles, J. R. (1977). Phosphorus-31 nuclear magnetic resonance of dihydroxyacetone phosphate in the presence of triosephosphate isomerase. The question of non-productive binding of the substrate hydrate. Biochemistry 16, 2738–2741.

Weber, J. P. & Bernhard, S. A. (1982). Transfer of 1,3-diphosphoglycerate between glyceraldehyde-3-phosphate dehydrogenase and 3-phosphoglycerate kinase via an enzyme-substrate complex. Biochemistry 21, 4189–4194.

Williams, J. F. (1980). A critical examination of the evidence for the reactions of the pentose pathway in animal tissues. Trends Biochem. Sci. 5, 315–320.

Williams, J. F. & Blackmore, P. F. (1983). Non-oxidative synthesis of pentose 5-phosphate from hexose 6-phosphate and triose phosphate by the L-type pentose pathway. Int. J. Biochem. 15, 797–816.

Williams, J. F., Rienits, K. G., Schofield, P. J., & Clark, M. G. (1971). The pentose phosphate pathway in rabbit liver. Studies on the metabolic sequence and quantitative role of the pentose phosphate cycle by using a system *in situ*. Biochem. J. 123, 923–943.

Williams, J. F., Clark, M. G., & Blackmore, P. F. (1978a). The fate of ^{14}C in glucose 6-phosphate synthesized from [1-^{14}C]ribose 5-phosphate by enzymes of rat liver. Biochem. J. 176, 241–256.

Williams, J. F., Blackmore, P. F., & Clark, M. G. (1978b). New reaction sequences for the non-oxidative pentose phosphate pathway. Biochem. J. 176, 257–282.

Williams, J. F., Gordon, R. C., Gerdes, R. G., Rienits, K. G., Arora, K. K., & Anderson, J. (1986). The metabolic significance of pentose cycle measurements in perfused liver. Biochem. Int. 13, 321–333.

Williams, J. F., Arora, K. K., & Longenecker, J. P. (1987). The pentose pathway: A random harvest. Impediments which oppose acceptance of the classical (F-type) pentose cycle for liver, some neoplasms and photosynthetic tissue. The case for the L-type pentose pathway. Int. J. Biochem. 19, 749–817.

Wood, T. (1974). Catalysis of pentose phosphate pathway reactions by cytoplasmic fractions from muscle, uterus and liver of the rat, and the presence of a reduced nicotinamide-adenine dinucleotide phosphate-triose phosphate oxidoreductase in rat muscle. Biochem. J. 138, 71–76.

Wood, T. (1985). The Pentose Phosphate Pathway, pp. 128–134, Academic Press, Inc., Orlando.

Wood, T. & Gascon, A. (1980). A lack of reactivity of D-arabinose 5-phosphate with enzymes of the pentose phosphate pathway. Arch. Biochem. Biophys. 203, 727–733.

Xu, A. S-L. & Kuchel, P. W. (1991). Difluorophosphate as a ^{19}F NMR probe of erythrocyte membrane potential. Eur. Biophys. J. 19, 327–334.

Xu, A. S-L., Potts, J. R., & Kuchel, P. W. (1991). The phenomenon of separate intra- and extracellular resonances of difluorophosphate in ^{31}P and ^{19}F NMR spectra of erythrocytes. Magn. Reson. Med. 18, 193–198.

Yeltman, D. R. & Harris, B. G. (1977). Purification and characterization of aldolase from human erythrocytes. Biochim. Biophys. Acta 484, 188–198.

York, M. J., Kuchel, P. W., Chapman, B. E., & Jones, A. J. (1982). Incorporation of labelled glycine into glutathione of intact human erythrocytes by enzyme catalyzed exchange—an NMR study. Biochem. J. 207, 65–72.

Young, J. D. & Ellory, J. C. (1977). Red cell amino acid transport. In: Membrane Transport in Red Cells (Ellory, J. C. & Lew, V. L., eds.), pp. 301–325, Academic Press, London.

STUDIES OF PHYSIOLOGICAL CONTROL OF ATP SYNTHESIS

K. F. LaNoue and C. Doumen

Advances in Molecular and Cell Biology
Volume 11, pages 207–232.

ISBN: 1-55938-844-7

ABSTRACT

Studies of the control of cardiac respiration using isolated heart mitochondria and submitochondrial particles are reviewed and compared with studies of cardiac respiration in perfused hearts and in intact organisms.

The studies of respiratory control in isolated mitochondria indicate that the rate of ATP synthesis is controlled by the substrates of the reaction, phosphate and ADP, and by the electrochemical potential gradient of protons across the mitochondrial inner membrane. The controversy about whether the ATP synthase itself or the adenine nucleotide translocase provides kinetic limitation to flux is discussed from the point of view that the more important kinetic control is provided by the synthase.

When rates of ATP synthesis are varied in intact organs other than the adult mammalian heart, NMR measurements suggest that free ADP also controls respiration and ATP synthesis *in vivo*. However, in the adult heart *in vivo* there appears to be an additional control mechanism which permits ATP synthesis to vary over a 5- to 6-fold range in the absence of appropriate increases in substrate level or increases in the electrochemical potential gradient of protons, a value recently measured in working hearts. Possible mechanisms are discussed in light of reports of allosteric control of the ATP synthase gathered using submitochondrial particles.

I. INTRODUCTION

ATP is the common currency for energy transactions in the cell. Thus, hydrolysis of the high energy phosphate bond of ATP drives the vast majority of diverse energy consuming reactions in cells. For example, the formation of ion gradients across intra- and extracellular membranes is dependent on ATP. ATP hydrolysis also provides the energy for biosynthetic processes, such as protein and nucleic acid synthesis for regulatory processes involved in the formation of cAMP and for phosphorylation of particular enzymes via ATP dependent protein kinases. In cells of higher organisms, most of the ATP consumed by these endergonic reactions is synthesized from ADP and phosphate in mitochondria by a process which uses the energy generated by the oxidative consumption of electron-rich substrates, such as fatty acids and pyruvic acid.

Downstream electron flow to oxygen drives ATP synthesis, but at the same time, electron flow and oxygen consumption are limited by the rate of consumption of ATP, thus conserving the energy available in electron rich carbon substrates (Chance and Williams, 1956; Lemasters, 1984). The link between electron flow and ATP synthesis was not fully appreciated until the details of the chemiosmotic hypothesis were set forth and accepted (Mitchell, 1961; Mitchell, 1979). A vast body of evidence now demonstrates that certain members of the electron transport chain act as proton pumps, thereby generating a gradient of protons across the mitochondrial membranes as electrons flow toward oxygen (Brand et al., 1976; Wikstrom et al., 1981). The proton gradient may then be used by the mitochondrial

ATP synthase as the source of energy for the synthesis of ATP (Alexandre et al., 1978; Kagawa, 1984). The mitochondrial ATP synthase is a multi-subunit protein which spans the mitochondrial membrane. The membrane spanning domain, frequently termed F_o, senses the electrochemical gradient of protons ($\Delta\mu H^+$) and transmits that information and energy contained in it via a series of conformational changes to F_1, the globular, hydrophilic domain which encompasses the catalytic sites (Boyer, 1989; LaNoue and Duszynski, 1992). Thus $\Delta\mu H^+$ is the connecting link between electron flow and ATP synthesis and this explains why one process cannot proceed without the other. The process of ATP synthesis acting through the electrochemical potential gradient is reversible and therefore ATP hydrolysis can drive electron flow in reverse, at least from cytochrome c back through NADH (Klingenberg, 1961).

The data supporting this concept, termed chemiosmotic coupling, has been gathered mostly using *in vitro* experiments with isolated mitochondria (LaNoue et al., 1986), preparations of inner mitochondrial membranes (Papa et al., 1981), isolated components of the electron transport chain (Prochaska et al., 1981) and isolated mitochondrial ATP synthase reconstituted into artificial liposomes (Knox and Tsong, 1983). Recently, studies of the mechanism of the synthesis of ATP *in vivo* have been initiated, in some instances to confirm concepts developed using simpler reconstituted *in vitro* systems (Matthews et al., 1981; Kingsley-Hickman et al., 1987) and in some instances to elucidate how disease processes may interfere with efficient energy coupling (Sako et al., 1988). These studies have brought to light additional layers of control not previously anticipated and as yet not fully explained. It is the goal of this review article to compare the *in vivo* and *in vitro* studies in order to demonstrate areas in which anticipated results have been obtained and areas in which *in vivo* results have been somewhat at odds with predicted results. Since discrepancies between predicted and observed data have been greatest in studies of cardiac metabolism, we will emphasize these studies. For the purposes of this review, the term *in vitro* will be used to describe subcellular studies and *in vivo* will be arbitrarily used to describe those studies which involve both intact organisms and intact organs, such as a perfused heart or liver.

II. STUDIES OF *IN VITRO* ATP SYNTHESIS

A. Chemiosmosis

Studies of mitochondria isolated by differential centrifugation from intact organs began in the early 1900s. It was recognized very early that they were the cellular structures responsible for consumption of oxygen (Lehninger, 1964). Mitochondria are ellipsoid intracellular organelles enclosed by two membranes. The outer membrane has large pores (composed of the protein porin) that allow passage of molecules up to 5 kDa. The membrane has a relatively high lipid to protein ratio and relatively low enzyme activity. The inner membrane, due to numerous invaginations termed cristae, has a much higher surface area than the outer membrane and a very low lipid to protein ratio; about 70% of the membrane is composed of membrane spanning proteins. Some of these proteins are involved in the respiratory electron transport chain and include complexes I, II, III and IV, plus ubiquinone and cytochrome c (Hatefi, 1985). Cytochrome oxidase (complex IV) catalyzes the final act in the flow of electrons by delivery of electrons to oxygen. The complex is responsible for adapting the arrival of electrons, one at a time to the smooth reduction of molecular oxygen to H_2O (which requires four electrons) without the release of a potentially destructive, intermediate reactive oxygen species (Chance et al., 1975).

The inner membrane is relatively impermeable to protons, K^+, and Na^+, and thus is well designed to hold the electrochemical potential gradient of protons which is developed due to the proton pumping activity of the electron transport chain. The ability of individual components of the electron transport chain, such as complex I, complex III (cytochrome bc_1 region) and cytochrome aa_3 to act as proton pumps, has been adequately demonstrated in systems where the isolated proteins have been incorporated into artificial liposomes (Ragan and Hinkle, 1975; Penttila, 1983).

The mitochondrial ATP synthase has also been isolated, its subunit structure determined, and each subunit cloned and sequenced (Walker et al., 1990; Lucken et al., 1990). The mammalian mitochondrial ATP synthase is one of a class of ATPases termed F type ATPases. These enzymes exist in plants, bacteria, and animal cells and are responsible for ATP synthesis which uses $\Delta\mu H^+$ as the energy source for formation of the high energy phosphate bond. When the purified protein is reconstituted into artificial liposomes, an artificially imposed proton gradient can drive ATP synthesis. A recent issue of the *Journal of Bioenergetics and Biomembranes* (Pedersen and Amzel, 1992) was devoted to this interesting enzyme. The issue contains a series of mini-reviews written by a variety of workers in the field. Current opinion, expressed in these, as well as other more comprehensive reviews (Hatefi, 1985; Futai et al., 1989; Penefsky and Cross, 1991), suggest that the effect of the proton gradient is indirect. In all likelihood, the opening and closing of a "gate" in a proton channel contained within the membrane spanning F_o, causes a conformational change in F_o subunits, which confers conformational change to the region of the protein (F_1) which projects into the mitochondrial matrix space (Senior and Wise, 1983; Fillingame, 1990). The gate movement caused by protons leaving the matrix space causes an increase in the binding energy of ATP at the catalytic site, releasing ATP from the enzyme, and allowing ADP and P_i to replace ATP at the catalytic site. Making and breaking high energy phosphate bonds within the catalytic site seems to require little energy. In fact, a recent publication from our laboratory shows that the rate of this process within the catalytic site is unaffected by a 60 mV decrease in the membrane potential which causes a 10-fold decrease in net ATP synthesis (Berkich et al., 1991).

B. Substrate Control of ATP Synthesis: Kinetic or Thermodynamic?

Since F_1, the portion of the ATP synthase molecule which includes the catalytic site extends into the mitochondrial matrix space, access to substrates is limited by the mitochondrial inner membrane (LaNoue and Schoolwerth, 1979). Two membrane spanning proteins, the phosphate transporter, and mitochondrial adenine nucleotide translocase are responsible for providing substrates to the synthase. The phosphate transporter catalyzes an electroneutral process which moves monovalent phosphate and a proton across the membrane (Coty and Pedersen, 1974). The physiological chemical gradient of H^+ favors phosphate influx. The adenine nucleotide transporter exchanges trivalent ADP for tetravalent ATP in an electrophoretic manner (LaNoue et al., 1978). This transport process is driven by the mitochondrial membrane potential in the direction of ATP efflux. Suggestions have been made that the adenine nucleotide transporter may limit the rate of ATP synthesis in isolated intact mitochondria and in intact organs (Davis and Davis-Van-Thienen, 1978; Lemasters and Sowers, 1979; Kunz et al., 1981; Wanders et al., 1981). On the other hand, little support exists for the concept that the phosphate transporter limits net synthesis since it has a very high V_{max} relative to physiological rates of ATP synthesis (Ligeti et al., 1985).

Data in support of the idea that the translocase limits flux are largely based on the fact that it is far from equilibrium and thus likely to operate unidirectionally and under kinetic control . The ATP synthase on the other hand is readily reversible and it has been generally assumed that it establishes an equilibrium between the adenine nucleotides inside the mitochondria and the $\Delta\mu H^+$ across the mitochondrial membrane. Several lines of evidence show that the translocase is not readily reversible in the presence of a high physiological $\Delta\mu H^+$. During net ATP synthesis, the ratio of ATP to ADP inside the mitochondria is usually lower than the ratio outside, but still the ratios do not conform to the Nernst equation (Klingenberg and Rottenberg, 1977). However, under conditions of zero net flux, this condition can be met (Kuster et al., 1981). Also, using liver mitochondria which may employ intramitochondrial ATP for the synthesis of carbamyl phosphate, a steady state situation can be set up involving net synthesis of ATP at the expense of $\Delta\mu H^+$, but involving no net transport of external nucleotide. Under these conditions, the transporter should be close to equilibrium. When this mitochondrial condition was compared to one with a similar rate of respiration where ATP was made inside the mitochondria but consumed outside, the external ATP/ADP ratio was quite different from the steady state situation involving no nucleotide transport. The data showed that the nucleotide transporter was far from equilibrium when it was involved in net transport (Kuster et al., 1981; Tager et al., 1981; 1983). These data do not prove that the translocase is rate limiting, but do strongly imply it is far from equilibrium.

Other more direct experiments have been carried out using both liver and heart mitochondria to assess the power of the translocase to limit ATP synthesis. Using a technique described by Kacser and Burns (1973), (see also Cornish-Bowden, this volume) the control strength of the translocase was determined by measuring respiration rates as a function of the concentration of a specific and noncompetitive inhibitor of the translocase. The rationale of this approach is that the steepness of the initial slope of the titration curve will reflect the degree to which the inhibited reaction determines overall flux in a given pathway. For example, if one adds sufficient inhibitor to block 20–30% of the translocase, with no effect on respiration, the control strength must be negligible.

The data suggested that the control strength is 20–30% in liver (Lemasters and Sowers, 1979; Groen et al., 1982), but negligible in heart (Forman and Wilson, 1983; Doussiere et al., 1984). These experiments have been criticized because substrate concentrations (ADP outside, ATP inside the mitochondria) were not controlled during the inhibitor titrations, thereby minimizing control strength (Gellerich et al., 1983; Davis and Davis-VanThienen, 1984). Thus, no clear consensus has been reached.

Moreover, the implicit assumption that the ATP synthase is always close to equilibrium and therefore not able to provide rate limitation has recently been called into question. Evidence in favor of this assumption is that the free energy of ATP hydrolysis calculated from the extramitochondrial mass action ration, (ATP/ADP)(P_i), is nearly equal to the free energy difference between NADH and cytochrome c under a variety of conditions.

A reaction in equilibrium is one in which the forward reaction is equal to the backward reaction. A "near equilibrium" reaction occurs when the difference between the forward and backward reactions (net flux) is small relative to the unidirectional forward and backward velocities. The first direct measurements of forward and backward velocity as a function of net synthesis of ATP in isolated liver and heart mitochondria were reported in 1986 by our laboratory working in conjunction with George Radda and colleagues (LaNoue et al., 1986). Unidirectional fluxes were measured isotopically using radiolabeled phosphate while net synthesis of ATP was determined enzymatically. When respiration was minimal and net ATP synthesis was zero because of lack of added ADP, isotopic exchange was rapid and forward and backward rates were equal. As the concentration of ADP was increased in the presence of saturating phosphate, the backward reaction decreased dramatically while the forward reaction increased only slightly. When respiration rates were maximal at saturating ADP levels, the back reaction had decreased to zero while the forward reaction had only doubled. Even at half maximal rates of ATP synthesis, the net synthesis of ATP was greater than the back reaction. Therefore, since the forward and reverse rates were not large relative to net synthesis, the conclusion could be drawn that at physiological rates of ATP synthesis, the overall sequence of reactions converting extramitochondrial ADP and phosphate to extramitochondrial ATP was far from equilibrium. The increase in ATP synthesis observed when external ADP increased was due in large measure to a decrease in the back reaction.

To determine whether the rates of the unidirectional reactions monitored were dependent on the translocase or the synthase activities, two independent types of experiments were carried out. In one type of experiment, isolated mitochondria were used to monitor the rate of transport of external ATP into the mitochondria in the absence of added ADP (no net ATP synthesis). Conditions were the same as those used to measure ATP/$^{32}P_i$ exchange but transport was monitored by using NMR saturation transfer between internal and external ATP. This experiment was possible because the β-phosphorus of the internal Mg^{2+}-chelated ATP has an NMR spectral peak separate from the β-phosphorus of external ATP, when a Mg^{2+} chelator such as CDTA is included in the external media. Since the rate of transfer thus measured was an order of magnitude more rapid than the observed ATP/^{32}Pi exchange, it was apparent that the isotopic ATP/^{32}Pi measurement monitored activity due to the ATP synthase (Masiakos et al., 1991). This conclusion was confirmed by additional experiments, in which oxygen^{18} exchanges between H_2O and ^{18}O-inorganic phosphate were monitored in isolated mitochondria and compared with the $^{32}P_i$ exchange rates (Berkich et al., 1991). The H_2O /^{18}O-phosphate exchange is a partial reaction of the ATP synthase, but since it does not involve nucleotides, it is not limited by the adenine nucleotide translocase. Over a wide range of ATP/^{32}Pi exchange rates, varied with substrate and temperature, the ^{18}O exchange rate followed a fixed 2:1 ratio with the $^{32}P_i$ exchange. The 2:1 ratio is reasonable based on the hypothesis that the molecular mechanism of the synthase involves an alternating catalytic site mechanism and that phosphorus exchange rates observed in intact mitochondria are a direct measure of synthase activity (Berkich et al., 1991; LaNoue and Duszynski, 1992).

When ATP synthesis is maximal, in the presence of saturating ADP levels, the backward reaction (which involves radiolabeled phosphorus transfer from ATP to inorganic phosphate) is zero. Under identical conditions, ^{18}O exchange measurements reveal that the synthase flux proceeds 80–90% in the forward direction. Thus while the translocase is completely unidirectional when ADP is saturating, the reversal of the synthase is also minimal. Taken together, the data have shown that the ATP synthase has a much lower V_{max} than the translocase and therefore may be a more likely site of flux control at intermediate physiological rates of respiration than the translocase (LaNoue and Duszynski, 1992).

C. Allosteric Control of the ATP Synthase

The data discussed thus far have concerned regulation of mitochondrial ATP synthesis by substrates. Since mitochondrial ATP synthesis is catalyzed by a proton pumping ATPase working in reverse, one of the substrates is $\Delta\mu H^+$; the other two are ADP and inorganic phosphate. Using intact mitochondria, the only means found thus far to increase the rate of ATP synthesis is to increase one of these substrates. An allosteric, regulatory subunit of the synthase has been isolated and characterized, but its activity has been demonstrated unequivocally thus far only in submitochondrial particles (van de Stadt et al., 1973). These submitochondrial particles, prepared from bovine hearts, exhibit slow rates of ATP hydrolysis until they are energized by addition of a respiratory substrate, after which rapid hydrolysis of ATP can be demonstrated. A small protein (10 kD) is associated with the particles during inhibition and the protein is released when inhibition is released. The inhibitor protein has been isolated and sequenced and when the purified protein is added back to submitochondrial particles depleted of inhibitor protein, ATP hydrolysis is inhibited (Pulman and Monroy, 1963; Power et al., 1983; Schwerzmann and Pedersen, 1986).

Despite these findings, it has been difficult to prove a physiological role for the inhibitor protein. Its function may be to place a brake on ATP hydrolysis in anaerobic or ischemic tissue. Evidence for this proposal comes from studies by Rouslin and co-workers (Rouslin, 1983; Rouslin, 1987; Rouslin and Pullman, 1987) who find that when hearts from larger mammals such as dogs are subjected to ischemia, submitochondrial particles subsequently isolated exhibit severely inhibited rates of ATP hydrolysis which is relieved by maneuvers designed to release the inhibitor protein from the particles.

There are also reports of a second inhibitor protein whose effect is minimized by Ca^{2+} (Yamada et al., 1980; Yamada and Huzel, 1988) but its Ca^{2+} sensitive effect has also been demonstrated directly only in submitochondrial particles (Yamada and Huzel, 1989). Indirect evidence obtained using isolated rat cardiac myocytes suggests that Ca^{2+} ions may regulate cardiac ATP synthase (Das and Harris, 1990a,b; Harris and Das, 1991). When cardiac cells were stimulated to contract with isoproterenol and electrical depolarizations, oligomycin sensitive ATP hydrolysis in sonicated cells was 50 percent faster than in similarly sonicated resting cells. The effect of isoproterenol could be blocked by the reagent, Ruthenium Red, which prevents entry of Ca^{2+} into the mitochondrial matrix. At the low concentrations used in the cell experiments, Ruthenium Red did not block Ca^{2+} entry into the cytosol. Since the ATPase assay of sonicated cells in these studies was carried out in the presence of Ca^{2+} chelators, the putative regulatory phenomenon must have been a stable one, and therefore not due to direct action of Ca^{2+} on the ATP synthase.

Unfortunately, this phenomenon cannot be demonstrated using isolated intact mitochondria. Although certain mitochondrial dehydrogenases including pyruvate dehydrogenase (PDH), α-ketoglutarate dehydrogenase (αKGDH) and isocitrate dehydrogenase (ICDH) are activated by Ca^{2+} in the physiologically relevant range, similar direct stimulation of the ATP synthase is not readily demonstrable. Indirect experiments suggest some slight stimulation but even these results are not routinely reproducible (Moreno-Sanchez and Hansford, 1988). Even the conditions for the Ca^{2+} activation of the dehydrogenases are somewhat stringent. Pyruvate dehydrogenase activation is dependent on inhibition by PDH kinase and stimulation by a Ca^{2+} dependent phosphatase (Reed, 1981). When intramitochondrial ATP/ADP is low, PDH is not phosphorylated and activation of the phosphatase by Ca^{2+} has little or no effect. Various other factors such as NADH levels and pyruvate concentration also alter phosphorylation/dephosphorylation. Only when ATP and NADH levels are high and pyruvate levels low, does Ca^{2+} effectively stimulate pyruvate dehydrogenase. In the case of αKGDH, the Ca^{2+} interacts directly with a subunit of the enzyme, lowering the K_m of α-ketoglutarate. Thus the effect of Ca^{2+} on the oxidation of α-ketoglutarate can be observed only at nonsaturating levels of α-ketoglutarate. Stimulation of ICDH by Ca^{2+} in the low physiological range occurs only if ADP levels are low. Therefore an increase in ADP could easily override inhibition of ICDH by insufficient Ca^{2+}. *In vitro* studies with isolated mitochondria show that when Ca^{2+} levels increase under conditions which do not favor stimulation of the Ca^{2+}-sensitive dehydrogenase, the effect on respiration and ATP synthesis is minimal or nonexistent (Wan et al., 1989). The overall conclusion can be drawn from the *in vitro* studies (as stated above), that ATP synthesis of isolated intact mitochondrial is altered only by substrate availability, the substrates being inorganic phosphate, ADP and $\Delta\mu H^+$.

III. *IN VIVO* STUDIES OF ATP SYNTHESIS

Over two decades ago, *in vivo* studies of the citric acid cycle and control of ATP synthesis were initiated by J. R. Neely et al. (1972). These workers studied perfused rat hearts and measured changes in high energy phosphates and citric acid cycle intermediates while increasing the work done by retrogradely perfused hearts. Respiration rates were measured and found to rise from 25 μmoles O_2/min·g dry wt at 50 mm Hg perfusion pressure to 61 μmoles O_2/min·g dry wt at 120 mm Hg perfusion pressure. In some cases hearts were perfused with glucose alone and in others with glucose plus radiolabeled acetate in order to more accurately trace carbon flow. This study was remarkable not so much for the work related changes in metabolites reported but rather for the relative lack of change in high energy phosphates and citric acid cycle intermediates in the face of large changes in ATP synthesis and citric acid cycle flux. Since rat hearts contain ~250 mg of mitochondria per g dry weight, the respiratory V_{max} of the tissue (based on the V_{max} measured with isolated mitochondria at 37 °C) can be estimated to be 150 μmoles/min·g dry wt. Thus, the low and high work rates were 17 and 41% of V_{max}, respectively. Assuming that phosphate levels were saturating and the K_m of ADP (20 μM) similar to that measured *in vitro*, the free ADP obtained from the theoretical Michaelis-Menten equation yielded values of 4 μM at low work and 14 μM at higher work. The measured values of ADP were 3–4 mM at both high and low work. The obvious discrepancy was dismissed at the time as due to technical problems in measuring phosphate, ADP, ATP, and creatine phosphate. Uncertainty concerning the measurements made were due to the possibility that rapid freezing and acid extraction of the tissue might lead to artifactual hydrolysis of ATP and creatine phosphate prior to enzymatic analysis. The measured ADP values were in the mM range, far higher than the K_m of the ADP translocase measured *in vitro*. It is clear now that most of the measured ADP is bound and therefore not available to the mitochondria. Free ADP may be calculated if one assumes the creatine kinase reaction is in equilibrium and if one knows the concentrations of the components of the creatine kinase mass action ratio (creatine, creatine phosphate and ATP) with reasonable accuracy. However, at that period in time, this method lacked precision due to uncertainty of the values of intracellular pH and Mg^{2+}, both of which influence the position of the creatine kinase equilibrium (Lawson and Veech, 1978; Veech, et al., 1978).

With the advent of the use of ^{31}P-nuclear magnetic resonance (NMR) spectroscopy it became possible to assess the tissue levels of phosphate metabolites, such as phosphate, ATP and creatine phosphate without subjecting the tissue to acid extraction. Since the method is noninvasive it has also been possible to use the same technique to monitor *in vivo* changes in the hearts of living mammals. ^{31}P-NMR spectroscopy not only provides a means to assess the quantity of ATP and creatine phosphate within the sensitive volume of the spectrometer, but it also provides a way to assess pH and free Mg^{2+}. The position of the spectral peak of ortho-phosphate shifts relative to creatine phosphate as a function of pH and the spectral peak of the β-phosphorous of ATP shifts relative to the α-peak as a function of free Mg^{2+} (Moon and Richards, 1973; Gupta and Moore, 1980). As this permits accurate assessment of the K_{eq} of creatine kinase, it also permits accurate assessment of free ADP. Using reasonable values of pH and free Mg^{2+}, free ADP can be calculated from the data reported by Neely et al. (1972). Their values for free ADP vary between 50–60 μM and do not increase with work. Technical improvements due to utilization of NMR techniques have led to a vast number of studies of changes in high energy phosphates not only in cardiac tissue but also in kidney (Freeman et al., 1983), brain (Cadoux-Hudson et al., 1989), liver (Cohen, 1983; Malloy et al., 1986), and skeletal muscle (Meyer et al., 1982; Kushmerick and Meyer, 1985; Brindle et al., 1989). The purpose of some of these studies has been to assess the physiological changes in high energy phosphates due to changes in substrate supply and work output. Most of the studies however focus on the effects of limiting the supply of O_2 to the tissue, with emphasis mainly on the reflow period following ischemia or anoxia to elucidate optimal conditions for recovery. This review, however, concerns only those studies which assessed the status of the high energy phosphates during normoxia.

With respect to this, the vastly improved technology of ^{31}P NMR has in fact confirmed the original puzzling results that ATP and creatine phosphate decrease very little with increased respiration and the small increases in ADP sometimes observed are inadequate to explain increases in respiration. It is interesting to note that some perfused hearts exhibit moderate decreases in creatine phosphate at very high work induced by adrenergic agonists and other inotropic agents (Matthews et al., 1982; Headrick et al., 1989). More recent ^{31}P NMR studies however show almost unanimously that in most *in vivo* adult mammalian hearts, acceleration of respiration in response to increased cardiac work is accompanied by little or no changes in CrP/P_i ratio (Balaban et al., 1986; Katz et al., 1989; Detre et al., 1990).

Notable exceptions are recent *in situ* rat and cat heart studies, showing small but definite decreases in PCr upon stimulation of cardiac work which may underscore species-specific differences in the mechanism of oxygen delivery (Bittl et al., 1987; Ligeti et al., 1987).

Similar studies carried out in skeletal muscle in intact animals yield data which are quite different, but in good accord with data obtained with isolated mitochondria. Skeletal muscles were monitored by NMR spectroscopy while rates of depolarization were increased inducing large increases in respiration. Plots of the data as respiration versus CrP/P_i or as respiration rates versus free ADP resulted in clear Michaelis–Menton kinetics yielding a K_m of free ADP for respiration of approximately 20 μM (Radda, 1986; Brindle et al., 1989). As this value is similar to that obtained with isolated mitochondria, these results suggested that the rate of ATP synthesis in skeletal muscle is controlled by ADP availability. The contractile elements consume ATP and produce phosphate and ADP which in series, stimulate the mitochondria to synthesize more ATP. By analogy, it would appear that this is not the case in the adult mammalian heart, but instead some parallel control mechanism stimulates ATP synthesis at the same time that its hydrolysis by the myosin ATPase is increased.

Evidently this putative parallel cardiac regulatory process develops after birth, not before, as neonatal hearts behave much like skeletal muscle. A close correlation between O_2 consumption and cytosolic free ADP exists in neonatal lamb hearts and neonatal rabbit hearts, while parallel studies carried out using adult rabbit hearts showed little or no change of ADP occurs with work (Perry et al., 1988; Portman et al., 1989).

These measurements of free ADP as a function of respiration are feasible only in relatively homogenous tissues and tissues which express creatine kinase. Until recently, such experiments could not be performed in liver tissue due to a lack of a functional creatine kinase system, making estimations of free ADP impossible. The development of a transgenic mouse line which expresses creatine kinase in the liver by Koretsky and co-workers (Koretsky et al., 1990) has removed this barrier and opened up new avenues in the bioenergetic studies of this organ. In this volume, Koretsky describes studies which show that when a transgenic mouse liver expressing creatine kinase is perfused and respiration increased by processes not involving increases in cytosolic free Ca^{2+}, creatine phosphate decreases, and free ADP increases according to Michaelis–Menten kinetic curves with a K_m of free ADP near 20 μM.

Therefore, the *in vivo* studies reveal that in most organs studied, increases in respiration can be accounted for by increases in ADP. However, in the *in vivo* adult mammalian heart the regulatory process for ATP synthesis does not appear to involve changes in free ADP, but appears to be uniquely controlled in a way which cannot yet be demonstrated *in vitro*.

IV. *IN VIVO* CONTROL OF ATP SYNTHESIS BY Ca^{2+}

The observation that work-induced acceleration of O_2 consumption occurs without a significant change in cytosolic ADP and P_i prompted a search for alternative regulatory mechanisms linking ATP utilization in the cytosol to production in the mitochondria. Since three mitochondrial dehydrogenases are activated by Ca^{2+} in the physiological range, the possibility suggested itself that respiration may be regulated by the generation of NADH. In specific terms, the build-up of NADH could potentially increase electron flow to oxygen, thereby increasing the proton gradient which represents the final link between substrate oxidation and ATP phosphorylation. As myosin ATPase is regulated by Ca^{2+} as well, an increase in respiration and ATP production could be achieved by this parallel mechanism without a change in free ADP.

Data both for and against this hypothesis of parallel activation have been collected. PDHa, the active dephosphorylated form of PDH, increases in perfused hearts after norephinephrine treatment. Presumably, Ca^{2+} stimulates PDH phosphatase. The presumption is supported by the observation that Ruthenium Red, an inhibitor of Ca^{2+} influx into mitochondria, prevents the norepinephrine induced increase in PDHa, without blocking the inotropic action of the hormone (McCormick and Denton, 1981; 1984; McCormick and England, 1983).

NMR studies by Balaban and co-workers (Katz et al., 1988) and others (Unitt et al., 1989) have provided information about the changes in high-energy phosphates as a function of work in the presence and absence of Ruthenium Red. In the absence of Ruthenium Red, no change in ATP or CrP was observed in glucose perfused rat hearts. Small, but significant changes were reported during perfusion with high (10 mM) pyruvate concentrations. The presence of Ruthenium Red, however, had no effect on the increase in respiration on addition of norepinephrine. In contrast to the control situation, free ADP increased in the presence of Ruthenium Red as respiration increased.

Although these data have been quoted as support for the Ca^{2+} activation hypothesis, certain aspects of the data are puzzling. Calculation of free ADP in the presence of Ruthenium Red suggests that it is always too high to control respiration. The Ca^{2+} activation hypothesis implies that respiration is low at low work due to inhibition of Ca^{2+}-sensitive dehydrogenases. Even though ADP is high, ATP synthesis is low because of a low $\Delta\mu H^+$. An increase in Ca^{2+} increases NADH and $\Delta\mu H^+$. Simultaneous activation of myosin ATPases and other cytosolic Ca^{2+} dependent ATPases would prevent a decrease in ADP. If Ruthenium Red prevents Ca^{2+} from entering the mitochondria, but allows the rise in cytosolic Ca^{2+}, muscle contractile activity would hydrolyze ATP to ADP, but the rise in ADP should be relatively ineffective in stimulating respiration since ADP was high prior to the increase in cytosolic Ca^{2+}; the translocase would remain saturated, and mitochondrial dehydrogenases would remain inhibited by lack of Ca^{2+}. However, experimental results show a normal increase in respiration with work. Other investigations of the effect of Ruthenium Red on respiratory control in intact hearts also report that Ruthenium Red fails to reduce work capacity or maximal rates of respiration (From et al., 1990).

Nevertheless, ample evidence indicates that the work capacity of the heart can be limited by substrate availability. When hearts are perfused with glucose in the absence of added insulin and in the presence of an inhibitor (4-bromocrotonic acid) of endogenous fatty acid oxidation, rate pressure product in excess of 50 RPP could not be achieved nor could the hearts respire at rates faster than 50 μmoles/min·gdw. It is also clear that some substrates provide more NADH to the electron transport chain than others. Using NMR spectroscopy, From et al. (1990) compared a variety of cardiac substrates (glucose, lactate, octanoate, pyruvates, and palmitate) for their ability to generate high energy phosphates at low work load and their ability to sustain high work levels. Pyruvate and octanoate produced higher levels of creatine phosphate and lower levels of free ADP than the other substrates.

Increasing cardiac work in hearts perfused with the latter substrates resulted in increases in free ADP possibly sufficient to cause observed increases in respiration. The free ADP levels in hearts perfused with glucose, palmitate, or lactate were already high at low work and remained high and constant after stimulation of cardiac work. The observations suggested the existence of two different mechanisms of respiratory control. With the more "potent" substrates, ADP appeared to control respiration. On the one hand, perfusion with substrates like glucose, palmitate and lactate resulted in only a moderate provision of NADH to mitochondria. This relative lack of reducing equivalents reduced the ratio of ATP/ADP enough to saturate mitochondria with ADP thereby precluding control by ADP. The authors presumed that control was provided by variable substrate delivery which appeared to be regulated by work. The natural link between work (muscle contraction) and NADH delivery to the electron transport chain is Ca^{2+}, but strangely, activation of pyruvate dehydrogenase by dichloracetate in glucose perfused hearts and glucose plus insulin perfused hearts, thereby replacing and bypassing Ca^{2+} control of the dehydrogenase, seemed to have little influence in the kinetics of ATP synthesis.

V. *IN VIVO* FLUX MEASUREMENTS OF THE ATP SYNTHASE

NMR is a powerful technique for the investigation of cellular metabolism *in vivo*. The technology allows detection of a wide range of nuclei in systems ranging from isolated cells and perfused organs to intact organisms, making the method extremely versatile. In particular, ^{31}P NMR has provided insight into energy metabolism since it permits continuous monitoring of phosphate metabolites during metabolic transitions. The technique has proven useful as well in the study of the flux of through individual enzymes *in vivo* e.g. the unidirectional flow of a phosphorus nuclei from one metabolite to another (Meyer et al., 1982; Brindle,

1988). The method relies on the selective perturbation of a particular phosphorus nuclei by electromagnetic pulsing, after which this electromagnetically labeled (or saturated) nucleus can be followed in its chemical exchange from one molecular species to another. Hence the nomenclature of saturation transfer. The limitations imposed are that the two phosphorus species must have separate resonances, a certain "critical" resonance signal in order to quantify the transfer and that the half time of exchange is comparable with the spin-lattice relaxation time of the exchanging nuclei (Meyer et al., 1982; Brindle, 1988). The experiments are, therefore, quite analogous to radioisotope techniques with the exception that saturation transfer permits instantaneous labeling of intracellular and intramitochondrial metabolites.

^{31}P Magnetic saturation transfer techniques have been used extensively to measure fluxes in the reaction catalyzed by creatine kinase (Matthews et al., 1982; Shoubridge et al., 1984; Bittl and Ingwall, 1985; Chance et al., 1986). Similar measurements have been employed to measure the flux between P_i and ATP in a variety of systems in an effort to evaluate the *in vivo* P:O ratio and the unidirectional flux through the ATP synthase. Both the execution and interpretation of the experiments have been susceptible to criticism. The interpretation of the comparison between rate of conversion of inorganic phosphate to ATP, as measured by NMR, to the rate of oxygen consumption (P:O) has varied. The ATP synthase is a reversible reaction, but the empirically measured ratio between O_2 consumption and ATP synthesis refers strictly to the net rate of ATP synthesis in relation to oxygen consumption. Because of the mechanisms of oxidative phosphorylation, this ratio is 3.0 even in well-coupled mitochondria and declines in proton "leaky" mitochrondria. Thus, P:O ratios provide a measure of the integrity of the mitochondrial membrane. The assumption that the rates of P_i conversion to ATP determined by saturation transfer experiments can be used to determine the *in vivo* P:O ratio is, therefore, not strictly valid as it only provides information about the NMR measured unidirectional rate of ATP synthesis, not net synthesis. Measurements of ratios of saturation transfer of P_i to ATP to O_2 consumption greater than 3 therefore demonstrate the presence of *in vivo* ATP/P_i exchange (or back reaction) equal to the P_i to ATP measurement minus the maximal rate of net synthesis or 3 times the site of oxygen consumption. If, for example, the ATP synthase catalyzed an equilibrium reaction, the NMR measured P:O ratio should be an order of magnitude higher than 3.0.

Although powerful, the execution of this technique is tied to certain implicit assumptions. In general, the resonance frequencies of mitochondrial and cytosolic ATP are assumed to be equal. Furthermore, presumptions are made that the saturation of the γ-phosphorus of ATP affects both mitochondrial and cytosolic compartments identically, thereby bypassing any possible limitations imposed by the translocase. Finally, the transport of cytosolic P_i across the mitochondrial membrane is assumed to be rapid relative to the conversion of P_i to ATP. This is especially important if the P_i resonance observed by NMR is solely cytosolic (Hutson et al., 1989; Garlick et al., 1992).

Notwithstanding the inherent limitations, significant advances have been made due to a combination of technical improvements with intellectual exchange of ideas. Although the first *in vivo* measurement of the ATP synthase reaction were made on *E.coli* (Brown et al., 1977), the first combination with measurements of oxygen consumption were initially carried out on yeast (Alger et al., 1982). Subsequently, identical studies were performed on perfused heart (Matthews et al., 1981) and kidney (Freeman et al., 1983) in order to evaluate mitochondrial oxidative phosphorylation *in vivo*. The first studies in yeast yielded a P_i to ATP rate of 3.5 μmoles/sec·g with O_2 consumption of 0.02 μmoles O_2/sec·g resulting in a P:O ratio of 87.5, indicating near-equilibrium of the ATP synthase. Later studies of yeast, however, revealed values close to the theoretical value of 3 (Brindle and Krikler, 1985). Similar agreements with the theoretical value of the P:O ratio were obtained from the heart, brain, and kidney studies (Matthews et al., 1981; Shoubridge et al., 1982; Freeman et al., 1983), suggesting that the mitochondrial ATP synthase reaction does not catalyze ATP/P_i exchange *in vivo*. As discussed above, isolated mitochondria data indicate that this is true only when ADP levels are saturating. These data were surprising considering the fact that the saturation transfer measurements were made on tissue consuming O_2 at rates well below the tissue specific V_{max} of O_2 consumption. The rat hearts for example were perfused with glucose in the low work Langendorff mode or with low concentrations of pyruvate (2 mM) at only 70 cm H_2O hydrostatic pressure . This suggests, as did previous studies, that ADP is saturating for respiration under these conditions, and that some additional mode of control of ATP synthesis operates.

In an attempt to explain these low rates of phosphate turnover, additional saturation transfer studies of ATP synthesis in the rat heart were pursued and expanded by examination of the effect of increased oxygen consumption (Kingsley-Hickman et al., 1987). At the same time, several studies implied that rates obtained by saturation transfer may be "contaminated" by the contribution of cytosolic P_i exchange with ATP due to the combined reactions of glyceraldehyde-3-phosphate dehydrogenase and phosphoglycerate kinase (Brindle and Krikler, 1985; Brindle and Radda, 1987). This would result in elevated P:O ratios, explaining the observed discrepancies between the reports of a P:O of 3 (Matthews et al., 1981; Freeman et al., 1983) in mammalian organs, and 82 in yeast (Alger et al., 1982). With this in mind, the new study revealed that the P_i turnover in glucose-perfused hearts was 12 times the oxygen consumption (P:O = 6). However, elimination of the glycolytic component by inhibition of glycolysis with iodoacetate or by perfusing with pyruvate as sole substrate reduced the P:O ratio to 2.4, giving additional support to the conclusion that glycolytic enzymes contribute to the measured exchange. Since no change in the P:O ratio was observed over a wide range of workload and O_2 consumption, the conclusion was drawn that the "operative" P:O ratio in the intact heart is less than the theoretical P:O with the mitochondrial ATP synthase proceeding only in the forward reaction (Kingsley-Hickman et al., 1987). In sharp contrast to the *in vitro* data with isolated mitochondria, the rate of net synthesis was not influenced by the back reaction, but instead net synthesis was equivalent to the forward reaction over the whole range of respiration tested, implicating once more the absence of respiratory control by ADP. Morever, recent studies of the partial reactions of the ATP synthase show that the forward reaction is a nearly linear function of $\Delta\mu H^+$ in the presence of saturating ADP levels (Berkich et al., 1991).

VI. *IN VIVO* STUDIES OF THE EFFECTS OF MITOCHONDRIAL NADH AND $\Delta\Psi m$ ON ATP SYNTHESIS

Since any factor which increases the supply of electrons to the electron transport chain can increase $\Delta\mu H^+$, numerous recent studies have attempted to measure the mitochondrial oxidation-reduction state as a function of respiration. As no redox couple exists for the cardiac mitochondrial space analogous to the cytosolic lactate/pyruvate couple, oxidation–reduction is most conveniently measured by monitoring NADH fluorescence. NAD^+ does not fluoresce and mitochondrial NADH fluoresces two orders of magnitude more than cytosolic NADH (Nuutinen, 1984). NADH surface fluorescence measurements can therefore provide a qualitative measure of mitochondrial redox state. Unfortunately, numerous technical difficulties caused by motion artifacts, interference by myoglobin and inherent difficulties in standardization have decreased the reliability of these measurements. The problems are more severe in *in vivo*, blood-perfused hearts. Still, most studies which have employed pyruvate as a substrate for isolated perfused rat hearts find that NADH surface fluorescence decreases with work (Steenbergen et al., 1978; Schaffer et al., 1978). Only one group has recorded an increase in NADH fluorescence when work was increased by doubling the rate of electrical pacing in glucose perfused hearts (Katz et al., 1986).

Due to the inherent difficulties associated with the measurement of NADH surface fluorescence, this laboratory has devised a method to estimate the mitochondrial electrical potential gradient ($\Delta\psi m$) in perfused rat hearts (Wan et al., 1993a,b). Since the gradient is the major component of $\Delta\mu H^+$ and the final link between the redox potential and ATP synthase, important information can be gathered from its vectorial displacement as cardiac work increases. The method used to determine $\Delta\psi m$ takes advantage of the property of tetraphenylphosphonium (TPP^+), a lipophilic cation, which at equilibrium distributes across lipid membranes according to the Nernst equation. Lipophilic cations have been used extensively to calculate $\Delta\psi m$ in isolated mitochondria and cells (Rottenberg, 1979; Ritchie, 1984). The cations bind passively to mitochondrial and other cell membranes. To obtain the correct equilibrium distribution, one must measure and correct for the cation binding. Although the cation equilibrates rapidly with isolated mitochondria, equilibration with perfused rat hearts is slow, making it impractical to wait for equilibrium to occur. Instead, since equilibrium is the situation where the rate of uptake is equal to the rate of efflux, the first order rate constants of uptake and release of TPP^+ were determined and used to calculate distribution at equilibrium. Thus at equilibrium:

$$K_{eq} = k_{out}/k_{in} = [TPP^+]_e/[TPP^+]_t \text{ and} \quad (1)$$

$$\Delta\psi m = 61.5 \log ([TPP]m/K_{eq}[TPP]t) - \Delta\psi c \text{ (Wan et al., 1993a).} \quad (2)$$

Measurements of $\Delta\psi m$ were carried out in hearts perfused with either glucose, lactate, or pyruvate at three widely different work loads spanning a fourfold change in respiration rates. As in previous studies, pyruvate perfused hearts at all work loads developed higher levels of ATP and creatine phosphate compared to hearts perfused with glucose alone. As one might expect from the higher phosphorylation potential at low work loads, total enzymatically measured NADH was approximately two times higher in pyruvate compared to glucose perfused hearts. Since cytosolic NADH is low in pyruvate perfused hearts, the data suggest that mitochondrial NADH is far higher in the presence of pyruvate. This led to the expectation that $\Delta\psi m$ would be higher in a pyruvate-perfused heart than a glucose-perfused heart and as expected, at low work, $\Delta\psi m$ was 27 mV more negative in the pyruvate perfusions. The absolute value obtained in the pyruvate perfusions was –140 mV, reassuringly close to values obtained *in vitro* using isolated rat heart mitochondria.

When hydrostatic pressure was increased, and when isoproterenol was added to the perfusate, increases in respiration were independent of perfusate substrate. Graded increases in respiration were accompanied by graded decreases not increases in $\Delta\psi m$. In no instance could increases in respiration or attendant decrease in $\Delta\psi m$ be accounted for by decreases in creatine phosphate or by calculated increases in free ADP. Maximal decreases in absolute $\Delta\psi m$ were obtained in pyruvate perfused hearts which exhibited a 30 mV change over the fourfold change in O_2 consumption.

The data are incompatible with the suggestion that regulation of the supply of substrate for the electron transport chain regulates ATP synthesis. Increases in ATP synthesis do not appear to be accompanied by increases in ADP, phosphate, or $\Delta\mu H^+$. Since these are the only factors known to increase ATP synthesis in mitochondria *in vitro*, it would appear that there is a regulatory element present in the intact cell which is not present in isolated mitochondria.

A similar result was obtained previously using a different method (Kaupinnen, 1983). Hearts were perfused with a different lipophilic cation, ^3H-trimethylphenylphosphonium ($TMPP^+$) and freeze clamped. The tissue was fractionated with nonaqueous solvents into mitochondrial and cytosolic fractions, and ^3H measured and the ratio of cytosolic to mitochondrial cation was used to calculate the value of $\Delta\psi m$. Values (–150 mV) similar to those obtained by Wan et al. (1993b) were obtained and when the heart rate was increased from 90 to 300 beats per minute the absolute value of $\Delta\psi m$ decreased to –125 mV. Since the principal of the measurement was similar to the one employed by Wan et al. (1993b), albeit differing in the methods of calculation, it is difficult to imagine the nature of an error that would produce an error in the directional change in $\Delta\psi m$. A more likely source of inaccuracy is in the estimate of free ADP. Free ADP, in a microcompartment near the mitochondrial inner membrane might influence respiration rather than the bulk phase (cytosolic) free ADP. The observed changes in free ADP are not in the wrong direction; they are only too small under most circumstances to account for observed changes.

VII. *IN VIVO* ESTIMATES OF THE CREATINE KINASE SHUTTLE

Data in support of such a microcompartment have been reported. The microcompartment may result from the activity of an isozyme of creatine kinase which is localized in mitochondria, between the inner and the outer membranes (Jacobs et al., 1964). Proposals have been made that the enzyme acts as part of a shuttle system for high energy phosphate which converts ATP coming out of the matrix space, to creatine phosphate which then diffuses to the muscle fibrils where creatine kinase bound specifically to the myofibrils converts the creatine phosphate to ATP (Bessman and Carpenter, 1985). Evidence in favor of operation of a shuttle system of this type comes from *in vitro* comparisons of ADP and creatine stimulation of mitochondrial respiration (Saks et al., 1978; 1980; 1985). Far lower levels of bulk phase ADP are needed to stimulate respiration in the presence of excess creatine. Likewise operation of a microcompartment near the muscle actomyosin ATPase can be demonstrated with skinned muscle fibers (Saks et al., 1984).

In vivo studies have not strongly supported the shuttle hypothesis in its strictest formulations. A strict formulation of the hypothesis would involve operation of the shuttle at approximately the same rate as ATP turnover. Measurements of the forward and reverse flux through creatine kinase in perfused hearts have shown, however, that flux through creatine kinase, as measured by NMR saturation transfer techniques, is about 10 times higher than ATPase flux (Bittl and Ingwall, 1985). Moreover, flux through creatine kinase varies only slightly with respiration rates. Before the proposal of the shuttle hypothesis, researchers believed that creatine kinase was a buffer for ATP/ADP ratios around an optimal ratio. This view has been altered somewhat in view of the obvious compartmentation of the isozymes so that the prevalent view now is that the important function of creatine kinase is to reduce diffusional barriers between the myofibrillar ATPase system and the mitochondrial ATP synthase (Meyer et al., 1984). Recent studies suggest that in mitochondria a multienzyme complex comprising the adenine nucleotide translocase, the mitochondrial creatine kinase isozyme, and porin of the outer membrane is involved in channeling high energy phosphates from the mitochondrial matrix to the cytosol of the cardiac cell (Schlegel et al., 1988; Schnyder et al., 1988). Porin is a large channel in the outer membrane of the mitochondria which allows passage of molecules under 5 kD from the cytosol to the intermembrane space. The porin is also known to be a part of the contact sites between the inner and outer membranes.

When mitochondria were subfractionated, a large portion (up to 50%) of mitochondrial creatine kinase was found in a fraction enriched for contact sites (Adams et al., 1989). This and a large body of additional data summarized in a recent review (Wyss et al., 1992) support the concept that mitochondrial creatine kinase is involved in facilitating the diffusion of high energy phosphates out and ADP into the mitochondrial intermembrane space.

Studies of the distribution and developmental patterns of isozymes of creatine kinase (Perry et al., 1988) also do not support a strict formulation of the creatine kinase shuttle hypothesis. A strict formulation would predict similar high activities at the two ends of the shuttle (mitochondria and muscle fiber) and lower activity in the bulk cytosol. Certain small mammals are born with no mitochondrial creatine kinase. As animals grow from 3 days to 18 days, the total creatine kinase activity remains constant, but the amounts associated with the mitochondria and with the myofibrils increase. This development parallels an increase in the work capacity of the heart although clearly, the compartmental pattern is not essential for function. Most recent data suggest that the mitochondrial creatine kinase is essential only for optimal function of the heart (Shoubridge et al., 1985; Yoshizaki et al., 1990; Kupriyanov et al., 1991) at very high workloads. This supports the concept, elegantly developed by Saks and co-workers (1980, 1985), that creatine kinase aids in the diffusion of ADP from the myofibrils to the mitochondria (Meyer et al., 1984; Jacobus, 1985).

These findings, however, cannot explain the lack of correlation between free ADP estimated from the bulk of the creatine kinase mass action ratio and respiration rates. If the myofibrils and the mitochondria were contained within the same microcompartment, direct transfer would be possible without involving the intervention of the bulk of the creatine kinase system which could independently perform its buffering function in the cytosol. In that case, the message of increased work could be transmitted directly to the mitochondria from the myosin ATPase. Relevant, but blunted changes of free ADP would be transmitted to the bulk phase creatine kinase system and measurements of free ADP from bulk phase creatine, creatine phosphate and ATP would correlate poorly with respiration.

However, this situation does not exist in heart muscle or skeletal muscle where there is a clear alternating arrangement of mitochondria and myofibrils. Free ADP must be transported over relatively long distances through the bulk of the cytosol to stimulate mitochondrial respiration. Therefore, the bulk phase creatine phosphate creatine ratio should reflect changes in the free ADP seen by the mitochondria. In fact, limitations in diffusion should produce larger changes in the bulk cytosol than in the mitochondrial intermembrane space. Therefore, the effect of mitochondrial creatine kinase should be to decrease the concentration gradient of free ADP from the cytosol to the mitochondria, not increase it.

VIII. CONCLUSIONS

The mechanism of control of ATP synthesis in the adult heart remains mysterious. Arguments presented throughout this review suggest that the increase in ATP synthesis which occurs in response to increased cardiac work is not caused by an increase in the ATP synthase substrates. This draws ones attention to allosteric control of the ATP synthase or the adenine nucleotide translocase. The low V_{max} of the synthase make it a particularly appealing target for control of flux through the pathway. Few current studies address the question of allosteric control of ATP synthesis, or how an allosteric message might signal the mitochondria of increased muscle fiber activity. Direct *in vitro* evidence for any control other than by substrates is lacking and therefore further advances in this area are likely to come from *in vivo* studies. Either Ca^{2+}, or possibly cAMP may activate a cascade of events in the cytosol which causes modifications of a regulatory protein in the mitochondria. Since calculated ATP synthesis based on *in vitro* measurements of the K_m's of ADP and phosphate and the relationship of $\Delta\psi m$ to the forward rates of ATP synthesis are higher than *in vivo* measured rates, the putative regulator, not present in isolated mitochondria, is likely to be an inhibitor.

REFERENCES

Adams, V., Bosch, W., Schlegel, J., Walliman, T., & Brdiczka, D. (1989). Further characterization of contact sites from mitochondria of different tissues: topology of peripheral kinases. Biochim. Biophys. Acta 981, 213–225.

Alexandre, A., Reynafarje, B., & Lehninger, A. L. (1978). Stochiometry of vectorial H^+ movements coupled to electron transport and to ATP synthesis in mitochondria. Proc. Natl. Acad. Sci. USA 75, 5296–5300.

Alger, J. R., den Hollander, J. A., & Shulman, R. G. (1982). *In vivo* phosphorus-31 nuclear magnetic resonance saturation transfer studies of adenosinetriphosphate kinetics in *Sacchoromyces cerevisiae.* Biochemistry 21, 2957–2963.

Balaban, R. S., Kantor, H. L., Katz, L. A., & Briggs, R. W. (1986). Relation between work and phosphate metabolites in the *in vivo* paced mammalian heart. Science 232, 1121–1123.

Berkich, D.A., Williams, G.D., Masiakos, P.T., Smith, M.B., Boyer, P.D., & LaNoue, K.F. (1991). Rate of various reactions catalyzed by ATP synthase as related to the mechanism of ATP synthesis. J. Biol. Chem. 266, 123–129.

Bessman, S. P. & Carpenter, C.L. (1985). The creatine-creatine phosphate energy shuttle. Ann. Rev. Biochem. 54, 831–862.

Bittl, J. A. & Ingwall, J.S. (1985). Reaction rates of creatine kinase and ATP synthesis in the isolated rat heart. A ^{31}P NMR magnetization transfer study. J. Biol. Chem. 260, 3512–3517.

Bittl, J. A., Balshi, J. A., & Ingwall, J. A. (1987). Effects of norepinephrine infusion on myocardial high energy phosphate content and turnover in the living rat. J. Clin. Invest. 79, 1852–1859.

Boyer, P. D. (1989). A perspective of the binding change mechanism for ATP synthesis. FASEB J. 3, 2164–2178.

Brand, M. D., Reynafarje, B., & Lehninger, A. L. (1976). Re-evaluation of the H^+/site ratio of mitochondrial electron transport with the oxygen pulse technique. J. Biol. Chem 251, 5670–5679.

Brindle, K. M. (1988). NMR methods for measuring enzyme kinetics *in vivo*. Prog. NMR Spectrosc. 20, 257–293.

Brindle, K. M. & Krikler, S. (1985). ^{31}P NMR saturation transfer measurements of phosphate consumption in *Saccharomyces cerevisiae*. Biochim. Biophys. Acta 847, 285–292.

Brindle, K. M. & Radda, G.K. (1987). ^{31}P-NMR saturation transfer measurements of exchange between P_i and ATP in the reactions catalyzed by glyceraldehyde-3-phosphate dehydrogenase and phosphoglycerate kinase *in vitro*. Biochim. Biophys. Acta 928, 45–55.

Brindle, K. V., Blackledge, M. J., Challis, R. A. J., & Radda, G.K. (1989). ^{31}P NMR magnetization-transfer measurements of ATP turnover during steady state isometric muscle contraction in the rat hind limb *in vivo*. Biochemistry 28, 4887–4893.

Brown, T. R., Ugurbil, K., & Shulman, R. G. (1977). ^{31}P nuclear magnetic resonance measurements of ATPase kinetics in aerobic *Escherichia coli* cells. Proc. Natl. Acad. Sci. USA 74, 5551–5553.

Cadoux-Hudson, T. A., Blackledge, M. J., & Radda, G. K. (1989). Imaging of human brain creatine kinase activity *in vivo*. FASEB J. 3, 2660–2666.

Chance, B. & Williams, G. R. (1956). The respiratory chain and oxidative phosphorylation. Advan. Enzymol. 17, 65–134.

Chance, B., Saronio, C., & Leigh, J. S. Jr. (1975). Functional intermediates in the reaction of membrane-bound cytochrome oxidase with oxygen. J. Biol. Chem. 250, 9226–9237.

Chance, B., Leigh, J. S. Jr, Kent, J., McCully, K., Nioka, S., Clark, B. J., & Maris, J. M. (1986). Multiple controls of oxidative metabolism in living tissues as studied by phosphorus magnetic resonance. Proc. Natl. Acad. Sci USA 83, 9458–9462.

Cohen, S. M. (1983). Simultaneous ^{13}C and ^{31}P NMR studies of perfused rat liver. Effect of insulin and glucagon and a ^{13}C NMR assay of free Mg^{2+}. J. Biol. Chem. 258, 14294–14308.

Coty, W. A. & Pedersen, P. L. (1974). Phosphate transport in rat liver mitochondria: kinetics and energy requirements. J. Biol. Chem. 249, 2593–2598.

Das, A. M. & Harris, D. A. (1990a). Control of mitochondrial ATP synthase in heart cells; inactive to active transition caused by beating or positive inotropic agents. Cardiovac. Res. 24, 411–417.

Das, A. M. & Harris, D. A. (1990b). Regulation of mitochondrial ATP synthase in intact rat cardiomyocytes. Biochem J. 266, 355–361.

Davis, E. J. & Davis-VanThienen, W. I. A. (1978). Control of mitochondrial metabolism by the ATP/ADP ratio. Biochim. Biophys. Res. Commun. 83, 1260–1266.

Davis, E. J. & Davis-VanThienen, W. I. A. (1984). Rate control of phosphorylation-coupled respiration by rat liver mitochondria. Arch. Biochem. Biophys. 233, 573–581.

Detre, J. A., Koretsky, A. P., Williams, D. S., & Ho, C. (1990). Absence of pH changes during altered work in the *in vivo* sheep heart. A ^{31}P NMR investigation. J. Mol. Cell. Cardiol. 22, 543–553.

Doussiere, J., Ligeti, E., Brandolin, G., & Vignais, P. V. (1984). Control of oxidative phosphorylation in rat heart mitochondria. The role of the nucleotide carrier. Biochim. Biophys. Acta 766, 492–500.

Fillingame, R. H. (1990). In: The Bacteria, pp. 345–391. Academic Press, New York.

Forman, N. G. & Wilson, D. F. (1983). Dependence of mitochondrial oxidative phosphorylation on activity of the adenine nucleotide translocase. J. Biol. Chem. 258, 8649–8655.

Freeman, M., Bartlett, S., Radda, G. K., & Ross, B. (1983). Kinetics of sodium transport in the kidney by saturation transfer ^{31}P NMR. Biochim. Biophys. Acta 762, 325–336.

From, A. H. L., Zimmer, S. D., Mishurski, S. P., Mohanakrishnan, P., Ulstad, V. K., Thoma, W. J., & Ugurbil, K. (1990). Regulation of the oxidative phosphorylation rate in the intact cell. Biochemistry 29, 3731–3743.

Futai, M., Nousini, T., & Maeda, M. (1989). ATP synthase (H^+-ATPase): Results by combined biochemical and molecular biological approaches. Ann. Rev. Biochem. 58, 111–136.

Garlick, P. B., Soboll, S., & Bullock, G. R. (1992). Evidence that mitochondrial phosphate is visible in ^{31}P NMR spectra of isolated, perfused rat hearts. NMR in Biomed. 5, 29–36.

Gellerich, F. N., Bohnensack, R., & Kunz, W. (1983). Control of mitochondrial respiration. The contribution of the adenine nucleotide translocator depends on the ATP-and ADP-consuming enzymes. Biochim. Biophys. Acta 722, 381–391.

Groen, A. K., Wanders, R. J. A., Westerhoff, H. V., Van der Meer, R., & Tager, J. M. (1982). Quantification of the contribution of various steps to the control of mitochondrial respiration. J. Biol. Chem. 257, 2754–2757.

Gupta, R .J. & Moore, R. D. (1980). ^{31}P-NMR studies of intracellular free Mg^{2+} in intact frog skeletal muscle. J. Biol. Chem. 255, 3987–3993.

Harris, D. A. & Das, A. M. (1991). Control of mitochondrial ATP synthesis in the heart. Biochem. J. 280, 561–573.

Hatefi, Y. (1985). The mitochondrial electron transport and oxidative phosphorylation system. Ann. Rev. Biochem. 54, 1015–1069.

Headrick, J., Clarke, K., & Willis, R. J. (1989). Adenosine production and energy metabolism in ischaemic and metabolically stimulated rat heart. J. Mol. Cell. Cardiol. 21, 1089–1100.

Hutson, S. M., Berkich, D., Williams, G. D., LaNoue, K. F., & Briggs, R. W. (1989). ^{31}P NMR visibility and characterization of rat liver mitochondrial matrix adenine nucleotides. Biochemistry 28, 4325–4332.

Jacobs, H., Heldt, H. W., & Klingenberg, M. (1964). High activity of creatine kinase in mitochondria from muscle and brain and evidence for a separate mitochondrial isoenzyme of creatine kinase. Biochem. Biophys. Res. Commun. 16, 516–521.

Jacobus, W. E. (1985). Theoretical support for the phosphocreatine energy transport shuttle based on the intracellular diffusion limited mobility of ADP. Biochem. Biophys. Res. Comm. 133, 1035–1041.

Kacser, H. & Burns, J. A. (1973). The control of flux. Symp. Soc. Exp. Biol. 32, 65–104.

Kagawa, Y. (1984). Proton motive ATP synthesis. In: Bioenergetics (L. Ernster, ed.), pp. 149–186, Amsterdam, Elsevier.

Katz, L. A., Koretsky, A. P., & Balaban, R. S. (1986). Respiratory control in the glucose perfused heart: a ^{31}P NMR and NADH fluorescence study. FEBS letters 221, 270–276.

Katz, L. A., Koretsky, A. P., & Balaban, R. S. (1988). Activation of dehydrogenase activity and cardiac respiration: a ^{31}P NMR study. Am. J. Physiol. 255, H185–H188.

Katz, L. A., Swain, J. A., Portman, M. A., & Balaban, R. S. (1989). Relation between phosphate metabolites and oxygen consumption of heart *in vivo.* Am. J. Physiol. 256, H265–H274.

Kaupinnen, R. A. (1983). Proton electrochemical potential of the inner mitochondrial membrane in isolated perfused rat hearts, as measured by exogenous probes. Biochim. Biophys. Acta 725, 131–137.

Kingsley-Hickman, P. B., Sako, E. Y., Mohankrishna, P., Robitaille, P. M. L., From, A. H. L., Foker, J. E., & Ugurbil, K. (1987). ^{31}P-NMR studies of ATP synthesis and hydrolysis kinetics in the intact myocardium. Biochemistry 26, 7501–7510.

Klingenberg, M. (1961). Zur reversibilitat der oxydativen phosphorylierung. Biochem. Z. 335, 253–272.

Klingenberger, M. & Rottenberg, H. (1977). Relationship between the gradient of the ATP/ADP ratio and the membrane potential across the mitochondrial membrane. Eur. J. Biochem. 73, 125–130.

Knox, B. E. & Tsong, T. Y. (1983). Voltage-driven ATP synthesis by beef heart mitochondrial F_0F_1-ATPase. J. Biol. Chem. 259, 4757–4763.

Koretsky, A. P., Brosnan, M. J., Chen, L., Chen, J., & Van Dyke, T. (1990). NMR detection of creatine kinase expressed in liver of transgenic mice: determination of free ADP levels. Proc. Natl. Acad. Sci. USA 87, 3112–3116.

Kunz, W., Bohnensack, R., Bohne, G., Kuster, U., Letko, G., & Schonfeld, P. (1981). Relations between extramitochondrial and intramitochondrial adenine nucleotide systems. Arch. Biochem. Biophys. 209, 219–229.

Kupriyanov, V. V., Lakomkin, V. L., Korchazhkina, O. V., Steinschneider, A. Ya., Kapleko, V. I., & Saks, V. A. (1991). Control of cardiac energy turnover by cytoplasmic phopshates: ^{31}P-NMR study. Am. J. Physiol. 261, 45–53.

Kushmerick, M. J. & Meyer, R. A. (1985). Chemical changes in rat leg muscle by phosphorus nuclear magnetic resonance. Am. J. Physiol. 248, C542–C549.

Kuster, U., Letko, G., Kunz, W., Duszynski, J., Bogucka, K., & Wojtczak, L. (1981). Influence of different energy drains on the interrelationship between the rate of respiration, proton-motive force and adenine nucleotide patterns in isolated mitochondria. Biochim. Biophys. Acta 636, 32–28.

LaNoue, K. F. & Schoolwerth, A. C. (1979). Metabolic transport in mitochondria. Ann. Rev. Biochem. 48, 871–922.

LaNoue, K. F. & Duszynski, J. (1992). Kinetic studies of ATP synthase: The case for the positional change mechanism. J. Bioenerg. Biomembr. 24, 499–506.

LaNoue, K. F., Mizani, S. M., & Klingenberg, M. (1978). Electrical imbalance of adenine nucleotide transport across the mitochondrial membrane. J. Biol. Chem. 253, 191–198.

LaNoue, K. F., Jeffries, F. M. H., & Radda, G. K. (1986). Kinetic control of mitochondrial ATP synthesis. Biochemistry 25, 7667–7675.

Lawson, J. W. R. & Veech, R. L. (1978). Effects of pH and free Mg^{2+} on the K_{eq} of the creatine kinase reaction and other phosphate hydrolysis and phosphate transfer reactions. J. Biol. Chem. 254, 6528–6537.

Lehninger, A. L. (1964). The mitochondrion, W.A. Benjamin, New York.

Lemasters, J. J. (1984). The ATP-to-oxygen stoichiometries of oxidative phosphorylation by rat liver mitochondria. An analysis of ADP-induced oxygen jumps by linear nonequilibrium thermodynamics. J. Biol. Chem., 259, 13123–13129.

Lemasters, J. J. & Sowers, A. E. (1979). Phosphate dependence and atractyloside inhibition of mitochondrial oxidative phosphorylation. The ADP-ATP carrier is rate limiting. J. Biol. Chem. 254, 1248–1251.

Ligeti, E., Brandolin, G., Dupont, Y., & Vignais, P. (1985). Kinetics of P_i-P_i exchange in rat liver mitochondria. Rapid filtration experiments in the millisecond time range. Biochemistry 24, 4423–4428.

Ligeti, L., Osbakken, M. D., Clark, B. J., Schnall, M., Balinger, L., Subramanian, H., Leigh, J. S., & Chance, B. (1987). Cardiac transfer function relating energy metabolism to workload in different species as studied with ^{31}P NMR. Mag. Res. Med. 4, 112–119.

Lucken, U., Gogol, E. P., & Capaldi, R. A. (1990). Structure of the ATP synthase complex (ECF_1F_o) of *Escherichia coli* from cryoelectron microscopy. Biochemistry 29, 539–5343.

Malloy, C. R., Cunningham, C. C., & Radda, G. K. (1986). The metabolic state of the rat liver measured by ^{31}P-NMR spectroscopy. Biochim. Biophys. Acta 885, 1–11.

Masiakos, P. T., Williams, G. D., Berkich, D. A., Smith, M. B., & LaNoue, K. F. (1991). ^{31}P NMR saturation-transfer study of the *in situ* kinetics of the mitochondrial adenine nucleotide translocase. Biochemistry 30, 8351–8357.

Matthews, P. M., Bland, J. L., Gadian, D. G., & Radda, G. K. (1981). The steady state rate of ATP synthesis in the perfused rat heart measured by ^{31}P-NMR saturation transfer. Biochem. Biophys. Res. Comm. 103, 1052–1059.

Matthews, P. M., Bland, J. L., Gadian, D. G., & Radda, G. K. (1982). A ^{31}P NMR saturation transfer study of the regulation of creatine kinase in the rat heart. Biochim. Biophys. Acta 721, 312–320.

McCormick, J. G. & Denton, R. M. (1981). The activation of pyruvate dehydrogenase in the perfused rat heart by adrenaline and other inotropic agents. Biochem. J. 194, 639–643.

McCormick, J. G. & England, P. J. (1983). Ruthenum red inhibits the activation of pyruvate dehydrogenases caused by positive inotropic agents in the perfused rat heart. Biochem. J. 214, 581–585.

McCormick, J. G. & Denton, R. M. (1984). Role of Ca^{2+} ions in the regulation of intramitochondrial metabolism in rat heart. Evidence from studies with isolated mitochondria that adrenaline activities the PDH and 2-oxoglutarate dehydrogenase complexes by increasing the intramitochondrial concentration of Ca^{2+}. Biochem. J., 235–247.

Meyer, R. A., Kushmerick, M. J., & Brown, T. R. (1982). Application of ^{31}P-NMR spectroscopy to the study of striated muscle metabolism. Am. J. Physiol. 242, C1–C11.

Meyer, R. A., Sweeney, H. L., & Kushmerick, M. J. (1984). A simple analysis of the "phosphocreatine shuttle". Am. J. Physiol. 246, C365–C377.

Mitchell, P. (1961). Coupling of phosphorylation to electron and hydrogen transfer by a chemi-osmotic type of mechanism. Nature 191, 144–148.

Mitchell, P. (1979). Vectorial chemistry and the molecular mechanics of chemiosmotic coupling: power transmission by proticity. Biochem. Soc. Trans. 4, 399–430.

Moon, R. B. & Richards, J. H. (1973). Determination of intracellular pH by ^{31}P magnetic resonance. J. Biol. Chem. 248, 7276–7278.

Moreno-Sanchez, R. & Hansford, R. G. (1988). Dependence of cardiac mitochondrial pyruvate dehydrogenase activity on intramitochondrial free Ca^{2+}-concentration. Biochem. J. 256, 403–412.

Neely, J. R., Denton, R. M., England, P. J., & Randle, P. J. (1972). The effects of increased heart work in the tricarboxylate cycle and the interactions with glycolysis in the perfused rat heart. Biochem. J. 128, 147–159.

Nuutinen, E. M. (1984). Subcellular origin of the surface fluorescence of reduced nicotinamide nucleotides in the isolated perfused rat heart. Basic Res. Cardiol. 79, 49–58.

Papa, S., Lorusso, M., & Guerrieri, F. (1981). Mechanism of respiration-driven proton translocation in the inner mitochondria membrane. Analysis of proton translocation associated with oxidation of endogenous ubiquinol. Biochim. Biophys. Acta 387, 425–440.

Pedersen, P. L. & Amzel, L. M. (1992). (Series eds.), Minireview series: F-type ATPases. J. Bioenerg. Biomembr. 24 (5), 427–521.

Penefsky, H. S. & Cross, R. L. (1991). Structure and mechanism of F_0F_1-type ATP synthases and ATPases. Advances in Enzymology 64, 173–213.

Penttila, T. (1983). Properties and reconstitution of a cytochrome oxidase deficient in subunit III. Eur. J. Biochem. 133, 355–361.

Perry, S. B., McAuliffe, J., Balschi, J. A., Hichey, P. R., & Ingwall, J. S. (1988). Velocity of the creatine kinase reaction in the neonatal rabbit heart: role of mitochondrial creatine kinase. Biochemistry 27, 2165–2172.

Portman, M. A., Heineman, F. W., & Balaban, R. S. (1989). Developmental changes in the relationship between phosphate metabolites and oxygen consumption in the sheep heart *in vivo*. J. Clin. Invest. 83, 456–464.

Power, J., Cross, R. L., & Harris, D. A. (1983). Interaction of F1-ATPase from ox heart mitochondria with its naturally occurring inhibitor protein. Studies using radio-iodinated inhibitor protein. Biochim. Biophys. Acta 724, 128–141.

Prochaska, L. J., Bisson, R., Capaldi, R. A., Steffens, G. C. M., & Buse, G. (1981). Inhibition of cytochrome C oxidase function by dicyclohexyl carbodiimide. Biochim. Biophys. Acta 637, 360–373.

Pulman, M. E. & Monroy, G. C. (1963). A naturally occurring inhibitor of mitochondrial adenosine triphosphatase. J. Biol. Chem. 238, 3762–3769.

Radda, G. K. (1986). The use of NMR spectroscopy for the understanding of disease. Science 233, 540–645.

Ragan, C. I. & Hinkle, P. C. (1975). Ion transport and respiratory control in vesicles formed from reduced nicotinamide adenine dinucleotide-coenzyme Q reductase and phospholipids. J. Biol. Chem. 250, 8472–8476.

Reed, L. J. (1981). Regulation of mammalian pyruvate dehydrogenase complex by a phosphorylation-dephosphorylation cycle. Curr. Top. Cell. Regul. 18, 95–106.

Ritchie, R. J. (1984). A critical assessment of the use of lipophilic cations as membrane potential probes. Prog. Biophys. Molec. Biol. 43, 1–32.

Rottenberg, H. (1979). The measurement of membrane potential and ΔpH in cell, organelles and vesicles. Methods Enzymol. 55, 547–569.

Rouslin, W. (1983). Protonic inhibition of the mitochondrial oligomycin-sensitive adenosine 5′-triphosphatase in ischemic and autolyzing cardiac muscle. Possible mechanism for the mitigation of ATP hydrolysis under nonenergizing conditions. J. Biol. Chem. 258, 9657–9661.

Rouslin, W. (1987). The mitochondrial adenosine 5′-triphosphatase in slow and fast heart rate hearts. Am. J. Physiol. 252, H622–H627.

Rouslin, W. & Pullman, M.E. (1987). Protonic inhibition of the mitochondrial adenosine 5′-triphosphatase in ischemic cardiac muscle. Reversible binding of the ATPase inhibitor protein to the mitochondrial ATPase during ischemia. J. Mol. Cell. Cardiol. 19, 661–668.

Sako, E. Y., Kingsley-Hickman, P. B., From, A. H. L., Foker, J. E., & Ugurbil, K. (1988). ATP synthesis kinetics and mitochondrial function in the postischemic myocardium studied by ^{31}P-NMR. J.Biol.Chem. 263, 10600–10607.

Saks, V. A., Rosenshtraush, L. V., Smirnov, V. N., & Chazov, E. I. (1978). Role of creatine phosphokinase in cellular function and metabolism. Can. J. Physiol. Pharmacol. 56, 691–706.

Saks, V. A., Kupriyanov, V. V., Elizarova, G. V., & Jacobus, W. E. (1980). Studies of energy transfer in heart cells. The importance of creatine kinase localization for the coupling of mitochondrial phosphoryl creatine production to oxidative phosphorylations. J. Biol. Chem. 255, 755–763.

Saks, V. A., Ventura-Clapier, R., Huchua, Z. A., Preobrazhensky, A. N., & Emilin, I. V. (1984). Creatine kinase in regulation of heart function and metabolism. I. Further evidence for compartmentation of adenine nucleotides in cardiac myofibrillar and sarcolemmal coupled ATPase-creatine kinase systems. Biochim. Biophys. Acta 803, 254–264.

Saks, V. A., Kutznetsov, A. V., Kupriyanov, V. V., Miceli, M. V., & Jacobus, W. E. (1985). Creatine kinase of rat heart mitochondria. The demonstration of functional coupling to oxidative phosphorylation in an inner membrane-matrix preparation. J. Biol. Chem. 260, 7757–7764.

Schaffer, S. W., Safer, B., Ford, C., Illingworth, J., & Williamson, J. R. (1978). Respiratory acidosis and its reversibility in perfused rat heart: regulation of citric acid activity. J. Am. Physiol. 234, H40–H51.

Schlegel, J., Zurbriggen, B., Wegmannn, G., Wyss, M., Eppenberger, H. M., & Walliman, T. (1988). Native mitochondrial creatine kinase forms octameric structures. I: Isolation of two interconvertible mitochondrial creatine kinase forms, dimeric and octameric mitochondrial creatine kinase: Characterization, localization and structure-function relationships. J. Biol. Chem. 263, 16942–16953.

Schnyder, T., Engel, A., Lustig, A., & Walliman, T. (1988). Native mitochondrial creatine kinase forms octameric structures. II: Characterization of dimers and octamers by ultracentrifugation, direct mass measurements by scanning transmission electron microscopy and image analysis of single mitochondrial creatine kinase octamers. J. Biol. Chem. 263, 16954–16962.

Schwerzmann, K. & Pedersen, P. L. (1986). Regulation of the mitochondrial ATP synthase/ATP complex. Arch. Biochem. Biophys. 250, 1–18.

Senior, A. E. & Wise, J. G. (1983). The proton-ATPase of bacteria and mitochondria. J. Membr. Biol. 73, 105–124.

Shoubridge, E. A., Briggs, R. W., & Radda, G. K. (1982). ^{31}P NMR saturation transfer measurements of the steady state rates of creatine kinase and ATP synthase in the rat brain. FEBS letters 140, 288–292.

Shoubridge, E. A., Bland, J. L., & Radda, G. K. (1984). Regulation of creatine kinase during steady state isometric twitch contraction in rat skeletal muscle. Biochim. Biophys. Acta 805, 72–78.

Shoubridge, E. A., Jeffrey, F. M., Kreogh, J. M., Radda G. K., & Seymour, A. M. (1985): Creatine kinase kinetics, ATP turnover, and cardiac performance in heart depleted of creatine with the substrate analogue β-guanidinopropionic acid. Biochim. Biophys. Acta 847, 25–32.

Steenbergen, C., DeLeeuw, G., & Williamson, J. R. (1978). Analysis of control of glycolysis in ischemic hearts having heterogeneous zones of anoxia. J. Mol. Cell. Cardiol. 10, 617–639.

Tager, J. M., Wanders, R. J. A., Groen, A. K., Van der Meer, R., Akerboom, T. P. M ., & Meijer, A. K. (1981). Control mechanisms of energy-dependent metabolic pathways in hepatocytes. Acta Biol. Med. Ger. 40, 895–906.

Tager, J. M., Wanders, R. J. A., Groen, A. K., Kunz, W., Bohnensack, R., Kuster, U., Letko, G., Bohme, G., Duszynski, J., & Wojtczak, L. (1983). Control of mitochondrial respiration. FEBS Letters 151, 1–9.

Unitt, J. F., McCormick, J. G., Reid, D., MacLachlan, L. K., & England, P. J. (1989). Direct evidence for a role of intramitochondrial Ca^{2+} in the regulation of oxidative phosphorylation in the stimulated rat heart. Biochem. J. 262, 293–301.

van de Stadt, R. J., de Boer, A. L., & van Dam, K. (1973). The interaction between the mitochondrial ATPase (F_1) and the ATPase inhibitor. Biochim. Biophys. Acta 292, 338–349.

Veech, R. L., Lawson, J. W. R., Cornell, N. W., & Krebs, H. A. (1978). Cytosolic phosphorylation potential. J. Biol. Chem. 254, 6538–6547.

Walker, J. E., Fearnley, I. M., Lutter, R., Todd, R. J., & Runswick, M. J. (1990). Structural aspects of proton-pumping ATPases. Philos. Trans. R. Soc. London B 326, 367–378.

Wan, B., LaNoue, K. F., Cheung, J. Y., & Scaduto, R. C. Jr. (1989). Regulation of citric acid cycle by calcium. J. Biol. Chem. 264, 13430–13439.

Wan, B., Doumen, C., Duszynski, J., Salama, G., & LaNoue, K. F. (1993a). A method of determining the electrical potential gradient across the mitochondrial membrane in perfused rat hearts. Am. J. Physiol., in press.

Wan, B., Doumen, C., Duszynski, J., Salama, G., Vary, C. T., & LaNoue, K. F. (1993b): The effects of cardiac work on the electric potential gradient across the mitochondrial membrane in perfused rat hearts. Am. J. Physiol., in press.

Wanders, R. J. A., Groen, A. K., Meijer, A. J., and Tager, J. M. (1981). Determination of the free energy difference in the adenine nucleotide translocator reaction in rat liver mitochondria using intra- and extramitochondrial ATP utilizing reactions. FEBS Letters 132, 201–206.

Wikstrom, M., Krab, K., & Sarate, M. (1981). Proton-translocating cytochrome complexes. Annu. Rev. Biochem. 50, 623–655.

Wyss, M., Smeitink, J., Wevers, R. A., & Walliman, T. (1992). Mitochondrial creatine kinase: a key enzyme of aerobic energy metabolism. Biochim. Biophys. Acta 1102, 119–166.

Yamada, E. W. & Huzel, N. J. (1988). The calcium binding ATPase inhibitor protein from bovine heart mitochondria. Purification and properties. J. Biol. Chem. 263, 11498–11503.

Yamada, E. W. & Huzel, N. J. (1989). Calcium-binding ATPase inhibitor protein of bovine heart mitochondria. Role in ATP synthesis and effect of Ca^{2+}. Biochemistry 28, 9714–9718.

Yamada, E. W., Schiffman, F. H., & Huzel, N. J. (1980). Ca^{2+}-regulated release of an ATPase inhibitor protein from submitochondrial particles derived from skeletal muscles of the rat. J. Biol. Chem. 255, 267–273.

Yoshizaki, K., Watari, H., & Radda, G. K. (1990). Role of phosphocreatine in energy transport in skeletal muscle of bullfrog studied by ^{31}P-NMR. Biochim. Biophys. Acta 1051, 14–150.

HEPATIC HIGH ENERGY PHOSPHATE METABOLISM IN TRANSGENIC LIVERS EXPRESSING CREATINE KINASE AS REVEALED BY ^{31}P NMR

Alan P. Koretsky, Kenneth R. Miller, and Jessica M. Halow

Advances in Molecular and Cell Biology
Volume 11, pages 233–255.

ISBN: 1-55938-844-7

ABSTRACT

There is no consensus for a detailed model of the regulation of mitochondrial ATP production in cells. The problem of metabolic control in intact tissues is extremely difficult for two reasons. First, it is not clear how best to measure the parameters believed to be important in control of mitochondrial metabolism. Second, it is difficult to make specific perturbations in order to test possible models of metabolic control. The ability to alter an animal's genome using transgenic techniques in combination with nondestructive spectroscopic tools, such as NMR, offers solutions to these two problems. In this paper, we describe recent work that has made use of a transgenic mouse model that expresses the enzyme, creatine kinase in liver. Creatine kinase is not normally expressed to any appreciable extent in liver. ^{31}P NMR studies of these transgenic livers have supplied information about the regulation of ATP catabolism, the role of creatine kinase in ATP metabolism, and the quantitative control of oxidative phosphorylation by ADP.

I. INTRODUCTION

Techniques for the identification and isolation of the protein components responsible for carrying out a biological process have become commonplace. Cloning genes and overexpressing proteins is now routine for biochemical studies (Sambrook et al., 1989). It remains a great challenge to advance from understanding the components responsible for catalyzing a process to quantitatively understanding the regulation of that process in the intact cell.

At least two major problems limit the development of quantitative models of biological processes *in vivo*. First, most proteins are regulated by a number of factors, including low molecular weight metabolites, protein modifications, and interactions with other macromolecules (see Serre et al., this volume). It is difficult to simultaneously monitor the relevant concentrations of all effectors of a biological process in the intact cell. Spectroscopic techniques, including nuclear magnetic resonance (NMR) (Gillies, 1992; Koretsky and Williams, 1992) and fluorescence (Wang and Taylor, 1989) offer approaches to "seeing" biological processes in a nondestructive manner. The second problem in gaining a quantitative understanding of a biological process is that it is difficult to make specific perturbations of the process. Molecular genetic techniques offer an approach to making directed changes in the amount, distribution, or behavior of a protein in an animal (Hogan et al., 1986; Palmiter & Brinster, 1986; Hanahan, 1989; Koretsky, 1992). The consequences of these changes on the process of interest can be studied using a variety of techniques.

The regulation of high energy phosphate metabolism in intact tissues remains an area of great interest with a number of outstanding questions. For example, there is no consensus on a detailed model explaining the increase in ATP production by oxidative phosphorylation that occurs when tissue work is increased (Hansford, 1985; Balaban, 1990). It is clear that a number of factors, such as levels of cytosolic ADP and inorganic phosphate (P_i) (Chance and Williams, 1955; Jacobus et al., 1982), levels of mitochondrial NADH (Erecinska et al., 1978; Koretsky and Balaban, 1987), and oxygen concentrations at cytochrome oxidase (Erecinska et al., 1978), can control oxidative ATP production. In addition, changes in mitochondrial ion concentrations, such as Ca^{2+} and Mg^{2+}, may regulate a number of enzymes, such as Krebs cycle dehydrogenases and the ATP synthase (Denton and McCormack, 1985; Hansford, 1985; Das and Harris, 1990; see also LaNoue and Doumen, this volume).

A number of questions concerning the delivery of ATP to sites of utilization and the regulation of ATP catabolism in the living cell remain unanswered. The importance of creatine kinase and the high energy phosphate compound, phosphocreatine (PCr) remains unclear (Bessman and Carpenter, 1985; Walliman et al., 1992). Creatine kinase is found at high levels in skeletal muscle, heart, and brain, with different isoforms being preferentially localized to sites of ATP utilization (myofibrils) and production (the space between the inner and outer mitochondrial membranes) (Walliman et al, 1992). It has been proposed that this specific localization is important for shuttling high energy phosphates from sites of production to sites of utilization (Bessman and Carpenter, 1985). However, it remains extremely difficult to determine the quantitative role of creatine kinase in energy metabolism (Meyer et al., 1984). Also, it is not clear how the balance between net synthesis and degradation of adenylates is maintained to keep ATP levels in the range found in particular tissues (Van den Berghe et al., 1977).

Questions concerning the regulation of high energy phosphate metabolism in mammalian tissues have been difficult to answer definitively due to obstacles in making accurate measurements of key regulatory metabolites and a lack of specific inhibitors to perturb metabolism in defined ways. Advances in NMR and fluorescence spectroscopy now make it possible to measure intracellular ATP, ADP, P_i, pH, free Mg^{2+}, and free Ca^{2+} levels in intact tissue. Many of these measurements depend on indirect methods. For example, because most cells contain a number of ADP binding sites, assaying total ADP gives an overestimate of the free level of ADP available to regulate metabolic processes (Sols and Marco, 1970). Similar problems exist for measuring mitochondrial NADH. One approach has been to use an enzyme system believed to be close to equilibrium as an indicator system for free metabolite levels. The two most frequently used systems are the creatine kinase catalyzed reaction to measure free ADP levels (Veech et al., 1979) and the β-hydroxybutyrate dehydrogenase system to measure mitochondrial NADH/NAD$^+$ ratios (Williamson et al., 1967). Liver contains sufficient β-hydroxybutyrate dehydrogenase activity to maintain equilibrium, while other tissues, such as heart, muscle and brain, do not. On the other hand, heart, skeletal muscle, and brain contain sufficient creatine kinase activity to measure free ADP levels and liver does not. Therefore, it is not possible to quantitatively measure both the mitochondrial NADH redox state and the cytosolic phosphorylation potential in any tissue.

The lack of reagents to specifically alter high energy phosphate metabolism has severely restricted experimental possibilities. For example, there are no known specific inhibitors of creatine kinase. While a number of exciting results that demonstrate the importance of creatine kinase in energy metabolism have been obtained using substrate analogues and nonspecific inhibitors (Shoubridge and Radda, 1984; Meyer et al., 1986; Fossel and Hoefeler, 1987), the fact that other processes may have been inhibited, or that the cell adapted in response to the inhibition, complicates the interpretation of results.

In this paper, we describe initial results of a program designed to combine molecular genetic techniques and noninvasive spectroscopic techniques in order to increase our understanding of high energy phosphate metabolism in intact tissue. In particular, we have produced transgenic mice expressing creatine kinase in liver (Koretsky et al., 1990). In combination with *in vivo* NMR results, this mouse model has provided information on the role of creatine kinase in high energy phosphate metabolism, the regulation of ATP catabolism, and the role of ADP in the regulation of oxidative phosphorylation.

II. EXPRESSION OF CREATINE KINASE IN LIVER

Creatine kinase is normally expressed at high levels in skeletal muscle, heart, and brain, but not to an appreciable extent in liver. The strategy used to express high levels of creatine kinase in liver was one that has been used to express a number of proteins in specific cells in the mouse that do not normally contain the protein of interest (Palmiter and Brinster, 1986; Hanahan, 1989; Koretsky, 1992). The promoter and enhancer region of the DNA that controls transcription of a liver specific gene, transthyretin, was placed 5′ of the coding region for the rat genomic clone of the B subunit of creatine kinase (Koretsky et al., 1990). This heterologous DNA construct was used to generate transgenic mice using standard microinjection techniques (Hogan et al., 1986). Briefly, embryos were isolated and the DNA construct was microinjected into the male pronucleus. The injected DNA integrates into the embryonic DNA approximately 10% of the time. Since this integration occurs before the embryo divides, the injected DNA is integrated in the DNA of all cells that arise from the embryo. The microinjected embryos were implanted into pseudopregnant foster mothers and developed to term. The resultant mice were screened for the presence of the transgene using polymerase chain reaction techniques performed on DNA isolated from the tail (Hogan et al., 1986; Sambrook et al., 1989).

Once founder mice containing the transgene were identified, they were bred with normal mice to establish a line of mice heterozygous for the transgene. For the transthyretin-creatine kinase construct, seven founder mice were identified. Five of the seven founders expressed creatine kinase in liver and three of these were used for further study. A homozygous line of transgenic mice was generated from one of the founder mice. The advantage of generating a homozygous line is that it can be maintained without the need for screening for the presence of the transgene. In this way, the mice can be sent to other laboratories without requiring them to have the expertise to screen for transgenic mice.

Transgenic mice expressing creatine kinase were characterized in a number of ways. Liver extracts were shown to have creatine kinase activity that ranged from 200–1000 μmoles/min/gm wet weight (Koretsky et al., 1990). These are levels of activity comparable to that found in heart, brain, and skeletal muscle. Furthermore, it was shown by agarose gel electrophoresis that the activity detected in liver was due to the BB isoform of creatine kinase. Figure 1 shows immunohistology performed on the liver from a control and a transgenic mouse. An antibody that recognizes the BB form of creatine kinase was used for immunohistology. The transgenic liver stains darkly compared to control indicating the presence of creatine kinase. The staining is detected throughout the section, indicating that expression has occurred throughout the liver.

To test that creatine kinase was active *in vivo*, ^{31}P NMR was performed on control and transgenic mice. Figure 2 shows representative examples of spectra obtained from these livers. Both the control and transgenic livers contain peaks due to phosphomonoesters (PME), P_i, and the three phosphorus atoms in ATP. The transgenic mouse spectrum contains a new peak that has been assigned to PCr. This assignment was based on the NMR chemical shift, HPLC retention time from extracts, and the fact that addition of ADP and creatine kinase to extracts resulted in the loss of the PCr (Brosnan et al., 1990; Koretsky et al., 1990). Thus, lines of transgenic mice have been generated that express high levels of creatine kinase. The enzyme is distributed throughout the liver and is able to catalyze the production of PCr.

A question that quickly comes to mind is whether or not expression of creatine kinase has affected the mice. We have not been able to detect any obvious phenotype. The transgenic mice grow as large as control mice, they breed productively, and we have not noticed any difference in longevity. There are no changes in liver ultrastructure as revealed by electron microscopy. In addition, the levels of two glycolytic and two mitochondrial enzymes are the same in control and transgenic mice expressing creatine kinase. Because the expression of creatine kinase does not appear to adversely affect the liver, these transgenic mice are an interesting and relevant model for studying hepatic high energy phosphate metabolism.

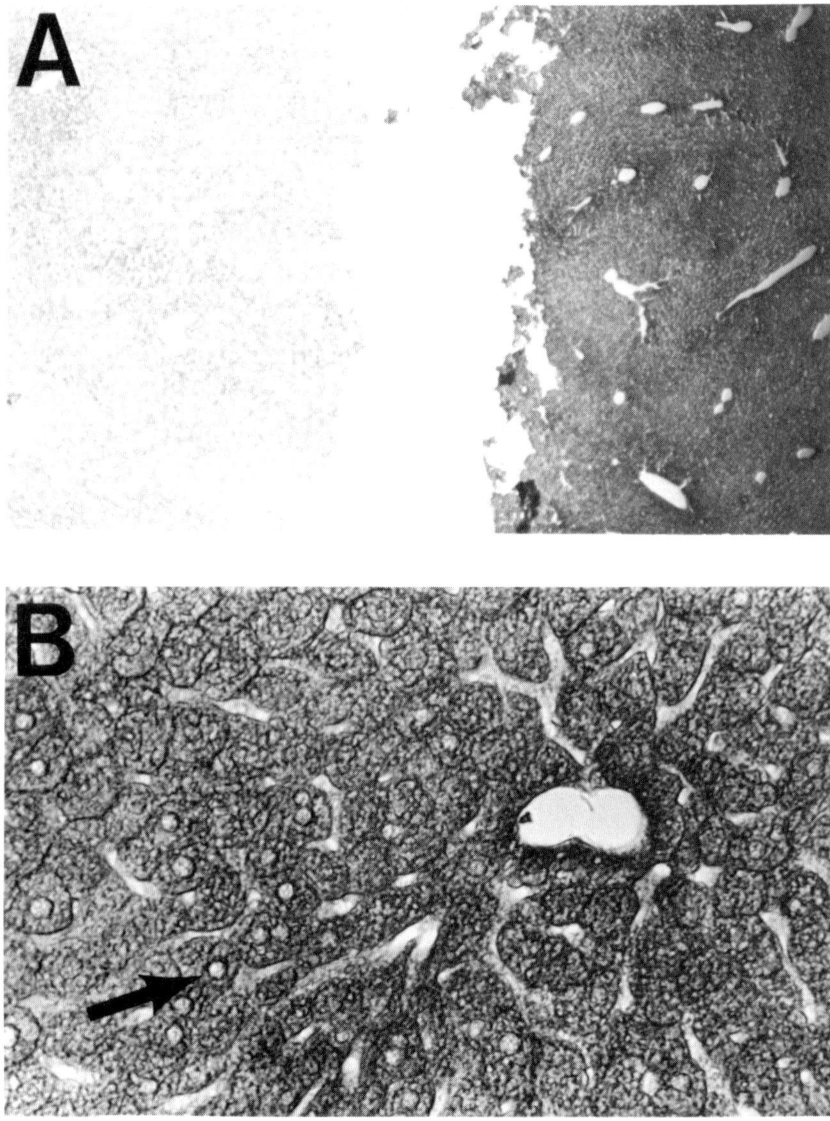

Figure 1. Immunolocalization of creatine kinase in liver of transgenic mice using anti-BB creatine kinase antibodies. (**A**) Low magnification view (x 36) showing normal liver (*left*) and transgenic liver expressing creatine kinase (*right*). Dark staining is seen throughout the transgenic liver section. (**B**) High magnification view (x 180) of immunohistology of the transgenic liver containing creatine kinase showing that the staining is predominantly cytoplasmic with little staining of nuclei (*arrow*). (Reprinted with permission from Koretsky et al., 1990.)

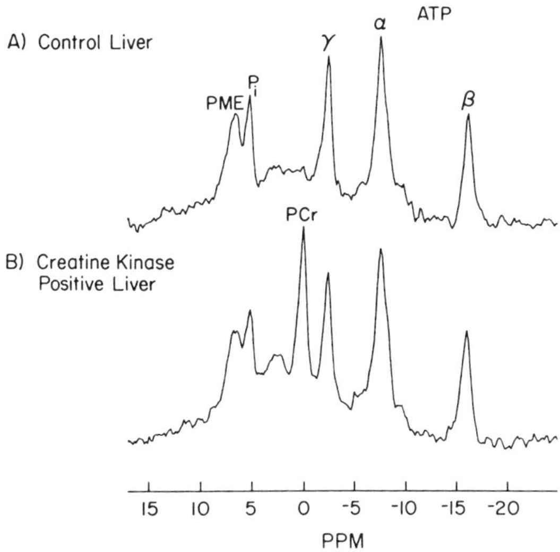

Figure 2. ^{31}P NMR spectra obtained from normal (**A**) and creatine kinase expressing liver (**B**). Both mice had been fed a diet supplemented with 2 percent creatine for five days prior to NMR experiments. Peaks are as labeled with PME, phosphomonoesters; P_i, inorganic phosphate; PCr, phosphocreatine; γ, α, β ATP, the three phosphorus atoms in ATP. Spectra were acquired on a Bruker 7 Tesla, Biospec II imaging NMR spectrometer and represent the accumulation of 600 free induction decays acquired from a surface coil placed on the surgically exposed liver *in situ.* A 90° pulse at the surface coil, a 7-sec recycle time, and 30 Hz linebroadening were used. (Reprinted with permission from Koretsky et al., 1990.)

III. THE ROLE OF PHOSPHOCREATINE AS AN ATP RESERVOIR

Creatine kinase is found at high levels in tissues with high and rapidly changing energy demands such as muscle and heart (Walliman et al., 1992). Based on the reaction catalyzed by creatine kinase, PCr + MgADP + H^+ = MgATP + creatine, it can be predicted that PCr can serve as a reserve for ATP during times when ATP utilization exceeds ATP production by oxidative phosphorylation or glycolysis. Indeed, during active exercise of skeletal muscle, or conditions where ATP production is inhibited, such as ischemia or anoxia, a fall in PCr is detected prior to a drop in ATP. The fact that the synthesis of ATP from PCr and ADP utilizes a proton makes it probable that the creatine kinase reaction plays a role in cellular pH regulation (Meyer et al., 1986).

The transgenic mice expressing creatine kinase in liver offer an opportunity to test the role of the creatine kinase reaction in buffering ATP and regulating intracellular pH. The fact that normal hepatocytes contain little or no creatine kinase

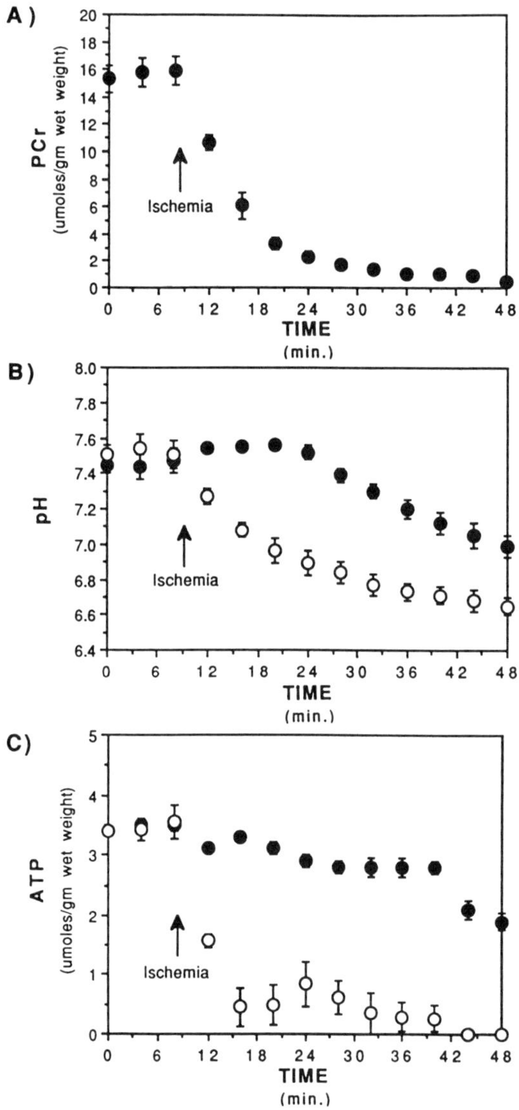

Figure 3. Time course of metabolites prior to and after ischemia as determined by ^{31}P NMR of perfused liver at 25°C from normal and transgenic mice expressing creatine kinase in liver. In each case, ischemia was initiated at the arrow by stopping the perfusate flow. (**A**) Time course of drop in PCr from transgenic liver expressing creatine kinase. (**B**) Time course of changes in pH from normal liver (○) and transgenic liver expressing creatine kinase (●). (**C**) Time course of drop in ATP from normal liver (○) and transgenic liver expressing creatine kinase (●). The symbols represent the mean ± a standard error with *n*=5.

allows the transgenic livers expressing creatine kinase to serve as a model for gain of function studies. Figure 3 shows the time course of changes in metabolites from normal and transgenic livers that were perfused for an interval and then subjected to ischemia. Metabolite levels were measured using ^{31}P NMR. At the onset of ischemia there is a fall in PCr levels in livers from the transgenic mice (Figure 3A). Figure 3B shows the changes in intracellular pH in normal and transgenic livers. The high, preischemic intracellular pH of 7.5 was due to the livers being perfused at 25 °C. Ischemia causes a large drop in pH to 6.6 in the normal livers. In the transgenic livers, the pH is maintained for a period after initiation of ischemia and then begins to fall at about one third the rate measured in controls. In normal livers, there is a rapid drop in ATP after ischemia, however, no drop in ATP is detected in transgenic livers until PCr levels get quite low (Figure 3C). Thus, the presence of creatine kinase and PCr in livers delays the fall in ATP and pH that normally occurs during ischemia. These results clearly demonstrate the buffering role that creatine kinase and PCr play in cellular energy metabolism.

In order to ask whether the maintenance of ATP levels by PCr in the transgenic livers protects hepatocytes, enzyme release was measured during hypoxic perfusion of control and transgenic livers. As with ischemia, the presence of PCr led to maintenance of ATP levels as compared to normal livers during low oxygen perfusion. The levels of the cytosolic enzyme, lactate dehydrogenase, were measured in the perfusate during the hypoxic perfusion. In normal livers levels of lactate dehydrogenase increased dramatically after 15 min of hypoxia. In transgenic livers with high levels of PCr, no significant increase in lactate dehydrogenase in the perfusate was detected until 90 min after hypoxia (K. R. Miller et al., 1993). The length of this protection was dependent on the starting level of PCr. Thus, maintenance of ATP in the transgenic livers expressing creatine kinase prevents hepatocellular damage caused by hypoxia.

IV. USE OF THE CREATINE KINASE CATALYZED REACTION TO DETERMINE FREE ADP LEVELS IN LIVER

ADP can regulate a number of cellular processes. Results from isolated mitochondria indicate that the K_m of ADP for oxidative phosphorylation is approximately 20 μM (Chance and Williams, 1955; Guylai et al., 1985). The reaction catalyzed by creatine kinase has been used to calculate free ADP levels in those tissues that contain high levels of creatine kinase (Veech et al., 1979). The idea is to assume the reaction is at equilibrium; then the amount of PCr, creatine, ATP, intracellular pH, and the concentration of intracellular free Mg^{2+} can be used to calculate free ADP levels using the equation:

$$[ADP]= [ATP][creatine]/[PCr]Kapp. \qquad (1)$$

Kapp represents the known apparent equilibrium constant of the reaction accounting for the concentration of free Mg^{2+} and intracellular pH (Lawson and Veech, 1979). [ADP] and [ATP] represent the sum of all ionic species present. At physiological levels of Mg^{2+}, the majority of the ATP and ADP is complexed with Mg^{2+} (Gupta et al., 1978). In heart, brain, and skeletal muscle, the levels of free ADP determined using the creatine kinase equilibrium range from 0.005–0.070 μmoles/gm wet weight depending on the tissue and physiological state (Veech et al., 1979; From et al., 1986; Detre et al., 1990; Kushmerick et al., 1992). These values are in the range where changes in free ADP can exert control over oxidative phosphorylation. The values of free ADP determined using the creatine kinase equilibrium contrasts with a value of approximately 0.5–1 μmole/gm wet weight ADP measured in perchloric acid extracts of tissue (Veech et al., 1979). The discrepancy between the value of ADP measured in extracts and that calculated with the creatine kinase equilibrium has been attributed to the large number of ADP binding proteins in cells (Sols and Marco, 1970). NMR estimates of ADP levels are below 0.4 μmoles/gm wet weight in most tissues consistent with the low value of free ADP determined using the creatine kinase equilibrium.

In liver, values reported for the ADP concentration range from approximately one μmole/gm wet weight down to 0.05 μmoles/gm wet weight (Morris et al., 1978; Veech et al., 1979; Iles et al., 1985). The high value comes from acid extracts while the low value comes from assuming a number of glycolytic enzymes are in equilibrium, measuring the relevant metabolites and calculating an ADP concentration. This is similar to the way the creatine kinase equilibrium has been used. It was of interest to use the transgenic mice expressing creatine kinase in liver to calculate hepatic free ADP levels. A combination of NMR data from the *in vivo* liver and HPLC analysis of tissue extracts allowed the quantitation of ATP, PCr, creatine, intracellular pH, and the concentration of free Mg^{2+} in the transgenic livers. Based on the creatine kinase equilibrium, a value of 0.059 μmoles/gm weight for the free ADP level in liver was determined (Koretsky et al., 1990). This value is in agreement with the value calculated from normal liver assuming equilibrium of five glycolytic enzymes (Veech et al., 1979). Determination of free ADP from creatine kinase is much simpler than using the glycolytic enzymes because only one equilibrium is involved and fewer metabolite levels need to be determined. In addition, all of the relevant metabolites necessary to determine free ADP levels can be measured using *in vivo* NMR techniques.

The calculation of free ADP levels using the creatine kinase reaction depends on the assumption that the enzyme is in equilibrium. The transgenic mice offer a unique opportunity to directly test this hypothesis. The different lines of mice all expressed different levels of creatine kinase and there was some variation in the level of creatine kinase from individual to individual within the same line. By using different mice it was possible to calculate free ADP levels as a function of creatine kinase activity (Brosnan et al., 1990). If the reaction was being maintained in equilibrium, there should be no dependence of calculated free ADP levels on

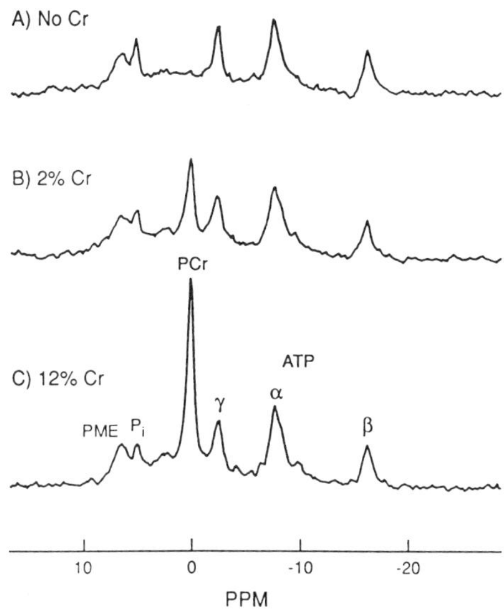

Figure 4. ^{31}P NMR spectra from transgenic mouse liver expressing creatine kinase. Mice had been fed diets supplemented with varying amounts of creatine (Cr) for five days with (**A**) No creatine, (**B**) 2% creatine, and (**C**) 12% creatine. Spectra were acquired from the surgically exposed liver *in situ* and represent the sum of 300 acquisitions using a 7-sec recycle time and 30 Hz exponential linebroadening. Peaks are as labeled with PME, phosphomonoesters; P_i, inorganic phosphate; PCr, phosphocreatine; γ, α, β ATP, the three phosphorus atoms in ATP. (Reprinted with permission from Brosnan et al., 1990.)

enzyme activity. No significant difference in free ADP levels was measured over a fourfold range in creatine kinase activity.

Another test of the equilibrium assumption is to alter the steady-state levels of creatine and PCr and determine if measured free ADP levels change. The ability to detect PCr by NMR in the transgenic livers expressing creatine kinase was dependent on having sufficient levels of creatine in the liver. The levels of PCr and creatine in the liver could be adjusted by altering the dietary intake of creatine (Brosnan et al., 1990). Figure 4 shows a series of ^{31}P NMR spectra obtained from different mouse livers that had been fed a different amount of creatine in their diet. As can be clearly seen there is an increase in the amount of PCr detected in proportion to the amount of creatine fed. For these experiments mice were fed different amounts of creatine for five days prior to measurements. The increase in PCr and creatine in the liver did not cause any changes in ATP, P_i, intracellular pH, or the concentration of free Mg^{2+} in the livers. The calculated free ADP levels were not significantly different over a 10-fold range of PCr concentration (Brosnan et al., 1990).

A final test of the equilibrium assumption used to determine free ADP levels in the transgenic livers could be performed due to the ability to easily manipulate creatine levels in the liver. This test relied on the use of an analog of creatine, cyclocreatine. The equilibrium of the creatine kinase catalyzed reaction with cyclocreatine lies thirty times more towards phosphocyclocreatine than does the equilibrium for creatine and PCr (LoPresti and Cohn, 1989). Due to the difference in equilibrium constants, a different level of free ADP should be calculated with the cyclocreatine/phosphocyclocreatine pair as compared to the creatine/PCr pair if the creatine kinase reaction were not in equilibrium. The free ADP level calculated from transgenic livers containing phosphocyclocreatine and cyclocreatine was not significantly different than that measured using PCr and creatine (Brosnan et al., 1990).

These tests all indicate that the equilibrium assumption used to calculate free ADP levels in the transgenic livers expressing creatine kinase is valid. The average level of free ADP in the transgenic livers is approximately 0.05 μmoles/gm wet weight. This is in a range where ADP can regulate a number of metabolic processes including oxidative phosphorylation. It also demonstrates that the liver is able to maintain low levels of ADP as do the brain and heart.

V. EFFECTS OF THE PRESENCE OF PHOSPHOCREATINE ON THE RESPONSE OF LIVER TO FRUCTOSE

The presence of PCr in the liver offers an additional source of high energy phosphate to phosphorylate ADP to ATP. Therefore, it is interesting to ask if the response of the liver to an agent that is known to drain high energy phosphates is altered in the presence of PCr and creatine kinase in the transgenic livers. The metabolism of fructose by liver has been studied by a number of investigators (Raivio et al., 1969; Woods et al., 1970; Van den Berghe et al., 1977; Morris et al., 1978). Fructose administration to an animal leads to accumulation of fructose-1-phosphate (F-1-P) in the liver due to the phosphorylation of fructose by ATP, catalyzed by fructo-kinase. Associated with this rise in F-1-P is a fall in P_i and ATP. The drop in Pi is presumably due to increased phosphorylation of ADP to ATP during the fructose load. The fall in ATP is primarily due to catabolism of adenylates to uric acid in humans, and to allantoin, in rodents. The rate limiting step in the breakdown of adenylates that occurs during a fructose load is the deamination of AMP to IMP catalyzed by AMP deaminase.

A number of questions remain concerning the mechanism of the observed changes that occur in liver metabolism during a fructose load. It is not clear whether free ADP levels rise during fructose loading. Based on liver extract data, investigators have reported either a rise, no change, or a fall in ADP levels (Woods et al., 1970; Van den Berghe et al., 1977; Morris et al., 1978). It is also not clear why AMP deaminase is activated to cause breakdown of adenylates during a fructose load.

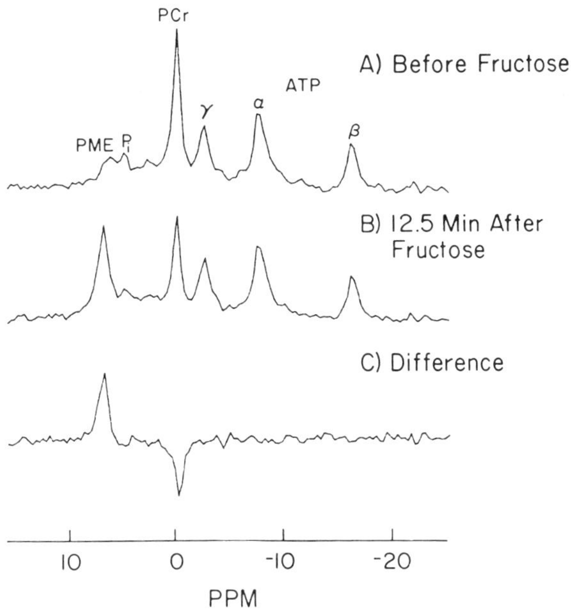

Figure 5. ^{31}P NMR spectra from a surgically exposed, transgenic liver *in situ* expressing creatine kinase prior to (**A**) and 12.5 min (**B**) after an intraperitoneal infusion of 0.15 mmole/gm body weight fructose. The spectrum shown in (**C**) is the difference spectrum (**B**-**A**). Spectra represent the sum of 300 acquisitions with a 1-sec recycle time and 30 Hz exponential linebroadening. Peaks are as labeled with PME, phosphomonoesters; P_i, inorganic phosphate; PCr, phosphocreatine; γ, α, β ATP, the three phosphorus atoms in ATP. (Reprinted with permission from Brosnan et al., 1991.)

Two possibilities are that a rise in free ADP leads to a rise in AMP which stimulates flux through AMP deaminase. Another possibility is that the fall in P_i associated with a fructose load activates flux through AMP deaminase. It is known that P_i is an inhibitor of AMP deaminase and a fall in P_i would be expected to stimulate breakdown of AMP to IMP (Wheeler and Lowenstein, 1980).

Figure 5 shows ^{31}P NMR spectra obtained from the *in situ* liver before (Figure 5A) and 12.5 min after (Figure 5B) an intraperitoneal injection of fructose (Brosnan et al., 1991). There is maximum accumulation of F-1-P 12.5 min after administering the fructose. In this case, the livers had been loaded with a medium amount of creatine by feeding the mice for five days prior to the experiment on a 5% creatine diet. The difference spectrum shown in Figure 5C clearly shows an accumulation of PME and a drop in PCr. No drop in ATP and only a very small drop in P_i is detected. Therefore, the presence of PCr and creatine kinase in the liver protected the ATP from a fructose load that was sufficient to cause a 50% drop in ATP in normal livers (Brosnan et al., 1991). Further evidence that the adenylate breakdown pathway was not activated in the transgenic liver during the fructose load was that there was no increase in IMP or allantoin compared to the large increase in these metabolites detected in normal livers. The protection of ATP was dependent on there being adequate levels of PCr. Transgenic livers that had not been loaded with creatine and PCr by prior feeding responded to fructose in the same manner as normal livers.

In transgenic livers, the level of free ADP was shown to increase 2.5-fold at 12.5 min after fructose administration (Brosnan et al., 1991). Because there was no loss of ATP or increase in IMP, it is clear that this rise in ADP is not sufficient to increase flux through AMP deaminase. The ability to vary PCr levels was very useful for studying the relation between ADP, P_i, and loss of ATP during a fructose load. At low levels of PCr, fructose caused a drop in P_i as well as a drop in PCr. As P_i levels began to fall so did ATP levels. This occurred even though free ADP levels rose to the same extent as they did in the presence of high levels of PCr. In all cases a drop in ATP was only detected if there was a fall in P_i levels (Brosnan et al., 1991). These results support the model that flux through AMP deaminase is regulated by P_i levels in the liver during a fructose challenge.

Another question that has been addressed about hepatic fructose metabolism is the reason for the accumulation of F-1-P in liver during a fructose load. Based on *in vitro* data comparing the rate of fructokinase and aldolase, the enzyme responsible for degradation of F-1-P, it was concluded that aldolase was inhibited *in vivo* during a fructose load (Woods et al., 1970). It was shown that increased IMP, which results from the breakdown of AMP during catabolism of adenylates, could be responsible for inhibition of aldolase. In the transgenic livers containing high levels of PCr there was an accumulation of F-1-P to the same extent as in normal livers, although there was no detectable increase in IMP (Brosnan et al., 1991). Therefore, IMP inhibition of aldolase does not explain the accumulation of F-1-P. Interestingly, the breakdown of F-1-P was twice as fast in the transgenic livers containing PCr as in the normal livers. This indicates that prevention of the rise in IMP may have resulted in higher aldolase activity in the transgenic livers as compared to normal livers, leading to a faster turnover of F-1-P.

VI. PHOSPHOCREATINE ACCUMULATION IN THE PERFUSED LIVER

The experiments described to calculate the value of ADP as a function of creatine (Figure 4) illustrates that total creatine levels (the sum of PCr + creatine) can be altered in the liver. Some work has been done characterizing creatine transport in a variety of cells (Loike et al., 1986). This work indicates that there are at least two creatine transport mechanisms, one that is not saturable and probably represents diffusion of creatine through the membrane, and another which is sodium dependent with a K_m of approximately 30 μM for extracellular creatine. In the experiments done on the *in situ* transgenic liver, creatine levels were typically 20- to 30-fold higher than levels measured in the blood (Brosnan et al., 1990), indicating that the liver can concentrate creatine. This observation, in combination with the known role of the liver as a site of creatine synthesis (Loike et al., 1986), indicates that the liver may effectively regulate blood creatine levels.

The fact that the amount of creatine in the liver can be altered by transport adds a complication to experiments with the livers expressing creatine kinase. It is possible that a fall in PCr could be due to a loss of creatine from the liver rather than from a rise in hepatic ADP levels. In the fructose experiments described earlier, care was taken to measure total creatine levels before and at the maximum change in PCr (Brosnan et al., 1991). In this case there was no significant change in total creatine indicating that transport of creatine was not a factor during those experiments. In order to more clearly define the effects of creatine transport on PCr levels in the transgenic livers, experiments were performed using perfused livers and manipulating creatine levels in the perfusate.

In order to examine creatine uptake, livers of transgenic mice that were not fed a creatine diet were perfused with no creatine in the perfusate and then switched to perfusate containing 10 mM creatine. The accumulation of PCr in ^{31}P NMR spectra from the perfused liver was monitored as a function of time. Figure 6A shows a ^{31}P NMR spectrum from the perfused liver prior to adding creatine to the perfusate. Figure 6B shows the spectrum after 40 min of perfusion with creatine and Figure 6C shows the difference spectrum. There is a large accumulation of PCr without any significant changes in any of the other phosphorus containing compounds detected by NMR. The half time of the PCr accumulation was 30 min. Similarly, when creatine was removed from the perfusate, there was a loss of PCr, with a half time of approximately 30 min. This loss in PCr was not accompanied by any changes in the levels of the other phosphorus containing metabolites detected by NMR (J. M. Halow, 1993).

It is evident from the difference spectrum shown in Figure 6C that there is a gain in PCr but no loss in phosphate from any other pool detected by NMR. Creatine is phosphorylated to PCr by creatine kinase at the expense of ATP. The ADP produced can be rephosphorylated to ATP at the expense of P_i. There are two possible explanations for why no loss in ATP or P_i was measured during the accumulation of PCr. First, the perfusate contained 0.5 mM P_i which could have been transported into the liver to prevent any fall in P_i. Second, there have been a number of studies that indicate that the P_i concentration measured by NMR is approximately one fourth that measured by freeze extraction and chemical assay of P_i (Iles et al., 1985). It could be that accumulation of the PCr occurred at the expense of this "NMR-invisible" pool of phosphate.

To test these possibilities, the creatine uptake experiments were done in perfusate that was free of P_i. Without phosphate in the buffer, there was only a very small accumulation of PCr, to approximately one third the levels illustrated in Figure 6B.

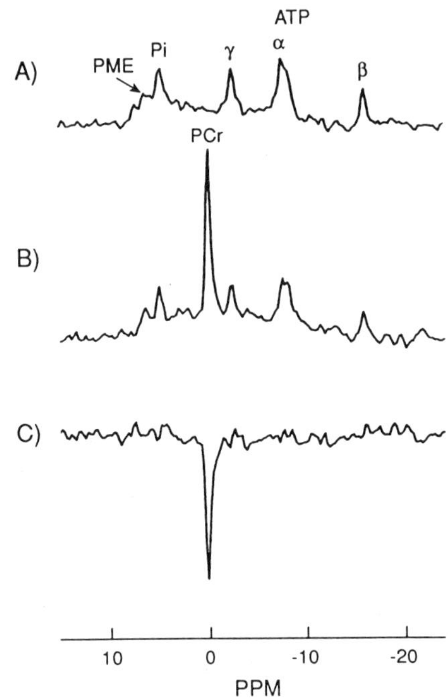

Figure 6. ^{31}P NMR spectra from a perfused transgenic liver expressing creatine kinase. (**A**) Spectrum from liver being perfused with buffer that lacks creatine and, (**B**) Spectrum from liver after being perfused with 10 mM creatine in the buffer for 50 min. The perfusion buffer also contained 0.5 mM inorganic phosphate. (**C**) The difference spectrum (**A-B**) clearly indicates a large accumulation of phosphocreatine without any changes in the other metabolites detected. Spectra represent the sum of 120 acquisitions with a 7-sec recycle time and 30-Hz exponential linebroadening. Peaks are as labeled with PME, phosphomonoesters; P_i, inorganic phosphate; PCr, phosphocreatine; and γ, α, β ATP, the three phosphorus atoms in ATP.

Associated with this small increase in PCr there was a decrease in P_i and a loss of 50% of the initial ATP (J. M. Halow, 1993). Thus, the accumulation of PCr in the transgenic livers is dependent on having phosphate in the buffer.

The loss in P_i and ATP associated with PCr accumulation in the P_i-free perfused transgenic liver is analogous to the experiments described that dealt with the mechanism for loss of ATP during fructose administration in normal livers. In the case of PCr accumulation, phosphorylation of the creatine which is being transported into the cell to PCr by ATP is predicted to lead to an increase in ADP. ADP is phosphorylated to ATP at the expense of P_i. The drop in P_i relieves inhibition of AMP deaminase and activates degradation of adenylates leading to a large loss in

ATP. Consistent with this model is the fact that the drop in P_i corresponds in time to the loss of ATP during creatine perfusion. Adding P_i to the perfusate prevents the fall in P_i and the loss of ATP and allows PCr to accumulate as long as net creatine transport into the cell occurs.

An interesting observation from these creatine accumulation experiments is that the liver is able to regulate P_i levels extremely well. Even though there was a large accumulation of PCr during creatine perfusion in the presence of P_i there was no detectable decrease in liver P_i. Furthermore, PCr could be washed out from the livers by perfusing with creatine free buffer. During the washout of PCr and creatine there was no detectable increase in liver P_i (J. M. Halow, 1993). Because these washout experiments were done without P_i in the perfusate, it will be interesting to see if similar results occur if there is P_i in the buffer.

Two important conclusions for interpreting experiments performed on the transgenic livers expressing creatine kinase grew out of the PCr accumulation experiments. First, the level of hepatic creatine can change due to transport with a half time of 30 min. Thus, it is important to be able to monitor levels of creatine as well as levels of ATP, PCr, intracellular pH, and the concentration of free Mg^{2+} for calculation of free ADP levels using the equilibrium established by creatine kinase. ^{1}H NMR spectroscopy can be used to measure the total creatine. Only the sum of creatine + PCr can be measured because the ^{1}H peaks due to PCr can not be resolved from those due to creatine. In the transgenic livers peaks due to creatine + PCr can be quantitated with ^{1}H NMR spectra obtained within three min (J. M. Halow, 1993). Therefore, the ^{1}H and the ^{31}P NMR data can be acquired in an alternating fashion every 5 to 10 min. The two spectra combined contain all the information necessary to calculate free ADP levels. The second conclusion is that stable levels of creatine and PCr can be maintained in the perfused livers by including creatine in the perfusate. Presently, we feed mice for three days on 2.5% creatine in a liquid diet and then include 1 mM creatine in the perfusate. This protocol maintains PCr to ATP ratios of 3 and a total creatine level of 27 µmoles/gm wet weight for at least four hours of perfusion. This level of creatine and PCr can be varied either by altering the amount of creatine in the diet and in the perfusate or by perfusing livers that had not previously been loaded with creatine with varying concentrations of creatine in P_i containing buffer until a steady-state level of PCr and creatine is reached.

VII. QUANTITATION OF THE ROLE OF ADP IN THE REGULATION OF OXIDATIVE PHOSPHORYLATION IN LIVER

One of the interesting outcomes of the fructose experiments described earlier is that the addition of fructose to livers with high levels of PCr led to an increase in free

ADP without any changes in ATP or P_i concentrations (Brosnan et al., 1991). This is an interesting observation because it has been difficult to separate the contributions of ADP and P_i to the regulation of oxidative phosphorylation in an intact tissue. Experiments in heart and skeletal muscle have correlated changes in ADP and oxygen consumption during increased contractile activity (From et al., 1986; Kushmerick et al., 1992). These experiments have measured apparent K_m's of ADP for oxidative phosphorylation that agree with values determined in isolated mitochondria. However, in these experiments, P_i was increasing along with ADP levels, complicating the interpretation of results. Administering fructose to the transgenic livers expressing creatine kinase offers a way to raise ADP without altering P_i and ATP and determine the response of oxidative phosphorylation to changes in free ADP.

In order to measure changes in oxidative phosphorylation, measurements of the oxygen consumption of the perfused liver were made. It is relatively straightforward to perfuse the transgenic livers and measure the oxygen tension of the perfusate before and after the liver. Along with a measurement of the perfusate flow rate, this data can be used to determine oxygen consumption. Fructose can then be added to the perfusate until a new steady-state in metabolite levels and oxygen consumption is reached. This can be repeated for various levels of fructose and the relation between the calculated level of free ADP and oxygen consumption can be compared.

Figure 7A shows time courses of metabolites from a perfused transgenic liver prior to and during addition of 5 mM fructose to the perfusate. As can be seen there was an increase in PME, due to an accumulation of F-1-P. This was accompanied by a decrease in PCr, and no change in ATP. In addition, there was no change in total creatine levels during this protocol as determined by ^{1}H NMR. There was a transient drop in P_i that recovered to pre-fructose levels. In this case calculated free ADP levels rose from 0.047 μmoles/gm wet weight prior to the fructose, to a steady-state level of 0.097 μmoles/gm wet weight during perfusion with fructose. Accompanying this increase in free ADP there was a 63% increase in oxygen consumption from 1.6 μmoles/min/gm wet weight to 2.6 μmoles/min/gm wet weight (Figure 7B).

A series of similar experiments with various levels of fructose allowed the relation between free ADP and oxygen consumption to be quantitated in the perfused transgenic liver (J. M. Halow, 1993). A Lineweaver-Burke analysis of the data yielded a K_m of 23 μmoles/gm wet weight of ADP for oxidative phosphorylation in the transgenic livers. This value is consistent with that measured in isolated mitochondria and consistent with the values measured in skeletal and cardiac muscles (Chance and Williams, 1955; Guylai et al., 1985; From et al., 1986; Kushmerick et al., 1992).

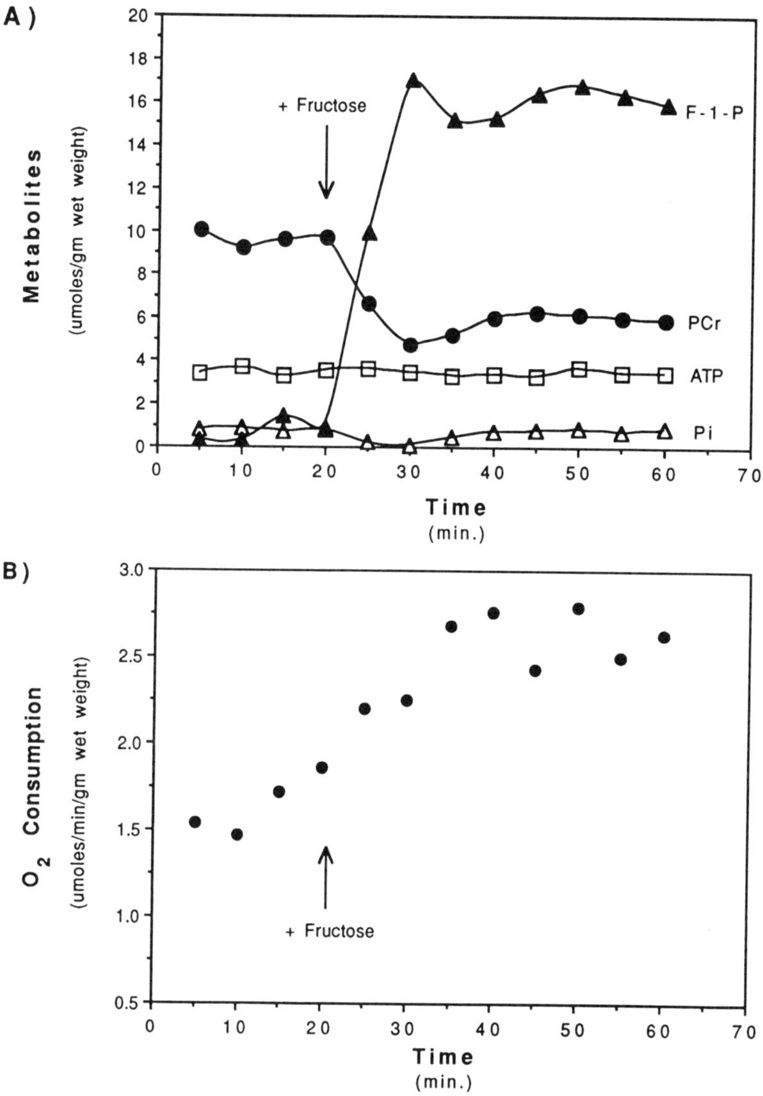

Figure 7. Time course of phosphorus containing metabolites (**A**) and oxygen consumption (**B**) prior to and during perfusion of a transgenic liver expressing creatine kinase with fructose. (**A**) Time course of phosphocreatine (PCr, ●), ATP (□), inorganic phosphate (Pi,), and fructose-1-phosphate (F-1-P, ▲) as determined by ^{31}P NMR. (**B**) Time course of oxygen consumption determined from the same liver as the metabolites shown in (**A**). The arrow indicates the time when 5 mM fructose was added to the perfusate.

VIII. SUMMARY AND FUTURE DIRECTIONS

The expression of creatine kinase in liver has led to a model that is useful for gaining insight into high energy phosphate metabolism. The presence of creatine kinase and PCr in the liver altered the response to ischemia and hypoxia by affording maintenance of pH and ATP levels. These results were anticipated based on the reaction that creatine kinase catalyzes. The fact that the addition of PCr and creatine kinase protected the liver against hypoxic damage opens up interesting possibilities for engineering tissues that are resistant to damage and thus more suitable for transplantation.

Furthermore, the presence of creatine kinase allowed for the calculation of free ADP levels based on the equilibrium reaction catalyzed by creatine kinase. The value of free ADP determined was in the range suitable for ADP to act as a key regulatory metabolite and comparable to free ADP levels found in heart and brain. The presence of PCr altered the response of the liver to fructose allowing the effects of a rise in ADP and a fall in P_i on the regulation of ATP catabolism to be separated. A rise in ADP is not sufficient to activate flux through AMP deaminase, rather a fall in P_i is necessary to relieve inhibition. Using the fact that the response to fructose was altered in transgenic livers expressing creatine kinase, the K_m of oxidative phosphorylation for ADP could be determined. This was accomplished without the complication of changing P_i or ATP levels which are usually encountered when attempting to analyze the role of ADP in regulating oxidative phosphorylation in intact tissues.

The expression of creatine kinase in liver, in combination with *in vivo* NMR techniques, has offered the possibility of measuring a key regulatory metabolite, ADP, in liver. In addition, the presence of creatine kinase and PCr in liver altered high energy phosphate metabolism in interesting ways, thereby allowing for new experimental approaches to apply to long standing problems. This one transgenic mouse model helped to solve both the problem of measurement of regulatory metabolites and the problem of making specific perturbations of cellular metabolism.

The ability to calculate free ADP in liver will allow for the quantitative understanding of the regulation of oxidative phosphorylation. This is because the liver also contains sufficient β-hydroxybutyrate dehydrogenase activity to determine mitochondrial NADH redox state (Williamson et al., 1967). Thus, for the first time, both of these key parameters can be quantitated in the same tissue. In addition, this model offers interesting possibilities for studying the quantitative role of other proteins that are important in energy metabolism. For example, myoglobin or the mitochondrial form of creatine kinase can now be added to these livers using transgenic mouse techniques. The effect of these proteins on the quantitative relation between free ADP, mitochondrial NADH redox state, and oxidative phosphorylation can then be determined.

The mice expressing creatine kinase in liver represent the result of the routine use of molecular genetic techniques to express a protein in a particular cell type (Hogan et al., 1986; Palmiter and Brinster, 1986; Hanahan, 1989; Koretsky, 1992). Advances in molecular genetic techniques over the past two years have made it possible to knock out a specific protein or replace a protein with a specific mutant form (Mansour et al., 1988). In addition, advances in NMR techniques continue to increase the sophistication with which it is possible to monitor the consequences of specific alterations in metabolism (Gillies, 1992; Koretsky and Williams, 1992). With this growing tool kit, our understanding of metabolic control in intact tissues will continue to improve rapidly.

ACKNOWLEDGMENTS

The authors acknowledge support from a NIH FIRST award to APK (HL-40354), a NIH Research Resource award to the Pittsburgh NMR Center for Biomedical Research (RR-03631), and a NIH predoctoral training grant award to KRM (GM-08067). This work benefitted from support and helpful discussions with Chien Ho, Terry Van Dyke, and M. Julia Brosnan. We are grateful to the Richard King Mellon Foundation, the Lucille P. Markey Charitable Trust, the Ben Franklin Partnership Program of the Commonwealth of Pennsylvania, and the Ralph M. Parsons Foundation for providing financial support for the establishment of the Pittsburgh NMR Center for Biomedical Research.

REFERENCES

Balaban, R. S. (1990). Regulation of oxidative phosphorylation in the mammalian cell. Am. J. Physiol. 258, C377–C354.

Bessman, S. P. & Carpenter, C. L. (1985). The creatine-creatine phosphate energy shuttle. Ann. Rev. Biochem. 54, 831–862.

Brosnan, M. J., Chen, L., Van Dyke, T. A., & Koretsky, A. P. (1990). Free ADP levels in transgenic mouse liver expressing creatine kinase: effects of enzyme activity, phosphagen type, and substrate concentration. J. Biol. Chem. 265, 20849–20855.

Brosnan, M. J., Chen, L., Wheeler, C. E., Van Dyke, T. A., & Koretsky, A. P. (1991). Phosphocreatine protects ATP from a fructose load in transgenic mouse liver expressing creatine kinase. Am. J. Physiol. 260, C1191–C1200.

Chance, B. & Williams, G. R. (1955). Respiratory enzymes in oxidative phosphorylation. J. Biol. Chem. 217, 409–427.

Das, A. M. & Harris, D. A. (1990). Regulation of the mitochondrial ATP synthase in intact rat cardiomyocytes. Biochem. J. 266, 355–361.

Denton, R. M. & McCormack, J. G. (1985). Ca^{2+} transport by mammalian mitochondria and its role in hormone action. Am. J. Physiol. 249, E543–E554.

Detre, J. A., Williams, D. S., & Koretsky, A. P. (1990). Nuclear magnetic resonance determination of flow, lactate, and phosphate metabolites during amphetamine stimulation of the brain. NMR Biomed. 3, 272–278.

Erecinska, M., Wilson, D. F., & Nishiki, K. (1978). Homeostatic regulation of cellular energy metabolism: experimental characterization *in vivo* and fit to a model. Am. J. Physiol. 234, C82–C89.

From, A. H. L., Petein, M. A., Michurski, S. P., Zimmer, S. D., & Ugurbil, K. (1986). ^{31}P NMR studies of respiratory regulation in the intact myocardium. FEBS. Lett. 206, 257–261.

Fossel, E. T. & Hoefeler, H. (1987). Complete inhibition of creatine kinase in isolated perfused rat hearts. Am. J. Physiol. 252, E124–E130.

Gillies, R. J. (1992). Special Topic: NMR. Annu. Rev. Physiol. 54, 733–846.

Gupta, R. K., Benovic, L., & Rose, Z. B. (1978). The determination of free magnesium level in the human red blood cell by ^{31}P NMR. J. Biol. Chem. 253, 6172–6176.

Guylai, L., Roth, Z., Leigh, J. S., & Chance, B. (1985). Bioenergetic studies of mitochondrial oxidative phosphorylation using ^{31}phosphorus NMR. J. Biol. Chem. 260, 3947–3954.

Halow, J. M. (1993). The role of ADP levels in regulation of oxidative phosphorylation in transgenic mouse liver expressing creatine kinase (Ph.D. thesis) Carnegie Mellon University, Pittsbugh, PA.

Hanahan, D. (1989). Transgenic mice as probes into complex systems. Science 246, 1265–1275.

Hansford, R. G. (1985). Relation between mitochondrial calcium transport and control of energy metabolism. Rev. Physiol. Biochem. Pharmacol. 102, 1–72.

Hogan, B., Costantini, F., & Lacy, E. (1986). Manipulating the mouse embryo: a laboratory manual. Cold Spring Harbor, NY.

Iles, R. A., Stevens, A. N., Griffiths, J. R., & Morris, P. G. (1985). Phosphorylation status of liver by ^{31}P-n.m.r. spectroscopy and its implications for metabolic control. Biochem. J. 229, 141–151.

Jacobus, W. E., Moredadith, R. W., & Vandegaer, K. M. (1982). Mitochondrial respiratory control. Evidence against the regulation of respiration by extramitochondrial phosphorylation potential or by [ATP]/[ADP] ratios. J. Biol. Chem. 257, 2397–2402.

Koretsky, A. P. (1992). Investigation of cell physiology in the animal using transgenic technology. Am. J. Physiol. 262, C261–C275.

Koretsky, A. P. & Balaban, R. S. (1987). Changes in pyridine nucleotide levels alter oxygen consumption and extra-mitochondrial phosphates in isolated mitochondria: a ^{31}P-NMR and NAD(P)H fluorescence study. Biochim. Biophys. Acta 893, 398–408.

Koretsky, A. P. & Williams, D. S. (1992). Application of localized *in vivo* NMR to whole organ physiology in the intact animal. Annu. Rev. Physiol. 54, 799–826.

Koretsky, A. P., Brosnan, M. J., Chen, L., Chen, J., & Van Dyke, T. A. (1990). NMR detection of creatine kinase expressed in liver of transgenic mice: determination of free ADP levels. Proc. Natl. Acad. Sci. USA 87, 3112–3116.

Kushmerick, M. J., Meyer, R. A., & Brown, T. R. (1992). Regulation of oxygen consumption in fast- and slow-twitch muscle. Am. J. Physiol. 263, C598–C606.

Lawson, J. W. R. & Veech, R. L. (1979). Effects of pH and free Mg^{2+} on the K_{eq} of the creatine kinase reaction and other phosphate hydrolyses and phosphate transfer reactions. J. Biol. Chem. 254, 6528–6537.

Loike, J. D., Somes, M., & Silverstein, S. C. (1986). Creatine uptake, metabolism, and efflux in human monocytes and macrophages. Am. J. Physiol. 251, C128–C135.

LoPresti, P. & Cohn, M. (1989). Direct determination of creatine kinase equilibrium constants with creatine or cyclocreatine as substrate. Biochim. Biophys. Acta 998, 317–320.

Mansour, S. L., Thomas, K. R., & Capecchi, M. R. (1988). Disruption of the proto-oncogene int-2 in mouse embryo derived stem cells: a general strategy for targeting mutations to non-selectable genes. Nature 336, 348–352.

Meyer, R. A., Sweeney, H. L., & Kushmerick, M. J. (1984). A simple analysis of the "phosphocreatine shuttle". Am. J. Physiol. 246, C365–C377.

Meyer, R. A., Brown, T. R., Knlowicz, B. L., & Kushmerick, M. J. (1986). Phosphagen and intracellular pH changes during contraction of creatine depleted rat muscle. Am. J. Physiol. 250, C264–C274.

Miller, K. R., Halow, J. M., & Koretsky, A. P. (1993). Phosphocreatine protects transgenic mouse liver expressing creatine kinase from ischemia and hypoxia. Am. J. Physiol. 265, C1544–C1551.

Morris, R. C., Nigon, K., & Reed, E. B. (1978). Evidence that the severity of depletion of inorganic phosphate determines the severity of the disturbances of adenine nucleotide metabolism in the liver and renal cortex of the fructose loaded rat. J. Clin. Invest. 61, 209–220.

Palmiter, R. D. & Brinster, R. L. (1986). Germ-line transformation of mice. Annu. Rev. Genet. 20, 465–499.

Raivio, K. O., Kekomaki, M. P., & Maenpaa, P. A. (1969). Depletion of liver adenine nucleotides induced by D-fructose. Biochem. Pharmacol. 18, 2615–2624.

Sambrook, J., Fritsch, E. F., & Maniatis, T. (1989). Molecular cloning: a laboratory manual. Cold Spring Harbor, N.Y.

Sols, A. & Marco, R. (1970). Concentrations of metabolites and binding sites. Implications in metabolic regulation. Curr. Top. Cell. Regul. 2, 227–273.

Shoubridge, E. A. & Radda, G. K. (1984). A ^{31}P-nuclear magnetic resonance study of skeletal muscle metabolism in rats depleted of creatine with the analogue β-guanidinoproprionic acid. Biochim. Biophys. Acta. 805, 79–88.

Van den Berghe, G. M., Bronfman, R., Vaneste, R., & Hers, H. (1977). The mechanism of adenosine triphosphate depletion in the liver after a fructose load. Biochem. J. 162, 601–609.

Veech, R. L., Lawson, J. W. R., Cornell, N. W., & Krebs, H. A. (1979). Cytosolic phosphorylation potential. J. Biol. Chem. 254, 6538–6547.

Walliman, T., Wyss, M., Brdiczka, D., Nicolay, K., & Eppenberger, H. M. (1992). Intracellular compartmentation, structure and function of creatine kinase isoenzymes in tissues with high and fluctuating energy demands: 'the phosphocreatine circuit' for energy homeostasis. Biochem. J. 281, 21–40.

Wang, Y.L. & Taylor, D. L. (1989). Fluorescence microscopy of living cells in culture. Methods in Cell Biol. 29, Parts A&B.

Wheeler, T. J. & Lowenstein, J. M. (1980). Effects of pyrophosphate, phosphate, and potassium chloride on adenylate deaminase from rat muscle. Biochemistry 19, 4564–4567.

Williamson, D. H., Lund, P., & Krebs, H. A. (1967). The redox state of free nicotinamide-adenine dinucleotide in the cytoplasm and mitochondria of rat liver. Biochem. J. 103, 514–527.

Woods, H. F., Eggleton, L. V., & Krebs, H. L. (1970). The cause of hepatic accumulation of fructose-1-phosphate on fructose loading. Biochem. J. 119, 501–510.

INDEX

J A I P R E S S

Advances in Molecular and Cell Biology

Edited by **E. Edward Bittar**, *Department of Physiology, University of Wisconsin-Madison*

Volume 7, Biology of the Cancer Cell
1993, 268 pp. $97.50
ISBN 1-55938-624-X

Edited by **Gloria H. Heppner**, *Director, Breast Cancer Biology Program, Michigan Cancer Center*

CONTENTS: Introduction. Cellular Genetic Alterations; Models of Breast and Colon Cancer, *S.R. Wolman, D. Visscher, Wayne State University School of Medicine.* Altered Expression of Transforming Growth Factor α and Transforming Growth Factor–β Autocrine Loops in Cancer Cells, *M.G. Brattain, Medical College of Ohio, K.M. Mulder, Pennsylvania State University, School of Medicine, S.P. Wu, G. Howell, L. Sun, Baylor College of Medicine, J.K.V. Wilson, Case Western Reserve University, B.L. Ziober, Baylor College of Medicine.* Altered Signal Transduction in Carcinogenesis, *Catherine A. O'Brian, Nancy E. Ward, Constantin G. Ioannides, Universty of Texas, M.D. Anderson Cancer Center, Houston.* The Significance of the Extracellular Matrix in Mammary Epithelial Carcinogenesis, *Calvin D. Roskelley, Ole W. Petersen, Mina J. Bissell, Lawrence Berkeley Laboratory.* Epithelial-Stromal Cell Interactions and Breast Cancer, *Sandra Z. Haslam, Laura J. Counterman, Katherine A. Nummy, Michigan State University.* The Tissue Matrix and the Regulation of Gene Expression in Cancer Cells, *Kenneth J. Pienta, Wayne State University School of Medicine, Brian C. Murphy, Robert H. Getzenberg, Donald S. Coffey, Johns Hopkins School of Medicine.* Tumor Cell Interactions in Cancer Growth and Expression of the Malignant Phenotype, *Fred R. Miller, Bonnie E. Miller, Michigan Cancer Foundation.* Effect of Class I MHC Gene Products on the Immunobiological Properties of BL6 Melanoma, *Misoon Kim, Elieser Gorelik, Pittsburgh Cancer Institute.* The Role of Angiogenesis in Tumor Progression and Metastasis, *Janusz W. Rak, Erick J. Hegmann, Robert S. Kerbel, Reichmann Research Institute.* Subject Index.

Volume 8, Organelles In Vivo
1994, 191 pp. $97.50
ISBN 1-55938-636-3

Edited by **Lan Bo Chen,** *Dana-Farber Cancer Institute, Harvard Medical School*

CONTENTS: Confocal Redox Imaging of Cells, *Barry R. Masters, Walter Reed Army Medical Center.* Calcium Channels and Vasodilation, *Alison M. Gurney and Lucie H. Clapp,*

UMDS, St. Thomas Hospital, London. Changes in DNA Supercoiling Status of Cells Treated with Antineoplastic Drugs, *William D. Wright and J.L. Roti Roti, Washington University School of Medicine.* Functional Morphology of the Golgi Region: A Lectino-Electromicroscopic Exploration, *Margit Pavelka and Adolf Ellinger, Universität Innsbruk.* Receptor Mediated Endocytosis of Plasminogen Activators, *Gouujun Bu, Phillipa A. Morton, and Alan L. Schwartz, Washington University School of Medicine.* Lipid Trafficking in Hepatocytes: Relevance to Biliary Lipid Secretion, *Kristien J.M. Zaal, Jan Willem Kok, Folkert Kuipers, and Dick Hoekstra, University of Gro;ningen, Faculty of Medicine.* Identification and Characterization of Functional Secretory Cells: Advantages of Multiparameter Flow Cytometry Kinetics, *Elizabeth R. Simons and Theresa A. Davies, Boston University.* An Outline of Neurosecretion, *Jane Somsel Rodman, Tufts University School of Medicine.*

Volume 9. Homing Mechanisms and Cellular Targeting
1994, 303 pp. $97.50
ISBN 1-55938-686-X

Edited by **Bruce R. Zetter,** *Harvard Medical School and Childrens Hospital*

CONTENTS: Introduction, *Bruce R. Zetter.* Directed Cell Migration in Embryonic Blood Vessel Assembly, *Thomas J. Poole, SUNY Health Science Center at Syracuse.* Leukocyte Interaction with Endothelium and Extracellular Matrix: The Role of Selectins and CD44, *Ivan Stamenkovic, Massachusetts General Hospital and Harvard Medical School.* Oligosaccharide-dependent Mechanisms of Leukocyte Adhesion, *John B. Lowe, Univesity of Michigan Medical School.* Cell Adhesion Molecules: Novel Therapeutic Targets for Chronic Inflammatory Diseases of the Central Nervous System, *Gregory N. Dietsch, Gary M. Peterman and W. Michael Gallatin.* Molecular Mechanism of Targeting of Hemopoietic Stem Cells to the Bone Marrow After Intravenous Transplantation, *Mehdi Tavassoli, University of Mississippi School of Medicine.* Tumor Cell Adhesion and Growth in Organ Preference of Tumor Metastasis, *Gareth L. Nicolson, The University of Texas M.D. Anderson Cancer Center.* Cancer Cell Chemotaxis: Mechanisms and Influence on Site-Specific Tumor Metastasis, *F. William Orr, McMaster University.* Experimental Orthotopic Models of Organ-specific Metastasis by Human Neoplasms, *Isaiah J. Fidler, The University of Texas, M.D. Anderson Cancer Center.* Inter and Intracellular Targeting of Drugs, *Smadar Cohen, Ben Gurion University of the Negev and Robert Langer, Massachusetts Institute of Technology.* Chimeric Molecules Constructed with Endogenous Substances, *Gregory T. Lautenslager and Lance L. Simpson, Jefferson Medical College.* Organ-Specific Targeting of Synthetic and Natural Drug Carriers, *S. Moein Moghimi, Lisbeth Illum and Stanley S. Davis, University of Nottingham.*

Advances In Cell and Molecular Biology of Membranes and Organelles

(Previosly published as Advances In Cell and Molecular Biology of Membranes)

Edited by **Alan M. Tartakoff,** *Institute of Pathology, Case Western Reserve University*

Volumes of Advances in Biochemistry and the Biology of Cells are intended to present interrelated reviews concerned with structure and function of membranes of prokaryotes and eukaryotes. Topics may include membrane protein and lipid biosynthesis, membrane-cytosketal relations, membrane permeability, signal transduction, etc. Individual volumes will have a common theme.

Volume 1, Endosomes and Lysosomes: A Dynamic Relationship
1993, 434 pp. $97.50
ISBN 1-55938-362-3

Edited by **Brian Storrie,** *Department of Biochemistry* and *Nutrition, Virginia Polytechnic Institute and State University* and **Robert F. Murphy,** *Department of Biological Sciences, Carnegie Mellon University*

CONTENTS: Preface, *Brian Storrie, Virginia Polytechnic Institute and State University and Robert F. Murphy, Carnegie Mellon University.* Models of Endosome and Lysosome Traffic, *Robert F. Murphy, Carnegie Mellon University.* Endocytic Receptors, *Michael G. Roth, University of Texas Southwestern.* Functions of the Mannose 6-Phosphate Receptors, *Bernard Hoflack, European Molecular Biology Laboratory, Germany and Peter Lobel, University of Medicine and Dentistry of New Jersey.* Chemistry of Lysosomal Cysteine Proteinases, *Robert W. Mason, Virginia Polytechnic Institute and State University and Donna Wilcon, Washington University School of Medicine.* Mechanism and Regulation of Autophagic Degradation of Cellular Proteins, *William A. Dunn, Jr., University of Florida.* Cell-Free Systems for Endocytosis, *William A. Braell, Harvard Medical School.* Genetic Analysis of Membrane Traffic in Mammalian Cells, *Penelope A. Colbaugh and Rockford K. Draper, The University of Texas at Dallas.* Plasma Membrane Lipid Transport in Cultured Cells: Studies Using Lipid Analogs and Model Systems, *Michael Koval, Washington University School of Medicine.* Endosomes, Lysosomes, and Trans-Golgi-Related Systems in Conventional Neurons and the Grof Retina: Shards and Suppositions, *Eric Holtzman, Eliene Augenbraun, Robert St. Jules, and Maria Santa-Hernandez, Columbia University.* The Role of Endocytosis in Epidermal Growth Factor Signaling, *Bryan K. McCune, Johns Hopkins University School of Medicine, William R. Huckle, and H. Shelton Earp, University of North Carolina at Chapel Hill.* Membrane Traffic Through the Late Stages of the Yeast Secretory, *Eric A. Whitters, Henry B. Skinner, and Vytas A. Bankaitis, University of Alabama at Birmingham.* Regulation of Lysosomal Trafficking and Function During Growth and Development of Dictyostelium Discoideum, *James A. Cardelli, Louisiana State University Medical Center.* Towards an Understanding of the Inheritance of Mammalian Lysosomes and Yeast Vacuoles, *Brian Storrie, Virginia Polytechnic Institute and State University.*

Volume 2, Membrane Transport in Protozoa
1993,483 pp. 2 Part Set $195.00
Set ISBN 1-55938-628-2

Edited by **Helmut Plattner,** *Fakultät für Biologie, Universität Konstanz*

PART A - CONTENTS: Preface. Involvement of the Trans-Golgi Network, Coated Vesicles, Vesicle Fusion and Secretory Product Condensation in the Biogenesis of Pseudomicrothorax Trichocysts, *Robert K. Peck, Barbara Swiderski and Anne-Marie Tourmel, Geneva, Switzerland.* Early Steps of the Secretory Pathway in Paramecium: Ultra-structural, Immuno-cyto-Chemical and Genetic Analysis of Trichocyst Biogenesis, *Nicole Garreau de Loubresse, Gif-sur-Yvette, France.* Calcium and Trichocyst Exocytosis in Paramecium: Genetic and Physiological Studies, *Jean Cohen and Daniel Kerboeuf, Gif-sur-Yvette, France.* Exocytotic Events During Cell Invasion by Apicomplexa, *Jean Francois Dubremeta and Rolf Entzeroth, Villeneuve dAsca, France/Bonn, FRG.* Pathways of Lysosomal Enzyme Secretion in Tetrahymena, Arno Tiedtke, *Thomas Kiy, Christian Vossk_hler and Leif Rasmussen, M_nster, FRG/Odense, Denmark.* Synchronization of Different Steps of the Secretory Cycle in Paramecium Tetraurelia: Trichocyst Exocytosis, Exocytosis-Coupled Endocytosis and Intracellular Transport, *Helmut Plattner, Gerd Knoll, and Regina Pape, Konstanz, FRG.* The Ciliary Membrane and its Engagement In Conjugation, *Jason Wolfe, Middletown, CT.* Ciliary and Plasma Membrane Proteins in Paramecium: Description, Localization and Intracellular Transit, *Yvonne Capdeville, Rene.e Charret, Claude Anthony, Julienne Delorme, Pierre Nahon and Andre. Adoutte, Orsay, Francea.*

PART B -- CONTENTS: Endocytosis and Intracellular Transport of Variant Surface Glycoproteins in Trypanosomes, *Michael Duszenko and Andreas Seyfang, Tu_bingen, FRG.* A Comparative Survey on Phagosome Formation in Protozoa, *Klaus Hausmann and Renate Radek, Berlin, FRG.* Endosomal Membrance Traffic of Ciliates, *Richard D. Allen and Agnes K. Fok, Honolulu, HI.* Membrane Flow in the Digestive Cycle in Paramecium, *Agnes K. Fok and Richard D. Allen, Honolulu, HI.* Signal Coupling During Endocytosis in Amoeba Proteus,